Pharmaceutical Formulation
The Science and Technology of Dosage Forms

Drug Discovery Series

Editor-in-chief:
David Thurston, *King's College, UK*

Series editors:
David Fox, *Vulpine Science and Learning, UK*
Ana Martinez, *Centro de Investigaciones Biologicas-CSIC, Spain*
David Rotella, *Montclair State University, USA*
Hong Shen, *Roche Innovation Center Shanghai, China*

Editorial advisor:
Ian Storer, *AstraZeneca, UK*

How to obtain future titles on publication:
A standing order plan is available for this series. A standing order will bring delivery of each new volume immediately on publication.

For further information please contact:
Book Sales Department, Royal Society of Chemistry, Thomas Graham House, Science Park, Milton Road, Cambridge, CB4 0WF, UK
Telephone: +44 (0)1223 420066, Fax: +44 (0)1223 420247,
Email: booksales@rsc.org
Visit our website at www.rsc.org/books

Pharmaceutical Formulation
The Science and Technology of
Dosage Forms

Edited by

Geoffrey D. Tovey
Geoff Tovey Associates, Harpenden, UK
Email: geofftovey@aol.com

ROYAL SOCIETY
OF **CHEMISTRY**

Drug Discovery Series No. 64

Print ISBN: 978-1-84973-941-2
PDF ISBN: 978-1-78262-040-2
EPUB ISBN: 978-1-78801-443-4
ISSN: 2041-3203

A catalogue record for this book is available from the British Library

The Royal Society of Chemistry is a charity, registered in England and Wales, Number 207890, and a company incorporated in England by Royal Charter (Registered No. RC000524), registered office: Burlington House, Piccadilly, London W1J 0BA, UK, Telephone: +44 (0)207 4378 6556.

For further information see our web site at www.rsc.org

Printed in the United Kingdom by CPI Group (UK) Ltd, Croydon, CR0 4YY, UK

Preface

Having spent a career in pharmaceutical development and more specifically pharmaceutical formulation it has always seemed strange to me that books on the subject of pharmaceutics do not lay more emphasis on formulation as a subject.

It is a fact that for any drug substance a medicine for administration to patients cannot exist without both a formulation and a process by which that formulation can be used to make the medicine. Hence, one of the aspects of early development of any new medicine looks at the formulation. A team of formulators is an important part of any development group and the question of how to formulate any potential candidate drug has to be considered at an early stage. For any medicine to be administered to a patient it needs to be in a form which the patient can take and in which it can reach the required target in the body effectively and safely. To devise a successful formulation the formulator has to gather information and take account of numerous factors that include the physical and chemical properties of the active ingredient in the medicine, its stability and its compatibility with other ingredients.

By way of a frequently encountered example, when considering formulating a solid oral dosage form there are several properties of an active ingredient which would be important to consider. These would include, but are not limited to, the bulk density and flow properties of the active ingredient as well as the particle size of the active ingredient. A low bulk density material has a large volume relative to its weight and there are negative implications from this when attempting to compress the material into a tablet using conventional tableting equipment. Related to low bulk density is poor powder flow. Poor flow refers to the low ability of a powder to flow in a desired manner and makes formulation difficult. In order to make a compressed

Drug Discovery Series No. 64
Pharmaceutical Formulation: The Science and Technology of Dosage Forms
Edited by Geoffrey D. Tovey
© The Royal Society of Chemistry 2018
Published by the Royal Society of Chemistry, www.rsc.org

dosage form the powder must flow into a die cavity on a tableting machine but materials with low bulk density typically do not flow well. At the same time, particle size is also important and can influence formulation method and techniques.

As part of a formulator's role decisions have to take account of preferred routes of administration and potential dosage forms. There are many possible routes including oral, parenteral, pulmonary and topical administration. Likewise there are numerous possible dosage forms available for drugs including tablets, capsules, granules powders and solutions. Decisions as to which route of administration and which dosage form is most suitable will depend on the properties of the active ingredient in the medicine, the disease to be treated and, potentially, the patient group which is to be treated; for example if the medicine is for the paediatric population an oral liquid is likely to be the required route and dosage form.

The intention of this book is mainly to emphasise the importance of formulation and the approach is essentially to focus on specific dosage forms such as tablets, capsules and liquids. Altogether eleven of the thirteen chapters have this focus. The starting point is a chapter on pre-formulation which is an essential need whatever dosage form is to be developed. The content of several of the chapters is applicable across a range of dosage forms; pre-formulation and excipients are examples. The need for modified release of the drug from the dosage form is also a commonly required feature of many products and separate chapters are included to cover the subject of coatings and controlled release. Other chapters concern the special needs of paediatric products and the different emphasis on how to formulate products acceptable to children and the elderly.

In addition to the main dosage forms there are others which are less commonly used although equally important when they are used. Some of these are included in a separate chapter on alternative dosage forms. A further special case concerns how the growing number of drugs emanating from biotechnology are formulated. Here the need for a different approach with different analytical techniques is essential and is covered in a separate chapter.

A unique inclusion is the chapter on intellectual property (IP). In this world where research-based companies operate side by side with generic companies it is not surprising that costs of development and pricing of products are key to economic success. The chapter on IP is intended to try to explain some of the issues and pressures of patents and other forms of protection which exist and how companies work through these complications. It is a fact of life that no company can afford to invest too much time and money into developing a product unless there is a reasonable chance of receiving an adequate return on their investment. Whether research-based or generic this has to be kept in mind. The IP status of the drug substance itself, its formulation and perhaps the process by which the product is made are all important factors in this regard; hence the inclusion of the chapter on IP in the book.

The final chapter is in some ways an over-arching chapter in which much of the content of the 'dosage form' chapters is seen to fit into a much larger picture of pharmaceutical development rather than the specifics of formulation. It is hoped that this chapter will add greatly to students' overall knowledge of the whole process of how new products are developed.

Another advantage of having worked in the area of formulation for a considerable time is that I have been able to get to know a large number of pharmaceutical scientists. I approached all of the authors to ask them to write chapters the subjects of which are within their own expertise and experience. All are world class in their fields and from your studies you may already recognise some of the names. I am confident in the quality of their writings. I am sincerely grateful to all of the authors for their commitment to contribute and particularly wish to acknowledge Dr Kendal Pitt for his advice and for helping me to keep the book on track. I am also grateful to the Royal Society of Chemistry for asking me to compile and edit the book and for the guidance of their editorial team.

I hope you will find the book a useful source of information whether you are pre-graduate, post-graduate, already working in the area of product development or just interested in pharmaceutical formulation and the development of medicines.

Geoffrey Tovey

For my father Lewis Thomas Tovey whose guidance led me to a career in pharmacy;
and to my wife Annie for her constant support and encouragement

Contents

Drug Discovery Series No. 64
Pharmaceutical Formulation: The Science and Technology of Dosage Forms
Edited by Geoffrey D. Tovey
© The Royal Society of Chemistry 2018
Published by the Royal Society of Chemistry, www.rsc.org

CHAPTER 1

Preformulation Studies

TREVOR M. JONES

King's College London, UK
*E-mail: trevor.m.jones@btinternet.com

1.1 Introduction

Discovering and developing new medicines is a long, complex and expensive process and the failure rate is high during the process. To minimise attrition it is essential, therefore, to understand the physicochemical characteristics of compounds or biological entities that are candidates for development into final products.

At various stages during the development of a new medical product the candidate drug must be formulated into a dosage form that is appropriate for the intended study *e.g. in vitro* screening using chemical, physicochemical or biological assays, pre-clinical *in vitro* laboratory safety tests, *in vivo* efficacy and safety studies in relevant animal species, first-in-human studies to determine the optimum drug to progress into clinical development, initial volunteer/patient studies and full-scale clinical trials (Figures 1.1 and 1.2).

The nature and composition of the formulations will be different for each stage of development but the formulation chosen for full-scale clinical trials must, as far as possible, be the same as the product that is intended for marketing. Otherwise extensive clinical comparative trials may be required to demonstrate the similarity between the formulation used in the clinical trials and that proposed for subsequent marketing.

Drug Discovery Series No. 64
Pharmaceutical Formulation: The Science and Technology of Dosage Forms
Edited by Geoffrey D. Tovey
© The Royal Society of Chemistry 2018
Published by the Royal Society of Chemistry, www.rsc.org

Drug Discovery

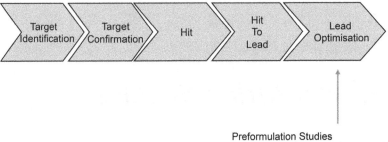

Preformulation Studies
To chose preferred compound
from a formulation/manufacturing perspective

Figure 1.1 Early stage preformulation studies.

Drug Development

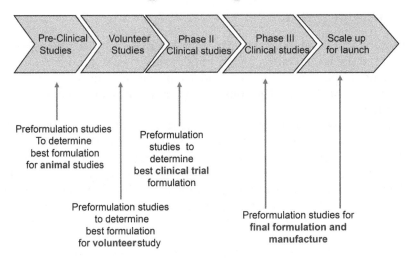

Figure 1.2 Preformulation studies at various stages of development.

To ensure that the various formulations are optimised for their intended use, pre-formulation studies should be conducted not only to evaluate the characteristics of candidate drugs but also potential formulation excipients, and their interactions with drug substances, in order to select appropriate formulation ingredients. In addition, preformulation studies should assess the effect of possible conditions of preparation, manufacture and storage on stability, so as to give confidence that a reliable assessment of the candidate drug has been performed during development and in regular, post-marketing, use.

Data acquired from preformulation studies also forms an important basis for understanding the potential pharmacokinetics of a drug in humans and animals.

In addition, as the chosen product is scaled up in manufacture and/or further process development is carried out *e.g.* to use alternative equipment or technologies; preformulation data can be a useful source of information to understand the opportunities for and limitations to process change.

Furthermore, a number of the characteristics measured in preformulation studies can be used to predict the stability of the formulation during manufacture, transport and storage so as to determine the shelf life of the marketed product.

Preformulation studies can therefore be defined as; Laboratory studies to determine the characteristics of active substance and excipients that may influence formulation and process design and performance.

It has been described as "Learning before doing".

1.2 Solubility

The aqueous and lipid solubility characteristics of a drug substance are of fundamental importance in determining whether it is capable of reaching sites of absorption, its interaction with putative therapeutic targets and its ultimate metabolism and excretion.

An assessment of solubility characteristics is, therefore, usually a starting point for preformulation studies.

1.2.1 Absolute (Intrinsic) Solubility

Using standard aqueous buffers the drug or excipient is vigorously stirred at a constant temperature, *e.g.* 37 °C, to achieve equilibrium, maximum (saturated) absolute solubility. For compounds with ionisable groups this equilibrium solubility of the unionised form is known as the intrinsic solubility.

Preformulation studies will start by measuring intrinsic solubility in a neutral, an acid and an alkaline environment; typically 0.1 M HCl, water and 0.1 M NaOH at 4 °C, 25 °C, 37 °C and an elevated temperature *e.g.* 50 °C.

These data can be recorded as the absolute (intrinsic) aqueous solubility at each pH and compared with data on known and related compounds.

The values obtained can provide insight into the state of the drug substance as it is subjected to a variety of different pH environment *e.g.* as it passes through the gastro-intestinal tract, circulates through various cellular, organ components, arterial and venous circulation and excretory fluids such as bile and urine.

In addition the solubility profile at different pH's can inform the type of the aqueous solvents that might potentially be used in formulations (*e.g.* parenteral injections, nasal or ophthalmic drops, oral solutions).

Furthermore, the information is useful to assess the possible effect that aqueous media used in dosage form manufacture, *e.g.* tablet wet granulation and film coating, may have on the compound.

1.2.2   Molecular Dissociation pK_a

The aqueous solubility of a compound is dependent, *inter alia*, on its state of ionization, including the ratio of ionised to unionised moiety.

The degree of ionisation can be estimated using the Henderson–Hasselbach equation which for weak acidic compounds (HA) is

$$pK_a = pH + \log[HA]/[A^-]$$

or in its rearranged form

$$pH = pK_a + \log[A^-]/[HA]$$

where Ka is the ionisation constant of the dissociation constant.

And for weakly basic compounds (BH)

$$pK_a = pH + \log[BH^+]/[B]$$

Or

$$pH = pK_a + \log[B]/[BH^+]$$

pK_a is obtained by measuring the pH changes of the substance in solution during potentiometric titration using either a weak base or a weak acid. When pH = pK_a the compound is 50% ionised.

The pK_a can be calculated from intrinsic solubility data; also measured using a variety of techniques *e.g.* conductivity, potentiometry and spectroscopy.

The pK_a value provides a useful indication as to the region of the gastrointestinal tract in which the drug will be in either the ionised or unionised state and, hence, some indication of its possible absorption characteristics.

Importantly, however, the chemical nature and concentration of the counter ion conferring solubility *e.g.* chloride or hydrochloride can have a significant influence on solubility and this should be examined during preformulation studies; so as to choose an optimum compound *e.g.* base or cation, for further development.

1.2.3   Solubility in Various Solvents

In addition to determining the solubility characteristics in an aqueous environment it is also useful to obtain preliminary data on the solubility of the drug/excipient in non-aqueous solvents that might be used in formulations, *e.g.* topical ointments/liniments or oily injections, and to provide data that can be used to select solvents for manufacture of the active ingredient, *e.g.* extraction or crystallisation, and for the final formulation, *e.g.* tablet granulation.

Since there are many organic solvents that might be employed, preliminary preformulation studies should focus on a selection of solvents such as:

For Formulation
- Ethyl alcohol
- Glycerin
- Propylene glycol
- Arachis oil
- Ethyl oleate
- Liquid paraffin

For Manufacture
- Industrial methylated spirits
- Isopropyl alcohol
- Benzyl alcohol
- Polyethylene glycol

1.2.4 Solubility Rate (Dissolution)

Whilst a knowledge of intrinsic solubility is essential, the rate at which a drug or excipient dissolves in any particular medium is also important.

Solubility rate will depend on many factors, such as particle size; particle size distribution and particle porosity—and, hence, the surface area available, which is changing as dissolution occurs—the wettability of the particle surfaces, the nature of the dissolution fluid, its polarity, rheological properties and the degree of stirring or agitation during dissolution.

Therefore, initial preformulation studies should focus on a model dissolution system, *e.g.* using pharmacopoeial paddle dissolution methodology. Studies should be performed at constant temperature and pH's using similar particle size fractions (sieve cut of powders) when comparing with reference materials.

More discriminating studies *e.g.* to examine surface area, pH or particle size can be performed as further development progresses.

1.3 Diffusion

Once in solution in an organ or cell in a biological fluid, *e.g.* synovial fluid, vitreous humour, mucous *etc.*, a drug will need to diffuse to the site of transfer or action.

The rate at which the drug can diffuse is dependent on a variety of physiochemical properties such as the viscosity of the fluid through which it is diffusing, the temperature of the fluid, the concentration gradient across the fluid—and hence the amount of drug in solution and the surface area with which it is in contact.

In a fluid with pure Newtonian rheological properties the rate of diffusion of a chemical entity can be calculated using the Noyes–Whitney Equation.

$$\mathrm{d}M/\mathrm{d}t = DS(\mathrm{Cs} - \mathrm{Cb})/H$$

where

$\mathrm{d}M/\mathrm{d}t$ is the rate of dissolution (*i.e.* the amount M diffusing in time t)

D is the diffusion coefficient from the saturated liquid layer adjacent to the crystal surface.

S is the surface area exposed.

Cs is the concentration in a saturated liquid layer directly adjacent to the crystalline solid surface.

Cb is the concentration in the bulk solution further out from the crystal, (Cs – Cb) is the concentration gradient.

H is the thickness of the liquid saturated layer.

Preformulation diffusion studies can be conducted using a Franz cell.[1]

In addition to determining the rate and quantity of drug that has permeated, the diffusion coefficient provides another means of comparing related compounds and those with known *in vivo* characteristics.

1.4 Partition Coefficient

Even when a drug substance is readily soluble at physiological pH's, its ability to transfer across membranes can be highly dependent on its capacity to partition into and cross lipophilic substrates, *e.g.* components of cell walls.

This lipophilicity can be quantified for comparative purposes by determining its partition coefficient P

$$P_{\mathrm{o/w}} = (C_{\mathrm{oil}})/(C_{\mathrm{water}}) \text{ at equilibrium}$$

which is a measure of the unionised drug distribution between an aqueous and an organic phase at equilibrium.

The technique used is to dissolve a known concentration of the compound in an aqueous solution and shake this together in a flask with an equal volume of the lipid. After the phases separate, the amount of drug remaining in the aqueous solution is determined, from which the amount that has partitioned into the lipid can be calculated.

Drug substances make contact with a variety of lipid substances in various compartments of the body so the choice of a lipid to determine the partition coefficient can be critical.

Over many years, *n*-octanol has been chosen as a model lipid in preformation studies since it has properties not too dissimilar to many biological short chain hydrocarbon lipids.

It is possible, therefore to build up a library of values for known drugs against which the new drug can be compared.

As the candidate drugs are being optimised to chose a lead compound for development, further studies can be established to examine their partition in solvents of increasing lipophilicity[2] (Figure 1.3).

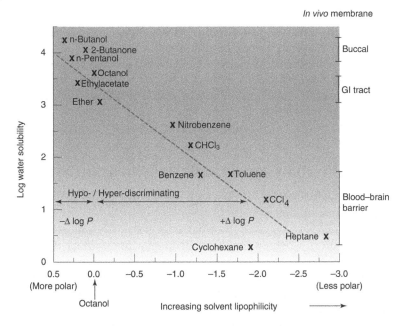

Figure 1.3 Partition coefficients in solvents of increasing lipophilicity. Reprinted with permission from *Aulton's Pharmaceutics*, Michale Aulton and Kevin Taylor, Chapter 23: Pharmaceutical preformulation, 367–395, Copyright 2013, with permission from Elsevier.

Since many factors determine the activity, absorption and permeation of drugs across membranes and into tissues and cells; the partition coefficient of itself is only a starting point for understanding the biopharmaceutical properties of a substance. Nevertheless it can provide valuable comparative data when examining a series of lead molecules to optimise efficacy and bioavailability.

1.5 Permeability

Once in solution in physiological fluids *e.g.* gastric juices or plasma, a drug must permeate cells and tissues to reach its target site of action. This will involve passive and/or active transport mechanisms. For passive diffusion the drug will need to partition with the lipid components of cells and/or diffuse through aqueous pores in tissues.

An index of its permeability can be obtained *in vitro* by measuring the permeability across a model membrane at a constant temperature, Typically the drug in solution is placed in one side of a two-compartment cell separated from the second compartment by a polymeric membrane, the second compartment containing a physiological representation fluid, *e.g.* normal saline.

The amount of drug permeating through the membrane can be measured at various time intervals. A variety of membranes may be chosen each differing in their lipid composition.

The data obtained permits the calculation of the diffusion rates and a comparison of permeability with that of drugs whose properties are known or comparison with related drug candidates.

Permeability is not therefore a single characterises but depends primarily on solubility, partition (aqueous:lipids), diffusion coefficient and the nature of the membrane (chemical and biological composition and thickness).

The rate of permeability will also depend upon other physicochemical properties of solutions (*e.g.* fluid temperature, viscosity, density).

1.6 The Biopharmaceutical Classification System

Combining knowledge of solubility with knowledge of permeability allows an initial estimate of bioavailability.

Amidon *et al.*[3] suggested a Biopharmaceutical Classification Scheme (Figure 1.4, Table 1.1) which has been used as a preliminary indication of bioavailability.

This categorisation can be used to establish whether candidate compounds possess physicochemical properties that are likely to be inferior in terms of bioavailability and hence suggest that further medical chemistry should be conducted to achieve potentially better bioavailability whilst retaining potency. It also forms the basis of FDA Regulatory Guidance on the need for bioavailability and bioequivalence studies.[4]

It is important to recognise that this classification is only an estimate of the likely bioavailability. Following oral administration of a drug, many other processes govern its pharmacokinetic properties. These include *in vivo*

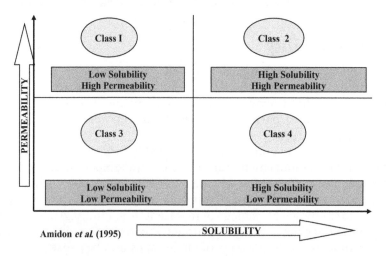

Figure 1.4   Biopharmaceutical classification scheme.

Table 1.1 Biopharmaceutical classification system.

Class 1	High permeability	High solubility	
Class 2	High permeability	Low solubility	Decreasing Bioavailability
Class 3	Low permeability	High solubility	
Class 4	Low permeability	Low solubility	
Highly soluble	Highest dose fully soluble in <250 ml over the pH range 1–7.5		
Highly permeable	>90% absorbed (humans)		
Rapidly dissolving	>85% dissolved in 30 min		
Dissolution rate limited : solubility rate limited			

stability and metabolism in various body fluids and compartments, receptor avidity and glomerular filtration rate.

Also, the biological fluids to which the drug is exposed may contain a variety of solutes that can affect solubility—*e.g.* surface active agents—so it is sometimes useful to measure solubility in "biorelevant" aqueous media in addition to standard pH buffer solutions.[5]

Furthermore, absorption and excretion—and hence bioavailability—can be affected by the biological nature of absorptive and efflux transporters.[6,7]

1.7   Moisture Uptake/Sorption

Chemical and biological materials have different capacities to adsorb and desorb water (called "hygroscopicity") depending on their chemical and physical state.

Drug substances and excipients will be stored in warehouses prior to manufacture, and exposed to various humidly environments during manufacture.

It is important, therefore, to determine their moisture sorption characteristics to establish those conditions that are acceptable and those that should be avoided.

Hygroscopicity information can be used to select packaging for the final dosage form that can protect the product from exposure to the many different humid environments to which it may exposed be during transport and storage. This is necessary to provide a maximum shelf life against chemical/microbiological degradation or, for example, in the case of tablets, physical degradation through disintegration or discoloration.

Laboratory evaluation consists of exposing thin layers of the drug or excipient on dishes at a variety of relative humidities (RH) and at different temperatures; measuring the weight gain or loss over a few days or weeks of exposure and hence the amount of water taken up under each specific temperature and humidity condition. The moisture content at

Hygroscopicity

Moisture sorption/uptake at different relative humidities

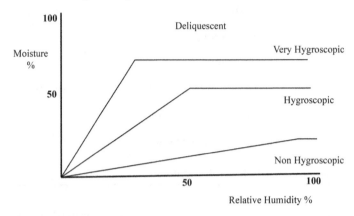

Figure 1.5   Hygroscopicity.

equilibrium at a specific relative humidity is called the equilibrium moisture content (emc).

The data can then be presented as a moisture sorption graph (Figure 1.5).

1.7.1   Classification of Hygroscopicity

Various attempts have ben made to standardise the terminology used in classifying hygroscopicity. The most widely used terms are:

- Deliquescent
- Very hygroscopic
- Hygroscopic
- Non-hygroscopic

However, there is no generally recognised classification.[8] Although, Callahan *et al.*[9] have proposed a useful definition (Table 1.2).

Some materials, *e.g.* maize, potato and corn starches, have the capacity to retain different amounts of water at the same relative humidity depending upon their moisture exposure history.[10]

For example if a starch powder is dried completely and then exposed to a humid environment it will adsorb and absorb moisture isothermally to a maximum emc at 100% RH. When the moisture saturated powder is placed in a low-humidly environment, desorption takes place more slowly due to the amylase chemical bonding that has occurred during sorption, which resists the rapid desorption of water molecules.

Table 1.2   Hygroscopicity.

Class I non-hygroscopic
Essentially no moisture increases occur at relative humidities below 90%. Further-more the increase in moisture content after storage for 1 week at above 90% relative humidity (RH) is less than 20%
Class II slightly hygroscopic
Essentially no moisture increases occur at relative humidities below 80%. The increase in moisture content after storage for 1 week at above 80% RH is less than 40%
Class III moderately hygroscopic
Moisture content does not increase above 5% after storage at relative humidities below 60%. The increase in moisture content after storage for 1 week at above 80% RH is less than 50%
Class IV very hygroscopic
Moisture increase may occur at relative humidities as low as 40–50%. The increase in moisture content after storage for 1 week above 90% RH may exceed 30%

Hygroscopicity
Moisture sorption/desorption at different relative humidities

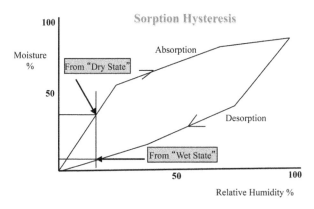

Figure 1.6   Sorption hysteresis.

The moisture sorption graph thus displays a hysteresis (Figure 1.6). At any particular relative humidity, starch powder may have a different moisture content depending upon its exposure history.

This can be important in the context of the use of materials with such moisture hysteresis properties. For example, starches are used as disintegrants in tablet formulations. Their capacity to initiate disintegration is dependent upon their swelling capacity which, in turn, is dependent upon their moisture content, which, in turn, is dependent on their sorption history. Thus a pre-dried starch is likely to be a more efficient disintegrant than a starch included in the final formulation having been pre-exposed to humid environments.

1.8   Polymorphism and Crystallinity

Drugs and excipients can exist in various crystalline or amorphous states depending on their chemical composition and method of isolation or crystallisation.

During crystallisation, molecules may arrange themselves in different geometric configurations such that the structure of the crystals formed has different packing arrangements or orientations. These different states are refers to as polymorphs.

Each polymorphic form may possess very different physicochemical characteristics (*e.g.* solubility, melting point), which can significantly affect the bioavailability of a drug as well as its stability (Figure 1.7).[11] In addition, polymorphism can affect the compression properties of drugs (*e.g.* paracetamol can exist in monoclinic or orthorhombic forms, the latter possessing preferable compaction properties).

In the amorphous state, crystal structures are generally disordered, such that the substance does not posses a sharp melting point but change its physical state slowly as temperature rises. The point at which this commences is called the glass transition temperature. This is another useful preformulation characteristic to consider when selecting processes for manufacture (including wet granulation for tableting and heat sterilisation for injectables) which might change the morphic structure and hence the physicochemical and biological properties of the product.

It is important, therefore, to establish whether a candidate for development has the propensity to exist in different polymorphic states, the

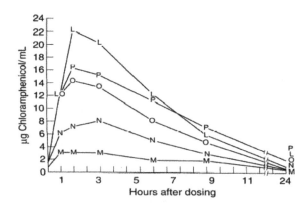

Figure 1.7   Comparison of mean blood serum levels after the administration of chloramphenicol palmitate suspensions using varying ratios of the stable (α) and the metastable (β) polymorphs. M, 100% α polymorph; N, 25:75 β:α; O, 50:50 β:α; P, 75:25 β:α; L, 100% β polymorph. Reprinted from *Journal of Pharmaceutical Sciences*, **56**, A. J. Aguiar, J. Krc, A. W. Kinkel, J. C. Samyn, Effect of polymorphism on the absorption of chloramphenicol from chloramphenicol palmitate, 847–853, Copyright 1967, with permission from Elsevier.[11]

properties of each polymorph (melting point, density, hardness, optical properties, hygroscopicity, solubility, stability *etc.*) and the conditions under which each may be formed; so as to provide guidance to establish a manufacturing process that ensures that the preferred polymorph is created and maintained.

Polymorphisms have been classified as:

1. Enantiotropic: one form changing into another form by varying temperature or pressure.
2. Monotropic: in which the polymorphic form is unstable at all temperatures and pressures.

Clearly it is highly desirable to chose a polymorph (where they exist) that is sufficiently stable at room temperature and to define the temperature conditions (during manufacture and storage) under which polymorphic change or instability could deleteriously affect the compound.

A useful staring point is to prepare samples of the drug or excipient using very different conditions *e.g.* a variety of solvents for crystallisation, different rates and temperature changes during crystallisation and drying. Samples can then be subjected to several analytical procedures to examine their possible polymorphism. Examples of these procedures are described in the following paragraphs.

1.8.1 Differential Scanning Calorimetry (DSC)

This technique measures the heat loss or gain that results from changes (whether physical or chemical or both) as a sample is subject to a programmed temperature change.

Changes in transitions—such as melting, desolvation and degradation—can be identified for different polymorphs to determine the preferred form for future use.

1.8.2 Thermogravimetric Analysis (TGA)

This technique measures changes in sample weight at either a constant temperature over time or when subject to a programmed temperature rise.

It provides additional data to DSC and is particularly of value in examination of solvation.

1.8.3 Powder X-ray Diffraction

This technique is very useful in establishing whether a compound exists as an amorphous state and for comparing the reproducibility of different batches of the chosen crystalline polymorph.

1.8.4   Crystallinity

Crystalline polymorphic materials can exist in a number of shapes or forms (sometimes called "habits") depending on the method and solvent used for final crystallisation.

This can range from highly angular crystals with an elongated shape, needle-like crystals and flat plate-like forms to more spherical habits.

Their shape can affect the "flowability" of the bulk powder *e.g.* during discharge from containers/hoppers *etc.* due to particle–particle mechanical and physical interactions or cohesion.

Each type of crystal may exist as well-formed, solid, structures or posses different degrees of internal stricture; sometimes leading to highly porous structures.

These differences can have profound effects on the rates of dissolution due to differences in surface area exposed to the solvent.

It is useful therefore in preformulation studies to crystallise the compound using different conditions of temperature, solvent, speed of crystallisation *etc.* to determine how critical the crystalline form may be and suggest preferred crystallisation conditions for further optimisation.

1.9   Stability

Clearly, the stability of a drug (and of formulation excipients) is critically important to ensure that the patient receives the correct dose of the active ingredient. Furthermore, for those drugs that can degrade to produce toxic materials, it is essential to determine the conditions under which this might occur so as to find methods of prevention or stabilisation and/or to determine limitations in terms of shelf life and storage conditions.

In addition, the stability of excipients and their stability in combination with other excipients and drug substances can be a critical factor in achieving a stable marketable product, *e.g.* the stability of antimicrobial or antioxidant preservatives in liquid formulations.

Degradation can occur through a number of chemical/physiochemical and biological pathways.

e.g.
- hydrolysis
- isomerisation
- oxidation
- polymerisation
- solid-state phase transformation
- dehydration or desolvation
- cyclization
- photolytic degradation
- microbial attack

Importantly, the kinetics of instability can vary according to the route and hence rate of degradation.

In addition, compounds, especially unsaturated fatty acids and oils, can degrade through different orders of reaction *e.g.* first, second, third *etc.* order kinetics. The chemical route of degradation depending on temperature.

Early preformulation studies should be designed to subject the drug or excipient to several "stress" conditions to identify key degradation pathways and the extent of degradation. From these studies it is possible to estimate the probable stability of the chemical or biological substance under the environmental conditions that it could be subject to during synthesis or extraction, manufacture, transport and storage.

The data might be used to provide feedback to the research team for modification of the labile groups to improve stability.

Alternatively, where such modification compromises the efficacy of the compound the data can be used to guide formulation stabilization strategies, to restrict the conditions to which it should be exposed during manufacture, transport and storage and to provide an early estimate of the potential shelf life of the final formulation.

Stability studies can be conducted on the materials in their solid state and in solution.

In addition to the value of such studies during preformulation, regulatory authorities require such data in submissions for product approval. This requirement is so that they can independently assess whether the product is likely to be adequately stable under the proposed conditions of manufacture, transport and storage until the shelf life or expiry date claimed on the label.

ICH (International Conference on Harmonisation of Regulatory Requirements) Drug stability test guideline Q1A (R2) requires that the drug substance be tested under different stress conditions.

It is suggested that stress testing include the effect of
- pH
- Temperature
- Humidity
- Light
- Oxidizing agents

1.9.1 Chemical Degradation in Solution

The chemical stability of a drug substance in the solid state can be evaluated under various temperature and humidity stress conditions.

Pre-weighed samples are stored in stability cabinets in open vials or thin layers for periods of up 8 weeks under conditions such as:

- 40 °C
- 60 °C
- 80 °C

- 25 °C 85% RH
- 40 °C 75% RH

At pre-determined time intervals, *e.g.* 2 weeks, 4 weeks, 8 weeks, samples are removed, dissolved in an appropriate solvent, and analysed using a robust, stability-indicating assay; typically a reverse-phase high-performance liquid chromatography (HPLC) assay that allows direct injection of stability samples.[12]

The exact temperature and humidity conditions and time intervals of storage that are chosen should take into account the chemical/biological nature of the substance and regulatory agency requirements.[13]

Ideally, the assay should allow detection of degradation peaks equivalent to 0.1% of the parent peak, but this is not always practical at the discovery stage.

Some techniques that perform and analyse multiple degradation experiments on drug substances under various stress conditions are amenable to high-throughput measurements in a 96-well format.

1.9.2   Hydrolytic Degradation

Hydrolysis can occur in many molecular species but particularly for carboxylic acid derivatives or substances containing a functional group based on carboxylic acid, *e.g.* ester, amide, lactone, lactam, imide and carbamate.

To identify and quantify potential degradation by this route, samples of the compound should be subject to stress testing in acidic and alkaline conditions, *e.g.* refluxing the drug in 0.1 N HCl and 0.1 N NaOH for 8–12 hours.

1.9.3   Stability in Solvents Used in Formulation and/or Manufacture

As with solubility evaluations at the preformulation stage, the stability of candidate drugs (and formulation excipients) in non-aqueous solvents that typically might be used in subsequent formulations or manufacturing processes should be examined.

1.9.4   Dimerization and Polymerisation

Similar molecules may interact to produce complex structures—including dimers and polymers of various lengths and orientations. The potential for such interactions should be evaluated by examining the polymeric state of samples during stress testing for heat, light and solution stability.

1.9.5   Photostability

For those substances that may degrade when exposed to light, a number of opportunities exist to prevent or minimise instability through the choice of specialised coatings or packaging.

Nevertheless, the propensity for compounds to degrade in this way should be evaluated at an early opportunity.

Solid-state photostability can be evaluated by exposing thin layers of samples to high-intensity light (HIL)/UV conditions initially at 25 °C (but subsequently at more elevated temperatures) in a photostability chamber.

The ICH guidelines recommend exposure at 1.2 million lux hours to visible light and 200 W hours m^{-2} to UV to represent the frequencies of light radiation in various geographical locations.

Since the drug may be required to be formulated as a solution (*e.g.* oral, parenteral or topical), photostability should also be evaluated in aqueous and, where appropriate, non-aqueous solution.

For both solid and solution photostability studies, samples protected from light are stored under the same conditions and used as controls.

1.9.6 pH-dependent Stability

Stability tests should also be performed under several physiological and formulation pH conditions in order to understand the characteristics of the drug candidate under physiological conditions and to provide key information for the formulation of solution dosage forms.

Typically this would involve measuring stability at 37 °C in a range of buffer solutions *e.g.* pH 1, pH 4, pH 7 and pH 9 at intervals from 1 day up to 1 month.

The studies should be designed using reasonable concentration of drug or excipient to detect even minor decomposition products in the range of detection.

1.9.7 Oxidative Stability

Drugs and excipients may be degraded by oxidation reactions of which there are two distinct types.

1. Oxidation through direct reduction reactions *via* atmospheric oxygen.
2. Oxidation by chain reaction involving the formation of peroxy free radicals.

This, latter, route of degradation is most likely to occur in compounds with double carbon bonds; especially long-chain unsaturated fatty acids and oils.

The oxidation process involves several steps *viz.*: initiation, propagation, and termination, and can be catalysed by heat, light, metals or free radicals.

Typically the reaction is as follows

$$\text{Initiation: } X^* + RH \rightarrow R^* + XH$$

$$\text{Propagation: } R^* + O_2 \rightarrow ROO^*$$
$$ROO^* + RH \rightarrow ROOH + R^*$$

Termination: ROO* + ROO* → stable product
ROO* + R* → stable product
R* + R* → stable product

If the chemical structure of the drug or excipient indicates that this route of oxidation is possible, then preformulation studies should subject samples of the substance to various elevated temperature stress conditions under open atmospheric conditions, *i.e.* in the presence of oxygen.

The route by which such oxidation occurs may be temperature-dependant *e.g.* the energetics and hence the site of peroxidation at low temperatures may be significantly different to those at high temperatures, giving rise to different orders of reaction at different exposure temperatures.

Thus, elevated temperature challenges may not reflect what happens at ambient temperatures.

In the solid state, oxidation can occur where molecular oxygen diffuses through the crystal lattice to the labile sites. These are called "oxygen" electron–transfer reactions.

1.9.8 Stability–Compatibility

Although early preclinical studies—and some animal studies on a lead candidate drug—may use simple solutions derived from preformulation studies on solubility and stability, as the candidate progresses to clinical trials, especially confirmatory large-scale trails, it will be required to be formulated with excipients.

Thus drug–excipient compatibility studies are required to determine the flexibility of choice available for various types of oral, parenteral, topical *etc.* formulation.

Based on a knowledge of the stability characterises of the drug substance, stability tests can be conducted on the drug in the presence of various excipients.

Clearly the range of excipients that might be eventually be chosen for the final, marketed, product can be extensive and, hence, a considerable number of possible combinations for evaluation can be identified. This is not usually justified at the early stages of development. Such initial studies should therefore be restricted to a few major potential excipients, *e.g.* lactose, sucrose, dextrose, magnesium stearate *etc.*, as a prelude to more extensive evaluation later in formulation design and development.

1.10 Solid-state Physico–Technical Properties

Most drug substances that are chosen for final product development will be in powder form. The technical properties of these solid-state materials will be important in formulation and manufacture, *e.g.* their compression characteristics for tablet formation, their flow properties in capsule and tablet production.

Such studies are required at later stages of development and, although still "pre" formulation studies, should be performed only when it is clear that a development candidate drug has ben identified and when the formulation of a final dose form is definitely required.

The physico–technical properties can be described as "fundamental" or "derived".[14-16]

Viz.: Fundamental. The inherent physicochemical properties of the compound (*e.g.* melting point, solubility, stability, taste, absolute density, hardness *etc.*).

Derived. Those characteristics which are dependant upon the physical state of the solid, which can vary according to how the substance is manufactured and processed, *e.g.* particle size, size distribution, surface area, specific surface, particle shape, bulk and tapped density, cohesiveness, dispersibility, flowability, compactability, including material tensile strength, stress relaxation and stress density; strength:pressure and force displacement profiles.

Further Reading

E. F. Fiese and T. A. Hagen, in *Chapter 8 Preformulation in The Theory and Practice of Industrial Pharmacy*, ed. L. Lachman, H. A. Lieberman and J. L. Kanig, Lea & Febiger Philadelphia, 1986.

S. Motola and S. N. Agharkar, in *Chapter 4 in Pharmaceutical Dosage Forms: Parenteral Medications*, ed. K. E. Avis, H. A. Lieberman and L. Lachman, Marcel Dekker, NY, 2nd edn, 1992, vol. 1.

J. T. Carstensen, in *Preformulation Chapter 7 in Modern Pharmaceutics*, ed. G. s. Banker and C. T. Rhodes, Marcel Dekker, 4th edn, 2002.

M. Gibson, *Pharmaceutical Preformulation and Formulation*, InterPharm/CRC, 2003.

A. T. Florence and D. Attwood, *Physicochemical Principals of Pharmacy*, Pharmaceutical Press, 5th edn, 2011.

S. Gatisford, in *Part 5 in Aulton's Pharmaceutics: The Design and Manufacture of Medicines*, ed. M. E. Aulton, K. Taylor, Churchill Livingstone Elsevier, 4th edn, 2013.

References

1. S. F. Ng, J. J. Rouse, F. D. Sanderson, V. Meidan and G. M. Eccleston, *AAPS PharmSciTech*, 2010, **11**, 1432–1441.
2. J. I. Wells, in *Pharmaceutical Preformulation Ellis Horwood (QV744) via Aulton's Pharmaceutics: The Design and Manufacture of Medicines*, ed. M. E. Aulton and K. Taylor, Churchill Livingstone Elsevier, 4th edn, 2013, 1998.
3. G. L. Amidon, H. Lennernas, V. P. Shah and J. R. Crison, A theoretical basis for a biopharmaceutical drug classification: The correlation of *in vitro* drug product dissolution and *in vivo* bioavailability, *Pharm. Res.*, 1995, **12**, 413–420.

4. FDA Guidance, *Waiver of In vivo Bioavailability and Bioequivalence Studies for Immediate-release Solid Oral Dose Forms Based on a Biopharmaceutical Classification System*, issued Aug 2000 see: www.fda.gov/order/guidance.

5. J. M. Butler and J. B. Dressman, *J. Pharm. Sci.*, 2010, **99**(12), 4940–4954.

6. C. Y. Wu and J. Z. Benet, *Pharm. Res.*, 2005, **22**, 11–23.

7. L. Z. Benet, *et al.*, *Adv. Drug Delivery Rev.*, 2016, **101**, 570–580.

8. S. M. Reutzel-Edens and A. W. Newman, in *Polymorphism: In the Pharmaceutical Industry*, ed. R. Hilfike, Wiley, N. Y., 2006, ch. 9.

9. J. C. Callahan, G. W. Cleary, M. Elefant, G. Kaplan, T. Kensier and R. A. Nash, *Drug Dev. Ind. Pharm.*, 1982, **8**, 355.

10. P. York, *J. Pharm. Pharmacol.*, 1981, **33**, 269–271.

11. A. J. Aguiar, A. W. Kinkel and J. C. Samyn, Effect of polymorphism on the absorption of chloramphenicol from chloramphenicol palmitate, *J. Pharm. Sci.*, 1967, **56**, 847.

12. M. Blessy, R. D. Patel, P. N. Prajapati and Y. K. Agrawal, *J. Pharm. Anal.*, 2014, **4**(3), 159–165.

13. H. Khan, M. Ali, A. Ahuja and J. Ali, *Curr. Pharm. Anal.*, 2010, **6**(2), 142–150.

14. T. M. Jones, The influence of physical characteristics of excipients on the design and preparation of tablets and capsules, *Pharm. Ind.*, 1977, **39**(5), 469–476.

15. T. M. Jones, The physico technical properties of starting materials used in tablet formulations, *Int. J. Pharm. Technol. Prod. Manuf.*, 1981, **2**, 17–24.

16. H. G. Brittain, *Physical Characterisation of Pharmaceutical Solids*, Marcel Dekker, 1995.

CHAPTER 2

Hard Capsules in Modern Drug Delivery

S. STEGEMANN*[a,b], W. TIAN[c], M. MORGEN[d] AND S. BROWN[c]

[a]Capsugel, Rijksweg 11, 2880 Bornem, Belgium; [b]Graz University of Technology, Inffeldgasse 13, 8010 Graz, Austria; [c]Capsugel, Oakbank Park Way, Livingston, West Lothian, UK EH53 0TH; [d]Capsugel, 64550 Research Road, Bend, Oregon 97701, USA
*E-mail: sven.stegemann@capsugel.com

2.1 Introduction

Hard capsules as a dosage form have been known since ancient Egyptian times and were mentioned in 1730 by the pharmacist de Pauli from Vienna, who produced oval-shaped capsules in the hope of covering up the unpleasant taste of the pure turpentine he prescribed for people suffering from gout.[1] The first patent for a capsule was granted in 1834 to the pharmacist Joseph Gérard Auguste Dublanc and the pharmacy student François Achille Barnabé Mothès.[2] Following several improvements and modifications, Jules César Lehuby had a patent granted in 1846 for his 'medicine coverings',[3] which formed the basis of his future inventions that lead to the two-piece capsules produced by dipping silver-coated metal pins into a gelatin solution and then drying them. However, the first commercial manufacturing of two-piece hard capsules started in 1931 when Arthur Colton, on behalf of Parke,

Drug Discovery Series No. 64
Pharmaceutical Formulation: The Science and Technology of Dosage Forms
Edited by Geoffrey D. Tovey

Davis & Co., succeeded in designing a machine which simultaneously manu-factured both bodies and caps and fitted them together to form a two-piece capsule.[4]

With the commercial manufacturing of empty hard capsules their use and application to pharmaceutical products grew very rapidly. In retail pharma-cies hard capsules were used to fill individual preparations for patients by hand using a manual capsule filling device. For the pharmaceutical industry, the pre-manufactured empty hard capsules were an important step forward in the commercial manufacturing of solid oral dosage forms using semi-automatic and fully automatic filling equipment. The continuous advance-ments in medical and pharmaceutical sciences has introduced more than 1500 new drug compounds over the past 70 years and capsule formulations have progressed from simple powder blends to more sophisticated delivery with multiple routes of administration.[5] Due to the flexibility of hard cap-sules to accommodate a variety of different drug delivery systems, such as multiparticulates, liquid and semi-solids, interactive powder blends for inha-lation or mini-tablets, hard capsules provide an effective option for modern drug product development. Through advancement in polymer sciences and engineering, the range of hard capsules has evolved substantially over the past years, providing functional features to the capsules which make hard capsules a drug delivery system on their own.

Delivering effective healthcare will continue to be driven by progress in medical and pharmaceutical sciences. In addition, personalization of ther-apies, demographic changes, demands from emerging markets, increasing global quality standards and access to affordable drug products are future challenges which can be addressed by the hard capsule dosage form and its drug delivery technologies.

2.2 Hard Capsules—Types, Characteristics and Applications

2.2.1 Hard Capsules as a Pharmaceutical Excipient

Hard capsules are considered as a pharmaceutical excipient, with mono-graphs in all major pharmacopoeias. Hard capsules are characterized by two pre-manufactured cylindrical sections with each having a hemispherical closed end and an open end whereby the open ends are slipping over each other until they are locked in the closed position. The longer part with the slightly smaller diameter is called the body and the shorter part slipping over the body is termed the cap (Figure 2.1).

Hard capsules are manufactured by a dip molding process. Two sets of dipping pins, one for the body and one for the cap, dip simultaneously into the polymer solution to form a consistent polymer film around the dipping pins, which solidifies after being pulled out from the solution. The dipping pins with the polymer solution pass through a drying section before being pulled off the pins, cut to the targeted length and joined to the pre-closed

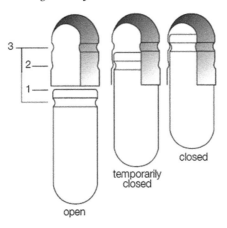

3
2
1

closed

temporarily
closed

open

1. The tapered rim prevents splitting and denting
   of the capsule
2. The notches prevent premature opening
   of the capsule
3. The rim closes the filled capsule safely
   (SNAP-FIT™ principle)

Figure 2.1   Two piece hard capsule design. Capsules in the open, temporarily closed (pre-closed) and closed position. Coustesy of Capsugel, Morristown, NJ, USA.

position (Figure 2.1). The aqueous polymer solution consists of the polymer, the dyes and in some cases a gelling system to support film formation on the dipping pins.

Hard capsules are manufactured in a variety of different sizes with the smallest size being size 5 with a fill volume of 0.13 ml and the largest size being a size 000 with a fill volume of 1.37 ml (Table 2.1). The empty capsules can be printed radially or axially on separate printing machines. Hard capsules should be stored in closed containers at 15–25 °C and 35–65% relative humidity. Under these storage conditions, hard capsule have demonstrated stability for at least five years.[6]

2.2.1.1   Capsules for Immediate Release (IR)

The traditional polymer for hard capsules is gelatin. Gelatin is a mixture of natural proteins derived from hydrolyzed collagen from bone and skin of bovine or porcine origin. Due to the sol–gel transformation of gelatin occurring within a very narrow temperature range, gelatin solidifies on the cold dipping pins immediately after dipping to form a capsule with a shell thickness of about 100 µm. Hard gelatin capsules dissolve quickly in aqueous media at 37 °C and rupture within 2–3 min to release the contents.

Hydroxypropylmethylcellulose, also referred to as HPMC or hypromellose, is another polymer that is used for hard capsules.[48] The HPMC capsules

Table 2.1   Sizes and dimensions of two-piece capsules.

Size	000	00el	00	0el	0	1el	1	2el	2	3	4el	4	5
Volume (ml)	1.37	1.02	0.91	0.78	0.68	0.54	0.50	0.41	0.37	0.30	0.25	0.21	0.13
Overall closed length (mm)	26.1	25.3	23.3	23.1	21.7	20.4	19.4	19.3	18.0	15.9	15.8	14.3	11.1
External cap diameter (mm)	9.91	8.53	8.53	7.65	7.64	6.91	6.91	6.36	6.35	5.82	5.31	5.32	4.91

manufactured with the traditional dipping process (gelling system HPMC (GS-HPMC)) of cold dipping pins dipping into a hot polymer solution require the addition of a gelling system to solidify the hot polymer solution on the cold dipping pins when being pulled out. Gelling systems are composed of either carrageenan or gellan gum and gelling promotors such as potassium chloride (*e.g.* Vcaps®, QualiV®). With the introduction of a dipping process with hot dipping pins dipping into a cold polymer solution the sol–gel transformation point of HPMC between 70 and 80 °C is used to form capsule shells on the dipping pins from pure HPMC solutions (thermogelation HPMC (TG-HPMC), *e.g.* Vcaps® Plus) that utilizes the thermal gelling properties of HPMC. The dissolution of HPMC capsules depends on the composition and process used for their manufacture. GS-HPMC capsule dissolution is dependent on the pH and the ionic strength of the dissolution media. In contrast to this, TG-HPMC capsules dissolve consistently and rapidly in all dissolution media and open within 5–10 min. The different dissolution profiles of a GS-HPMC and a TG-HPMC are shown in Figure 2.2.

Hard capsules can also be made of pullulan, a starch derivative. Pullulan-based hard capsules contain a gelling system and are manufactured with the traditional process. Pullulan-based hard capsules (*e.g.* Plantcaps®) disintegrate faster than gelatin or HPMC capsules and are mainly used for health and nutrition products.[7]

2.2.1.2   *Capsules for Modified Release (MR)*

Capsules with modified release characteristics can be prepared from HPMC and HPMC derivatives. Delayed-release characteristics are provided by HPMC capsules containing gelling systems that stay intact at low pH for up to two hours (*e.g.* DRcaps™). Using the thermal-gelling process and a blend of HPMC and hydroxypropyl methylcellulose acetate succinate (HPMCAS) leads to a capsule that provides enteric properties according to the pharmacopoeia specification for gastro-resistant dosage forms (*e.g.* Vcaps® Enteric). The release profile of an enteric HPMC capsules is shown in Figure 2.3.

The characteristics of the different types of capsules are summarized in Table 2.2.

2.2.2   Hard Capsules for Special Applications

Hard gelatin capsules are one of the most flexible dosage forms and can be used in a variety of different applications.

2.2.2.1   *Capsules for Liquid and Semi-solid Formulations*

Hard capsules are frequently used for liquid and semi-solid formulations. Liquid-filled capsules are normally sealed after filling by banding or fusion sealing (*e.g.* LEMS®). Capsules for fusion sealing are specifically designed to

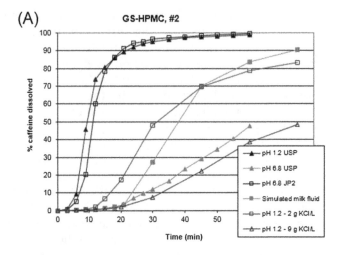

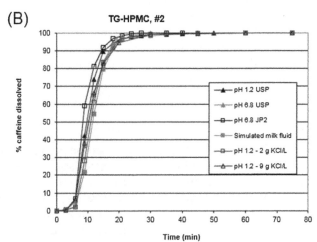

Figure 2.2 Dissolution profiles of caffeine in (A) GS-HPMC (*e.g.* QualiV®) and (B) TG-HPMC (*e.g.* Vcaps® Plus). Coustesy of Capsugel, Morristown, NJ, USA.

ensure that there is a tight fit between the cap and the body after closing of the capsules (*e.g.* Licaps®).

2.2.2.2 Capsules for Pulmonary Delivery

Special capsules have been developed for orally inhaled products to optimize the delivery of dry powder inhalation therapy (DPI). Hard capsules for DPI are based on gelatin or HPMC and are normally customized for the inhalation product formulation. The main feature of DPI capsules is to provide

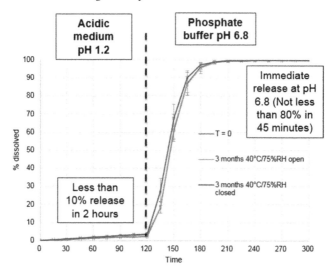

Figure 2.3 Release profile of acetaminophen in enteric HPMC capsules (Vcaps® Enteric). Coustesy of Capsugel, Morristown, NJ, USA.

effective mono-dose packaging of the inhalation formulation, secure the release from the capsules through ease of opening in the device (*e.g.* piercing, shearing) and complete release of the fine particles from the capsules. Capsule-based DPI systems are gaining interest due to their ease and economy of manufacturing as well as the increasing demand from the emerging markets.[22]

2.2.2.3 Sprinkle Capsules

For patients who cannot swallow larger solid oral dosage forms, *e.g.* young children or older patients, hard capsules can be formulated with multiparticulates that are sprinkled on food or beverages for administration. The main feature of this capsule design is the ease of opening, which can be done by the patients or their care givers (*e.g.* Coni Snap® Sprinkle).

2.2.2.4 Hard Capsules for Clinical Trials

The double blinding of medications for clinical trials is a major challenge in drug product development. Over-encapsulation of the study medications is a straightforward technology to achieve double blinding. Special capsules have been developed that prevent re-opening of the capsules after closing as well as the ability to incorporate of a variety of solid oral dosage forms. The capsules (*e.g.* DBcaps®) are available in sizes ranging from size E, with a volume of 0.21 ml, up to size AAA, with a volume of 1.47 ml.

Table 2.2 Comparison of characteristics of different types of capsules.

	Hard gelatin	GS-HPMC	TG-HPMC	Delayed release HPMC	Enteric-HPMC	Pullulan
Polymer	Gelatin	HPMC	HPMC	HPMC	HPMC/HPMCAS	Pullulan
Gelling systems	No	Carrageenan/potassium chloride Gellan gum/potassium acetate	No	Gellan gum/potassium acetate	No	Carrageenan/potassium chloride
Release	IR	IR	IR	MR–Delayed release	MR–Enteric	IR
Origin	Animal	Plant	Plant	Plant	Plant	Plant
Global regulatory acceptance	EP, USP/NF, JP	EP, USP/NF, JP	EP, USP/NF, JP	EP, USP/NF, JP	EP, USP/NF, JP	GRAS
Moisture content	13–16%	4–9%	2–9%	2–9%	4–9%	11–13%
Chemical interaction	Cross-linking	No	No	No	No	No
Dissolution	Consistent	pH- and ionic strength- dependent	Consistent	Slow at low pH	Above pH 6.8	pH- and ionic strength- dependent
Mechanical stability	Dependent on water content	Robust	Robust	Robust	Robust	Dependent on water content

2.3 Selection of Capsules in Formulation Development

The development of a pharmaceutical drug product starts with the definition of the target product profile (TPP) describing the desired pharmaceutical, technical and performance requirements as well as patient need. Based on these targets, prior knowledge and physicochemical properties of the active pharmaceutical ingredient (API), a formulation strategy, including the appropriate manufacturing process, is defined. The choice of capsule type will reside within the overall formulation strategy, taking into account the API characteristics and the TPP. For example, moisture-sensitive formulations are more stable in HPMC capsules as the moisture contents of these capsules are lower compared with gelatin (<9% *vs.* 13–16%) and can be reduced further to *e.g.* 2% without losing the mechanical flexibility. HPMC capsules should also be selected for formulations that contain residual aldehydes, which can cross-link gelatin affecting the disintegration of capsule shells. HPMC-based capsules have been shown to aid in solubilization of poorly water soluble drugs compared with performance in gelatin capsules.[8] Another criteria is the expected release profile of the capsule. For modified release formulations, the capsule itself can be designed to provide this functionality. For example, for products that need to be protected from the gastric juices, an HPMC capsule with enteric properties can be chosen. In situations where other release targets are required HPMC and gelatin capsules can be filled with the formulation and then coated with functional coatings to target the release of the API in different areas of the gastrointestinal tract.[9] Other criteria that have to be included in the selection process are the route of administration, needs of the targeted patient population, *e.g.* swallowing capabilities, but also considerations regarding marketing and identification of drug product.[10]

Following the selection of the formulation and manufacturing process, a risk assessment may be performed and experiments designed [design of experiments (DoE)] to define the relationship between the critical quality attributes (CQA) and critical process parameter (CPP)/critical material attribute (CMA). Investigations into the CQA of the empty capsules have shown that the variabilities of the empty capsules are well within the specification and reproducible over different manufacturing locations and a two year period, supporting quality by design (QbD)-based drug product development.[11]

2.4 Hard Capsule Drug Delivery Addressing Pharmaceutical Needs

2.4.1 Hard Capsule Powder Blend Formulation and Processing

Hard capsules are often selected as the dosage form for first-in-human clinical trials as the drug substance can be filled alone or in a blend with very few functional excipients to achieve the desired drug dissolution and absorption.

It has also been demonstrated that powder-filled capsules are ideal for drugs approved through the fast track, breakthrough therapies, accelerated development or priority review designation process of the FDA[12] or through the European Medicines Agency (EMA) accelerated assessment and conditional approval programs.[13,14] These products take advantage of the existing scientific knowledge and expertise on hard capsules as well as relative simplicity of formulation and processes that de-risk and speed up the product development cycle.

Powder blends for hard capsule formulation are composed of the drug substance and possibly components from up to six different functional excipient classes. Diluents are added to the formulation to improve formation of a powder plug upon compression at 20–30 N in the capsule filling machine that is required to achieve a clean filling operation. The most used diluents are lactose monohydrate or mannitol, which is preferred for hygroscopic drugs. Due to its plastic deformation and compactability microcrystalline cellulose (MCC) is used to achieve the required powder plug formation for poorly compressible drug compounds.[15] Lubricants are added to the powder blend to improve the flow properties of the powder, the plug ejection and fill mass uniformity in manufacturing. Lubricants are generally hydrophobic and can affect dissolution based on their concentration and processing time.[16] Glidants are added to improve flow properties, reducing electrostatic charging and adhesion to metal surfaces in the filling process. Disintegrants are added to the formulation to facilitate the disintegration of the powder plug and disperse the formulation in the media or gastric juice after capsule rupture. In capsule formulation, superdisintegrants, including sodium croscaramellose and crospovidone, are considered, but moderate disintegrants such as sodium glycol starch or corn starch may often be sufficient for hydrophilic formulations. Wetting agents might also be added to provide hydrophilic properties to the plug and assure wetting and water penetration into the plug supporting drug dissolution. In situations where dry or wet granulation is required for the drug substance due to low drug dose, poor flow properties (*e.g.* needle-shaped drug particles) or content uniformity issues (*e.g.* powder segregation) binders might be added. For powder blend formulation the five different classes of functional excipients consist of 21 different excipients that are globally accepted. The excipients including the recommended percentages (w/w) in the formulation are listed in Table 2.3.

As for any pharmaceutical product, process design and development are integral to the pharmaceutical development process. Powder-filled hard capsules contain a blend of the active ingredient with one or more functional excipients, which are mixed in a dry blender and filled on automated filling machines. To achieve content uniformity, product stability and performance according to the defined critical quality attributes, the mixing operation must provide a homogeneous blend with sufficient flow properties.

The powder flow of the formulation can be measured by different methods, like tapped and bulk density, angle of repose and others. The flowability

Table 2.3 Excipients for hard capsule formulation. 0 = center of gravity (standard percentage in formulation), − = lower amount and + = higher amount to modify functional excipient effect and optimize product performance.

Disintegrant	−	0	+
Alginic acid	7.0	8.0	9.0
Croscarmellose	1.5	2.0	2.5
Crospovidone	2.0	3.0	4.0
Maize starch	7.5	10.0	12.5
Pregelatinised starch	7.5	10.0	12.5
Sodium starch glycolate	3.5	5.0	6.5

Glidant	−	0	+
Colloidal silica	0.25	0.50	0.75
Purified talc	4.50	5.00	5.50

Lubricant	−	0	+
Glyceryl monostearate	1.0	2.0	3.0
Magnesium stearate	0.75	1.00	1.25
Purified talc	4.5	5.0	5.5
Stearic acid	0.5	1.0	1.5
Sodium stearyl fumerate	0.5	1.0	1.5

Binder	−	0	+
Alginic acid	1.0	2.0	3.0
Gelatin	4.0	5.0	6.0
Hydroxy propyl cellulose	2.0	3.0	4.0
Hydroxyl propyl methylcellulose	2.0	3.0	4.0
Povidone	1.0	2.0	3.0
Pregelatinised starch	7.5	10.0	12.5

Diluents
Lactose monohydrate
Maize starch
Micro crystalline cellulose
Mannitol
Pregelatinised starch

and compressibility of a powder for capsule filling can be characterized by the Carr's Index (CI):

$$CI = \frac{(TBD - LBD)}{TBD} \times 100$$

whereby TBD is the tapped bulk density and LBD is the loose bulk density. In Table 2.4 the CI values are related to the flow characteristics. The CI of formulations for capsule filling should target a CI of between 18 and 35 to allow sufficient fill homogeneity and plug formation.[17] The CI also serves for the calculation of the required capsule size, taking into account that

Table 2.4   Carr's Index and flowability classification of powders.

Carr's Index	Flowability
1–10	Excellent
11–15	Good
16–20	Fair
21–25	Passable
26–31	Poor
32–37	Very poor
>38	Extremely poor

additional slight densification in the filling machine might occur. It should also be noted that the capsule body should only be filled up to about 90% to prevent powder leakage and other mechanical issues during capsule closing.

Besides powder flow, the homogeneity of the blend throughout the process needs to be investigated. As a rule of thumb, particle size of all components should be between 10 and 150 μm and anisometric particles (*e.g.* needle-shaped) should be avoided or addressed by grinding or wet granulation.[18] Attention should also be given to the potential risk of powder segregation, which might occurs as a result of differences in particle density, size, shape, surface properties, particle friction and other physical attributes of the particle and the blend.[19] The investigation into powder homogeneity starts with determining mixing time required for optimal homogeneity.[20] Powder segregation during capsule filling can be evaluated by using near-infrared (NIR) spectroscopy and multivariate analysis.[21]

A special application of hard capsules filled with an interactive powder mixture is their use for pulmonary drug delivery. Dry powder inhalation products are characterized by API with a small particle size of 2–5 μm that need to be delivered to the deep lung through oral inhalation. As these small particles are cohesive, they are blended with coarse carrier particles (*e.g.* lactose) with a particle size of 90 μm or a mixture of smaller and larger carrier particles (*e.g.* a fraction of 30 μm and 90 μm). The fine drug particles attach to the coarse particles during the mixing to form a sufficiently fluid powder blend for capsule filling. The development of dry powder formulations for inhalation focuses on the interaction between the drug and the carrier to achieve the right balance between surface adhesion required for manufacture and the detachment forces required to separate the small drug particles from the carrier during inhalation. Formulations for dry powder inhalation products are typically in the mg dose range and are filled as a free flowing powder, without densification into the capsules. Compared with alternative non-metered inhalation systems, capsule-based DPI systems can be manufactured on standard capsule filling machines and integrated into commercially available capsule inhalation devices.[22]

2.4.2   Highly Potent and Low-dose Drug Formulations

According to a market report published by Transparency Market Research[23] the global highly potent drug market was worth USD 9.1 billion in 2011 and

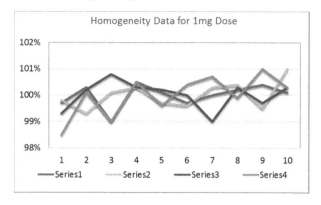

	Batch			
Sample No.	1	2	3	4
% RSD	0.43%	0.52%	0.52%	0.76%

Figure 2.4 Filling accuracy for low-dose drugs based on a 1 mg liquid formulation product.[24] Coustesy of Capsugel, Morristown, NJ, USA.

is expected to reach USD 17.5 billion in 2018, growing at a compound annual growth rate of 9.9% from 2012 to 2018, with the prevalence of oncology products in the pharmaceutical pipeline being a key driver. The liquid-filled hard capsule (LFHC) process renders it ideal for high potency drug manufacture. Once the API has been wetted following addition to a liquid excipient in the mixer, the potential for airborne and accidental exposure is greatly reduced. This is in contrast with other oral solid forms, where much more stringent and costly control is required during the preparation of the final dosage forms directly from powders, such as tableting and powder filling.

Uniformity of dose poses the most significant challenge for the formulation of low-dose products (5 mg or less) throughout development, validation and commercial manufacture. However, once a liquid solution is prepared, dose uniformity is assured through the excellent weight control attainable during filling (mostly <1% RSD) (Figure 2.4). This provides a simple solution for assuring uniformity for low-dose applications which, aligned with highly potent compounds, are increasingly prevalent in pharmaceutical development pipelines.

2.4.3 Enhancing Bioavailability of Poorly Aqueous Soluble Drugs

According to the Developability Drug Classification System (DCS), which is an evolution of the original Biopharmaceutical Classification System[25] regarding pharmaceutical drug development, drugs are classified based on their aqueous solubility and permeability.[26] Poor aqueous solubility can result in insufficient oral bioavailability of DCS 2a/b (poor solubility/high permeability) and DCS 4 (poor solubility/poor permeability) drugs. Poor aqueous solubility can be attributed to either a high lattice energy, characterized by a high melting point, or a high lipophilicity indicated, *e.g.*, by a high Log P value.

For compounds that are dissolution-rate-limited (DCS class 2a), drug particle size reduction can increase the dissolution rate, and therefore bioavailability, through an increase in the surface area to mass ratio. The dissolution rate of a compound is described by the Noyes–Whitney equation.

$$\frac{\mathrm{d}C}{\mathrm{d}t} = \frac{AD\left(C_s - C\right)}{h}$$

where $\mathrm{d}C/\mathrm{d}t$ is the dissolution rate, A is the surface area of the solute, D is the diffusion coefficient of the compound, C_s is the solubility of the compound in the solvent, C is the concentration of the compound in the solvent at a time t and h is the thickness of the diffusion layer. Based on this equation, increasing surface area and interaction with the solvent (*e.g.* wetting agent) can increase the dissolution rate.

For compounds in DCS class 2b or 4, improvement in dissolution rate alone is not sufficient to achieve adequate bioavailability. For such compounds, formulation of the drug compound in a high-energy form, such as a high-energy salt, a liquid lipid solution, or a solid amorphous dispersion, can markedly improve the bioavailability by increasing the dissolved drug levels above the crystalline solubility.

2.4.4   Physical Modification of the Drug Substance

2.4.4.1   *Particle Size Reduction*

As mentioned above, dissolution rate enhancement by particle size reduction is based on increased surface area of the drug crystals and corresponding enhanced exposure to the dissolution media. Particle size reduction can be achieved by milling using a variety of milling techniques (*e.g.* jet milling, pin milling, hammer milling, colloid milling, wet milling *etc.*) and particle classification (*e.g.* air swept screening, pneumatic screening or vibratory screening).

The micronized particles are formulated similarly to the standard powder formulation. To stabilize the supersaturated solution of the compound in the media, crystallization inhibitors can be added to the formulation (*e.g.* HPMC). Crystallization inhibition can also be achieved by selection of an HPMC capsule to enhance the bioavailability of a drug compound. For example, dabrafenib is a poorly water-soluble drug that is formulated as a high-energy crystalline mesylate salt form in TG-HPMC capsules containing the micronized drug, microcrystalline cellulose, magnesium stearate and silicon dioxide (Tafinlar®). A supersaturated solution of the high-energy salt form is observed during an *in vitro* dissolution test, which correlates with an enhanced bioavailability of the drug *in vivo*.[8] Wetting agents can also be used in conjunction with precipitation inhibitors to increase the dissolution of the compound. Aprepitant (Emend®) is micronized to the 0.12 µm particle size and incorporated into a formulation of sucrose, microcrystalline cellulose, HPMC and sodium dodecyl sulfate.

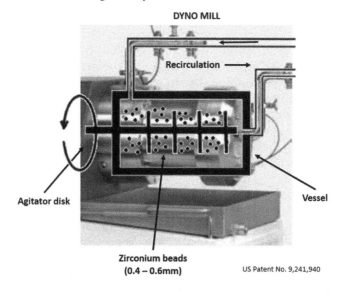

Figure 2.5   Nano milling process of a typical Dyno® Mill. Coustesy of Capsugel, Morristown, NJ, USA.

Milling in the presence of a non-aqueous medium provides a further method of size reduction. This has the advantages of reduced propensity for particle re-agglomeration, maintaining homogeneity of the mix, in particular direct filling of the milled content together with the non-aqueous medium into hard capsules, thereby minimizing process steps. *In situ* size reduction followed by direct capsule filling provides an effective means of enhanced drug bioavailability with minimal cost[27] (Figure 2.5).

2.4.4.2   Solid Amorphous Dispersions (ASDs) by Spray Drying

Amorphous solid dispersions are typically mixtures of drug molecularly dispersed within an amphiphilic or hydrophilic polymer matrix. Such dispersions enhance the dissolution rate and increase the dissolved drug levels above the drug's crystalline solubility by removing the strong cohesive forces of the drug crystals. A common method of manufacturing ASDs is to spray dry drug and excipient(s) from a common good solvent. Typical dispersion polymers may include HPMC, HPMCAS, poly(vinylpyrrolidinone) (PVP), poly(vinylpyrrolidinone co-vinyl acetate) (PVPVA) and methacrylates (various Eudragits™), as well as alternative or additional excipients (*e.g.* amino acids). The spray drying process results in rapid drying of the drug–excipient mixture, giving a homogeneous molecular dispersion of the drug and excipient(s).[28] The solubility enhancement of a drug compound in its amorphous form is based on the high free energy of the dispersed form. This high free energy can make amorphous dispersions prone to nucleation and the growth of the more stable crystalline form, either in the solid state, or in the liquid use environment. Through rational selection of the polymer

and spray conditions, and taking into account the physicochemical properties of the compound, stable amorphous dispersions can be designed and
developed.[28,29] Spray-dried particles can be filled into hard capsules. The
properties of the ASD, driven by the nature of the drug, and the dispersion
excipient, dictate what additional excipients are used, and how they are optimally incorporated for best performance. For example, for the BCS 2 drug
tacrolimus, a diluent (*e.g.* lactose) a disintegrant (*e.g.* croscarmellose sodium)
and a lubricant (*e.g.* magnesium stearate) (Prograf®) are added. For other
formulations an osmogen, such as salt or sugars, can be added to prevent
gelation of the ASD particles prior to capsule shell dissolution and dispersal
of the ASD particles.[30]

HPMC has favorable drug solubilization properties, and HPMC comprising
the shell of HPMC-based capsules can aid in solubilization of poorly aqueous
soluble drugs through specific drug–polymer interactions to sustain supersaturated concentrations of drug.[31–33] This drug supersaturation can result from
use of high-energy drug forms, such as ASDs, but also from high energy salts,[8]
and lipid-based formulations, such as self-emulsifying drug delivery systems
(SEDDS).[34] Likewise, HPMC-based capsules can also sustain supersaturated levels of drug that result from ionization and extensive dissolution of weakly basic
drugs in the low-pH gastric environment as they transit to the higher pH of the
intestine and revert to the lower solubility neutral form. By taking advantage of
the solubilizing characteristics of the HPMC in the capsule shell of HPMC-based
capsules, the solubilizing performance requirements for the drug formulation
to be encapsulated can be relaxed, perhaps allowing omission of solubilizing
excipients in the formulation, thus allowing a higher drug loading in the capsule.

Spray drying has been successfully applied in pulmonary delivery of large
molecules to the systemic circulation as well as of high-dose drugs to the lung.
Pulmonary delivery takes advantage of the high drug loading and the aerodynamic particle shape and size achieved by the spray drying process. Spray
dried insulin in a glassy sugar matrix was formulated using a mixture of mannitol, glycine, sodium citrate and sodium hydroxide to achieve the desired
drug stability, particle and surface morphology and dispersibility.[35] The
insulin formulation was approved (Exubera®) and came to the market, but
was discontinued later on for marketing reasons. Using an emulsion-based
spray drying process, inhalation particles were developed to deliver a high
dose of tobramycin to patients with cystic fibrosis (TOBI Podhaler®).[36] With
the evaporation of the oil phase (Perflubron) a porous particle morphology
was achieved, providing the desired aerodynamic properties to the particles.[37] Meanwhile successful development of spray dried formulations for
vaccines (*e.g.* H1N1) has been reported and these are expected to advance to
the commercial stage.[37]

2.4.4.3   Hot-melt Extrusion (HME)

To increase solubility and dissolution rate of poorly water-soluble compounds, hot-melt extrusion is employed to embed the drug in an amorphous
form or as a nano-dispersion in a hydrophilic carrier matrix. In contrast to

spray drying, hot-melt extrusion is a solvent-free technology based on the miscibility of drug and carrier in their molten stage at higher temperatures followed by rapid cooling to form solid amorphous dispersion.[38] The hot-melt extrusion is a continuous process where the drug and excipients are fed into an extruder which convey, meter, mix and melt the components and form a solid matrix.[39] Passing the molten material through a narrow die orifice and cutting the strands at high speed leads the formation of round pellets that solidify rapidly.[40] Beside the drug and the carrier, HME formulation can include other excipients, like disintegrants, plasticizer, antioxidants and thermal lubricants, to increases stability, processing or release.[39] Due to the pelletized form of these extrudates, they can be filled in hard capsules without any additional excipients. Nabilone, a synthetic cannabinoid for the treatment of chemotherapy-related nausea and vomiting has been formulated with PVP as a carrier and corn starch as a swelling agent to support fast dissolution of the active component (Cesamet®).

2.4.4.4   Lipid-based Formulations

Size reduction and amorphous dispersions *via* spray drying and HME proved to be effective in overcoming poor solubility due to the high lattice energy of drug crystals. Lipid-based formulations are most appropriate in providing solutions to overcome the challenge of poor solubility due to moderate to high lipophilicity. In addition to enhancing intestinal solubility of the drug, lipid based formulations can additionally increase drug exposure through increased permeability and the avoidance of first pass metabolism.

Feeney *et al.*[41] provides a comprehensive overview of the development of lipid-based formulations over the last 50 years. In particular, the Lipid Formulation Consortium advanced the science and practice of lipid-based formulations through the classification of these formulations and their evaluation by standardizing the *in vitro* evaluation process (Table 2.5).

Fundamentally lipid based formulation for drug delivery take advantages of the ability of human body to digest and absorb lipids as a nutrient. Triglycerides are hydrolysed into mono- and di-glycerides and free fatty acids on digestion and through the passage to the lower gastrointestinal tract. The digested species are then combined with biliary secretions to form a multitude of colloidal species from coarse/micro-emulsions, multilamellar/unilamellar vesicles and mixed micelles. It is these colloidal species that are utilized to facilitate drug solubilisation and absorption across the biological barriers.

The lipid formulation classification alongside the standardized digestive modelling of the formulations provide some clarity for the formulators in formulation design and selection. The lipid-rich Type 1 formulations contain the highest amount of triglycerides and have a low hydrophilic–lipophilic balance (HLB) value, and require digestion to increase amphilicity and dispersion into the intestinal fluids, whereas Type IV formulations contain high HLB and hydrophilic co-solvents which disperse into micellar solution. Because of the higher hydrophilicity, Type III (especially IIIB) and IV are

Table 2.5 The lipid formulation classification system [adapted from Pouton 2000 [42] and 2006 [43]].

| Formulation classes | Type I | Type II | Type III | | Type IV |
			Type IIIA	Type IIIB	
Phase	Oil	SEDDS	SEDDS	SMEDDS[a]	Oil-free
Composition	Pure oil, no surfactant	Water-in-soluble components	Includes water-soluble surfactants and possible co-solvents		Comprises only water-soluble surfactants and co-solvents
Oil, glycerides	100	40–80	40–80	<20	0
HLB < 12 (%)	0	20–60	0	0	0–20
HLB > 12 (%)	0	0	20–40	20–50	30–80
Co-solvents	0	0	0–40	20–50	0–50
Phase structure	Limited or no dispersion	Rapidly dispersing	Rapidly dispersing	Transparent dispersion	Micellar solution
Particle size (nm)	Coarse	100–250	100–250	50–100	1–50
Metabolism	Requires digestion	Likely to be digested	Digestion may not be necessary		Limited digestion

[a]SMEDDS, self-micro-emulsifying drug delivery systems.

Table 2.6 List of selected excipients used in LFHC formulation.

Medium chain trigelycerides	Labrafac, Miglyol, Crodamol
Long chain triglycerides	Soya bean oil, maize oil, sesame oil, hydrogenated vegetable oil
Medium and long chain monoglycerides and diglycerides	Capmum, Peceeol, Maisine, Compritol
Pegylated glycerides, medium and long chain, monoglycerides, diglycerides and triglycerides	Labrasol, Labrafil M-1944; Labrafil M-2125, Gelucire 44, 48, 50
Pegylated castor oil derivatives	Kolliphor RH40, Kolliphor EL
Sorbitan esters	Polysorbates (Tweens) and Spans

preferred vehicles for increased drug loading. For the same reasons, there is a greater propensity for the API to recrystallize when dispersed into the gastrointestinal fluids.

The modern day pharmaceutical formulators have both the luxury of the availability of numerous lipid excipients and, at the same time, the challenge of designing and selecting the optimal formulations for the drug molecule from the large range of excipients. Lipids excipients can be categorized based on their chemical classes: natural oils and fats, fatty acids, mono- and di-glycerides; polyethylene glycol derivatives of glycerides and fatty acids; sorbitan derivatives; polyglycerol fatty acid esters; cholesterol and phospholipids (Table 2.6).

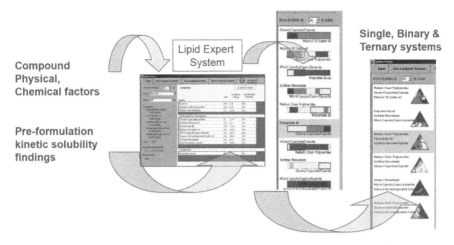

Figure 2.6 Structure of the Lipidex software for developing liquid formulations. Coustesy of Capsugel, Morristown, NJ, USA.

It is worth noting that the preparation of the semi-synthetic excipients involves esterification and trans-esterification from natural oils and substrates such as polyethylene glycol, sorbitol, glycerol and propylene glycol. Therefore, the final composition of the excipient is as much dependent on the input materials for synthesis as on the actual manufacturing process. Therefore, the end products from different suppliers that comply with the same monograph standards often exhibit different functional performance because of the different manufacturing processes employed. To aid formulation design and optimization, a computer-aided expert system that is based on extensive excipient database and a decision tree giving access to a database of experimentally generated phase diagrams is often employed. One such system is the Lipidex® system that uses the physicochemical properties of the API and the kinetic solubility of the API in a range of excipients as the input parameters and produces a range of formulation options for further evaluation through digestive modelling (Figure 2.6).

In addition to solubility enhancement, lipid-based formulations can enhance drug exposure through the induction of fed state, efflux and drug metabolism inhibition or saturation and lymphatic transportation.

2.5 Targeted Drug Delivery

2.5.1 Capsules with Solid Formulation

2.5.1.1 Coating of the Capsules

To target the release of a drug in the gastrointestinal tract, hard capsules can be coated with different functional coatings. The release is targeted through the selection of coating polymer and their dissolution within the physiological pH range. Alternatively timed or controlled release can be achieved through the use of an erodible coating.

A typical modified release coating consists of the coating polymer, mainly polymethacrylates (*e.g.* Eudragit®), cellulose-based coatings (*e.g.* Aquateric®) or polyvinyl derivatives (*e.g.* Coateric®), a plasticizer, such as triacetin, triethylcitrate or others, as well as talc, magnesium stearate or silicon dioxide as antiadhesion agents. The coating of gelatin capsules might require an undercoat to improve the adhesion of the coating to the capsule surface as well as to prevent the uptake of water from the aqueous coating solutions or suspensions.

Using a film composition of Eudragit L 30 D-55 and triethylcitrate for intestinal targeting and a film composition of Eudragit FS 30 D, triethylcytrate, glycerol monostearate and Tween 80 for delivery in the ilium or proximal colon, a human *in vivo* study confirmed that the *in vitro* dissolution results (dissolution at pH > 6, 8 and pH > 7.2 respectively) correlated with the *in vivo* disintegration in the small intestine and past the mid small intestine in the ilium region.[9]

2.5.1.2 Modified-release Formulation in Hard Capsules

Modified-release multiparticulate forms like pellets, granules or mini-tablets are traditionally filled into hard capsules without any additional excipients. Filling multiparticulates in hard capsules provides some important advantages over alternative dosage forms like tablets:

- Avoiding the compression of multiparticulates
- Reduced formulation and process complexity
- Flexibility of dosing and dose strength development
- Product and dose differentiation by color and imprint
- Ease of fixed dose combination or dual-release products

Omeprozole was one of the first such products launched in 1988 in Europe and in 1989 in the USA. Omeprazole required enteric protection and fast release in the upper intestinal tract. Due to the pH-sensitivity and fast degradation of omeprazole below pH 4 (<10 min), a multilayer pellet was developed that contained the drug, a buffering layer and the enteric protective coat. This concept was further extended by a dual release formulation of dexlansoprazole, a drug with a biological half-life of about 1 hour, to achieve a once daily dosing regimen (Dexilant®). The product contains two populations of pellets, one population releasing the drug at a lower duodenal pH and one population releasing the drug at a higher distal intestinal pH. The *in vivo* studies confirmed two plasma peaks, one at 1–2 h and one at 4–5 h post dosing.[44]

Diltiazem is a drug with a high first-pass metabolism, a biological half-life of 3.0–4.5 h and a bioavailability of about 40%. To reduce the complexity of developing multiple dose strengths of an extended-release product, hard capsules have been used to accommodate different doses of diltiazem. Mini tablets containing 60 mg of diltiazem were developed and filled in hard capsules to achieve 120 mg, 180 mg and 240 mg doses (Dilacor XR®). The dosage form provides an extended release over 24 h *in vivo* with a time to maximum plasma concentration (T_{max}) after 4–6 h. The biological half-life under steady-state conditions shifted to 5–10 h, providing the desired plasma profile for once-daily dosing.[45]

Especially for more sophisticated pharmaceutical product design, hard capsules offer the possibility of combining different types of formulations, two or more active drugs and various release profiles. For example, modified-release pellets of tamsulosin have been combined with a soft gelatin capsule of dutasteride (Duodart®) or modified-release pellets of dipyridamole have been combined with an immediate-release tablet of acetylsalicylic acid (Aggrenox®).

2.5.1.3 Modified-release Formulations Based on Liquid-filled Hard Capsules

Modified-release profiles can be attained using the liquid fill formulation approach. The formulation design typically consists of formulating a matrix structure by combining a thermal softening or waxy component that is typically a solid or semi-solid at room temperature but becomes a liquid at a temperature above its melting point; and a hydrophilic polymeric material that is very soluble and/or gelling when in contact water.

The mechanism of release can either be erosion-driven, if an intermediate-HLB waxy material and a highly soluble polymer are employed, or diffusion-driven, if a low-HLB waxy material and a highly soluble polymer are employed. If a medium and intermediate waxy material and a gelling polymer are combined, then a diffusion-driven process is also attained. The selection of the most appropriate matrix system will depend on the dose strength and the physicochemical properties of the active agent. For example, to achieve the sustained release of a highly water soluble active agent, a low-HLB waxy material (HLB < 6) with a soluble polymer would be most appropriate. On the other hand, a gelling polymeric matrix, an intermediate- or high-HLB material (HLB > 8) or waxy material may be most appropriate for a less soluble active substance.

The incorporation of the waxy matrix in such formulations can minimize either accidental or intentional alcohol dose dumping of a sustained-release formulation. There is a regulatory expectation that potential issue of alcohol dumping is studied and reduced during the development of modified-release products.

Given the increased recognition of the circadian and other rhythmic cycles in diseases such as psychiatric and somatic illness, it is desirable to schedule drug administration, taking into account the pharmacokinetics of the active agent to maximize effectiveness and to minimize side effects. One approach is to vary the release profiles of a single active agent to maximize pharmaceutical effects. A capsule-in-capsule technology (Duocap™) has been successfully commercialized. The technology allows the insertion of a pre-filled smaller capsule containing either a liquid or semi-liquid formulation into a larger liquid-filled hard capsule. The capsule-in-capsule approach provides a simple solution to achieve the variable release profiles. The following examples use two active substances to illustrate the concept more clearly. If a single active substance is used, then the release profile will be cumulative of the two. Other profiles can be designed to achieve the target product profiles.

Example 1, the larger LFHC is released immediately, whereas the small inner capsule is released with a sustained release profile to maintain a desired therapeutic level. Example 2, the larger LFHC is designed to release immediately, whereas the small inner capsule is delayed. The capsule-in-capsule approach incorporating LFHC technology is ideal to derive such release profiles.

2.5.2 Abuse Deterrent

Pharmaceuticals that affect the central nervous system with indications for pain, anxiety, depression and hyperactivity are often the targets of misuse and abuse, and are increasingly regulated. Efforts to reformulate some opioids have been ongoing for more than two decades. Recently, the U.S. Food and Drug Administration issued a guidance document on the evaluation and labeling of abuse-deterrent opioids in recognition of the problem of widespread opioid abuse.[46]

One development approach is to focus on formulations that physically limit the ability of drugs to be mechanically or chemically modified for the purpose of injection, insufflation or rapid oral absorption. For example, this includes sustained pain relief formulations that can be extracted with alcohol or other solvents using common kitchen chemistry techniques in order to induce a high. This type of extractability can be substantially reduced with a high-melting-point wax-based matrix formulation, which maintains the slow release characteristics (for example, 85% release over 24 hours). Liquid-filled hard capsule technology is a viable route for processing such a formulation since mixing and filling can be completed at a temperature exceeding the melting point of the designated waxy component.

In a single-center, randomized, analytically masked, fasted five way cross-over study under naltrexone block in the USA, Levorphanol immediate release and four extended release prototype formulations were administered to 15 healthy volunteers during each study period. Twenty-one sequential blood samples were obtained during each dosing period over 48 h. There was a 7–14 day washout period between each of the five dosing-periods. Plasma samples were analyzed using a fully validated and robust LC-MS–MS method. All four Levorphanol extended release capsule formulations demonstrated robust extended release characteristics suitable for once-a-day dosing. For the four extended-release formulations, the mean T_{max} ranged from 9.15 to 12.29 h *vs.* 2.40 h for the immediate-release formulation; the ratio (%) of dose-normalized extended-release to immediate-release C_{max} ranged from

26.7% to 40.9%; and the ratio of dos- normalized extended-release to immediate-release $AUC_{infinity}$ ranged from 82.16% to 99.27% 9.[47]

Since many abuse deterrent formulations are used in second- and third-generation generic products, bioequivalence to the Reference Listed Products that are in most cases, immediate-release formulations will be required (for example, 85% release over 60 min). To achieve this, formulations incorporating a liquid carrier and hydrophilic polymers are designed to achieve a careful balance of high viscosity (to reduce syringeability, for example) and sufficient mobility to allow high-speed filling in the liquid fill hard capsule process.

2.6 Manufacture of Commercial Hard Capsules Products

The filling of hard capsules is an established technology, with equipment available ranging from that for very small scale manual filling (1–100 capsules), through intermediate-scale semi-automatic filling (10 000–35 000 capsules) to large-scale fully automatic filling (>200 000 h).

The principles of filling hard capsules manually as well as using automatic equipment include the orientation of the capsules, opening of the capsule from its pre-locked position, the filling of the body, the final closing with the cap and the ejection of the filled and closed capsule. The fully automated capsule filling machines can be equipped with multiple filling stations to manufacture hard capsule products with more than one fill formulation. At commercial scale, the principles of the filling operation depend on the type of fill materials:

- Powder and granule blends
- Multiparticulates and mini-tablets
- Tablet filling
- Liquid and semi-solid formulations
- Powder micro-dosing

2.6.1 Powder and Granule Blends

The simplest form of filling powder or granules in hard capsules is through gravimetric filling of the powder into the body part. While this principle is used by the manual filling devices, semi-automatic machines support the gravimetric filling by an active powder transport using a screw-tube conveyor. This principle of the auger filling process is shown in Figure 2.7.

For fully automatic filling machines two types of filling principles are being used. The dosing disk type filling is based on the gravimetric flow of powder into dosing chambers that are tamped five times to build up a plug, which is transferred into the capsule body (Figure 2.8).

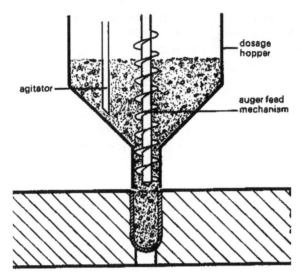

Figure 2.7   Auger filling principle for hard capsules. Coustesy of Capsugel, Morristown, NJ, USA.

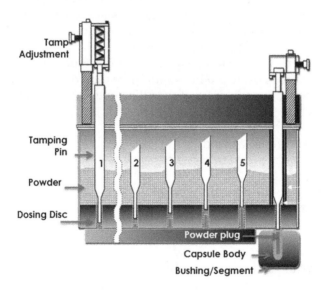

Figure 2.8   Capsule filling with the dosing disk filling principle. Coustesy of Capsugel, Morristown, NJ, USA.

The fill weight is determined by the height of the dosing disk as well as by powder densification. The dosator type filling is based on a powder bed of a set height in which dosators strike in and densify the powder through a piston movement to form a powder plug which is transferred and ejected into the capsule body (Figure 2.9).

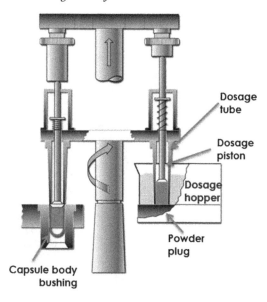

Figure 2.9   Capsule filling with the dosator filling principle. Coustesy of Capsugel, Morristown, NJ, USA.

Both filling principles can handle a wide range of formulations with less strict requirements compared with tablets. As mentioned above, the formulation flow characteristics should have a CI in the range of 18–35, should form a stable plug upon slight densification (plug strength approximately 1 N) and sufficient lubrication to reduce the ejection force of the plug from the dosator or dosing disk.[17]

2.6.2   Multiparticulates and Mini-tablets

Multiparticulates are free-flowing materials that can be filled *via* volumetric filling (Figures 2.10 and 2.11) and dosator filling using aspirational air flow (Figure 2.12).

Volumetric filling can be achieved in a number of ways. The dosing disk method can be adapted in a way that the filled dosing chambers are not tamped and the filled dosing chamber is transferred above the capsule body, where a sliding plate opens the chamber to release the multiparticulates. Similarly, the dosator filling principle is adapted, whereby the dosator strikes in the multiparticuate bed to vacuum transfer the multiparticulates by air and transfer them to the capsule body.

In the double-slide method a chamber is opened and closed by a moving plate. A second plate then opens and closes at the bottom of the chamber to fill the metered pellet dose into the capsule.

The piston dosing systems are based on the gravimetric principle where a dosing piston is underneath the multiparticulate bed. The dose

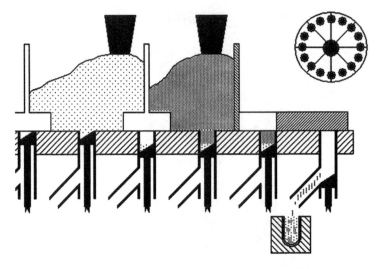

Figure 2.10   Multiparticulate volumetric chamber filling principle for multiple products. Coustesy of Capsugel, Morristown, NJ, USA.

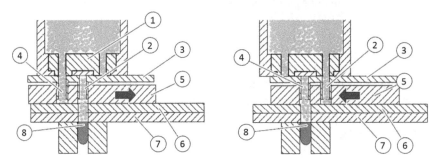

1. Dosing head
2. 1st Dosing chamber
3. Upper dosing disk
4. 2nd Dosing chamber
5. Lower dosing disk
6. Transfer disk
7. Filling disk
8. Capsule body

Figure 2.11   Pellet dosing based on volumetric filling by the moving dosing disk principle. Courtesy of Bosch Packaging, Waiblingen, Germany.

is adjusted by the moving piston and a mechanical closure before the piston moves further down to reach to open a channel through which the multiparticulates flow into the capsule body. This principle is also used for the filling of two different multiparticulates whereby the first filling and closing of the piston is followed by lowering of the piston to collect second filling before moving to the release stage of the multiparticulates into the capsule body (Figure 2.10). In the dosing disk principle the dosing chamber is closed at the bottom and gravimetrically filled with the pellets.

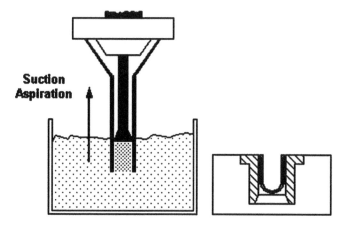

Figure 2.12 Multiparticulate dosator capsule filling principle.

When the dosing disk moves the upper end closes und the dosing chamber moves directly above the capsule body to release the pellets (Figure 2.11). Filling of multiparticulates by dosator-type machines is achieved by the dosator driving into the pellet bed and sucking in the pellets by vacuum (Figure 2.12).

2.6.3 Tablet Filling

To fill tablets into capsules, the tablets are fed in a tube to align in rows. The tablet on the bottom enters into a sliding chamber that is moved sidewise to the open body part and inserted by gravimetric methods or assisted by a downwards moving pin.

2.6.4 Liquid and Semi-solid Formulation

Liquid and semi-solid formulation are filled through piston pump systems. The filling of a liquid or semi-solid formulation is dependent on the viscoelastic properties of the formulation and the need to fulfill certain characteristics at the filling temperature. As a general rule, the formulation should have a viscosity of between 50 and 1000 Centipoise (cP) (although formulations of much higher viscosity can be suitable for manufacturing) and should not exceed 70 °C. The particle size in suspension should be ideally be less than 20 µm and formulations should be such that no stringing, dripping, splashes or solidification of the formulation should occur at the dosing nozzle. Unless a hot-melt is filled that completely solidifies below 40 °C, hard capsules are recommended to be band or fusion sealed using separate band sealing or LEMS® sealing equipment. For research purposes, a machine that is capable of filling and sealing 1500 capsules an hour (*e.g.* CFS® 1500) has been developed.

2.6.5   Powder Micro-dosing

Filling of micro-doses of powder into hard capsule is a challenge, especially when these powders are cohesive and cannot be densified to maintain their dispersibility, as is required for inhalation products.

Dosator type filling machines can be equipped with special dosators that can fill accurately doses of 5–15 mg upwards into hard capsules, depending on powder characteristics. To increase the accuracy and micro-dosing capability, especially of cohesive powder mixtures, capsule filling machines with vacuum drum filling can achieve fill weights of 1 mg. The vacuum drum filling consists of a drum with bores of fixed or adjustable volume and the powder bed above. A vacuum is applied to confine the targeted dose into the bores before the drum turns downside to release the powder into the capsule body.

For research purposes micro-dosing equipment exists that can fill lab-scale quantities. The automated machines use gravimetric vibrational technology to fill the capsule until the targeted weight is reached starting in the microgram range with a fill weight accuracy of less than ±1% (*e.g.* Xcelodose®). Such micro-dosing systems are used for first-in-human and phase 1 clinical trials whereby the drug substance is filled into the capsule without the addition of any supplementary excipients.

Filling machines for commercial hard capsule products are provided by several companies and are constantly being improved and advanced in their capability. These filling machines offer a broad range of possibilities in terms of capacity, speed, PAT controls and flexibility. In Table 2.7 the major filling machine manufactures are listed with a selected list of their machine types.

2.7   Conclusions

Hard capsule formulations, in their simplest form, contain the drug and a lubricant (*e.g.* Amoxil® Capsules 500 mg). Manufacturing of such hard capsule products consist of weighing, sieving, mixing and filling. Additional components, like diluents, disintegrants or wetting agents, might be added to the blend to optimize product performance. The hard capsule is an important dosage form in advanced drug delivery. Hard capsules have been developed that provide modified release properties that do not require additional coating. In addition, hard capsules can be filled with a range of targeted or bioavailability enhancing formulations in the form of a powder blend (*e.g.* micronized particles), multiparticulates (*e.g.* spray dried particles, pellets, granules), solid dispersions (*e.g.* hot-melt amorphous dispersions), small solid dosage forms (*e.g.* mini-tablets) or liquid fill formulations that provide a wide range of applications in terms of modified and targeted delivery and bioavailability enhancement as well manufacturing of highly potent active agents. For pulmonary drug delivery, hard capsules provide an economically viable and worldwide accessible mono-dose delivery platform, which is used for local lung therapy as well as for the systemic administration of large molecules.

Table 2.7 List of major filling machines for commercial manufacturing of capsule products.

Machine	Dosing principle	Output (cps h⁻¹)	Products to be filled
Bosch (http://www.bosch.com)			
GKF 701	Dosing disk	42 000	Powder, pellets, tablets, liquids
GKF 1400	Dosing disk	84 000	Powder, pellets, tablets, liquids
GKF 2500	Dosing disk	150 000	Powder, pellets, tablets
GKF 3000	Dosing disk	175 000	Powder, pellets, tablets
GKF 705	Dosing disk	42 000	Powder, pellets, granules
GKF 1505	Dosing disk	92 000	Powder, pellets, granules
GKF 3005	Dosing disk	175 000	Powder, pellets, granules
Harro Höfliger (http://www.hoefliger.com)			
Modu-C high speed	Multi functional	200 000	Powder, pellets, granules, tablets, paste, liquids, micro-dose, micro-tablets, capsule-in-capsule
Modu-C Mid speed	Multi functional	100 000	Powder, pellets, granules, tablets, paste, liquids, micro-dose, micro-tablets, capsule-in-capsule
Modu-C low speed	Multi functional	25 000	Powder, pellets, granules, tablets, paste, liquids, micro-dose, micro-tablets, capsule-in-capsule
IMA (http://www.ima.it)			
Zanasi 6/12 25/40	Dosator	40 000	Powder, pellets, liquids, tablets
Zanasi 8/16	Dosator	16 000	Powder, pellets, liquids, tablets, micro-tablets
Zanasi plus	Dosator	85 000	Powder, pellets, liquids, tablets, micro-tablets
Adapta	Multi functional	100 000	Powder, pellets, liquids, tablets, micro-tablets
Imatic 100–200	Dosator	200 000	Powder, pellets
MG2 (http://www.mg2.it)			
Alterna	Dosator	70 000	Powder, pellets, tablets
Suprema	Dosator	48 000	Powder, pellets, micro-tablets
MG Compact	Dosator	48 000	Powder, pellets, tablets, micro-tablets, micro-dose
G70 140	Dosator	140 000	Powder, pellets, tablets, micro-tablets, micro-dose
Planeta 100	Dosator	100 000	
G 100	Dosator	90 000	Powder, pellets, liquids, tablets, micro-tablets, capsule in capsule, micro-dose
G 250	Dosator	200 000	Powder, pellets, tablets, micro-tablets, micro-dose
Romaco-Macofar (http://www.romaco.com)			
CD40	Dosator	40 000	Powder, granules, pellets, tablets
CD60	Dosator	60 000	Powder, granules, pellets, tablets

For the increasing demand for patient-centered drug product design, hard capsules provide the necessary features to overcome swallowing issues (*e.g.* sprinkle capsules), medication errors (*e.g.* product differentiation through bi-chromatic color and imprint) as well as the increasing personalization of medicines (*e.g.* fixed dose combinations and sub-doses). In addition, the recent and continuing focus in medical and pharmaceutical sciences to address unmet medical needs and therapeutic areas that only affect a small number of patients as well as the trend to strengthen local production, flexibility and transferability of pharmaceutical manufacturing are becoming important criteria in future healthcare provision. Due to the relative simplicity of the manufacturing process for hard capsule products that require just mixing and filling unit operations, and the large range of fill formulations that can address the numerous challenges of modern pharmaceutics, such as low drug solubility, poor bioavailability, sustained and targeted release profiles, abuse deterrent formulations, highly potent drugs, low dose drugs, speed to clinic and speed to market, the hard capsule dosage form will play an increasingly important role in future drug product development.

Acknowledgements

The authors would like to thank Jane Fraser for her valuable scientific input and review of the book chapter.

References

1. F. M. Feldhaus, *Dtsch. Apoth. Ztg.*, 1954, **94**, 321.
2. L. A. Planche and F. Gueneau de Mussy, *Bull. Acad. R. Med. Belg.*, 1837, 442–443.
3. S. L. M. Dorvault, *L'officine ou répertoire général de pharmacie pratique*, 1923, p. 504.
4. A. Colton, U.S. Patent 1 787 777, 1931; A. Colton, British Patent 360 427, 1931.
5. S. Stegemann, in *Developing Drug Products in an Aging Society. AAPS Advances in the Pharmaceutical Sciences Series*, ed. S. Stegemann, Springer, New York, 2016, ch. 4.1, vol. 24, pp. 191–216.
6. *Technical Reference File*, http://www.capsugel.com/media/library/Coni-Snap_brochure_full.pdf.
7. http://www.capsugel.com/consumer-health-nutrition/products/plantcaps-capsules.
8. D. Ouellet, *et al.*, *J. Pharm. Sci.*, 2013, **102**, 3100–3109.
9. E. T. Cole, *et al.*, *Int. J. Pharm.*, 2002, **231**, 83–95.
10. S. Stegemann, *et al.*, *Int. J. Pharm.*, 2017, **517**, 112–118.
11. S. Stegemann, *et al.*, *AAPS PharmSciTech*, 2014, **15**(3), 542–549.
12. FDA, *Guidance to Industry – Expedited Programs for Serious Conditions – Drugs and Biologics*, 2014, UCM 358301, https://www.fda.gov/downloads/Drugs/Guidances/UCM358301.pdf.

13. *Guideline on the Scientific Application and the Practical Arrangements Necessary to Implement Commission Regulation (EC) No 507/2006 on the Conditional Marketing Authorisation for Medicinal Products for Human Use Falling within the Scope of Regulation (EC) No 726/2004; EMA/CHMP/509951/2006, Rev.1*, 23 July 2015.
14. *Guideline on the Scientific Application and the Practical Arrangements Necessary to Implement the Procedure for Accelerated Assessment Pursuant to Article 14(9) of Regulation (EC) No 726/2004; EMA/CHMP/697051/2014-rev. 1*, 23 July 2015.
15. G. Shlieout, *et al.*, *AAPS PharmSciTech*, 2002, **3**(2), 45–54.
16. K. S. Murthy, *et al.*, *J. Pharm. Sci.*, 1977, **66**, 1215.
17. P. K. Heda, PhD Thesis, University of Maryland, 1998.
18. J. Hogan, *et al.*, *Pharm. Res.*, 1996, **13**, 944–949.
19. R. Hogg, *KONA Powder Part. J.*, 2009, **27**, 1–15.
20. H. Wu and M. A. Khan, *J. Pharm. Sci.*, 2009, **98**, 2784–2789.
21. O. Scheibelhofer, *et al.*, *AAPS PharmSciTech*, 2013, **14**, 234–244.
22. S. Stegemann, *et al.*, *Eur. J. Pharm. Sci.*, 2013, **48**, 181–194.
23. http://www.transparencymarketresearch.com.
24. P. Chong, *et al.*, *AAPS Annual Meeting*, 2010, poster.
25. G. L. Amidon, *et al.*, *Pharm. Res.*, 1995, **12**, 413–420.
26. J. M. Butler and J. B. Dressman, *J. Pharm. Sci.*, 2010, **99**, 4940–4954.
27. S. Brown, WO WO 2013057518 A1, 2013.
28. D. T. Friesen, *et al.*, *Mol. Pharmaceutics*, 2008, **5**, 1003–1019.
29. A. Paudel, *et al.*, *Int. J. Pharm.*, 2013, **453**, 253–284.
30. M. Morgen, *et al.*, *AAPS Annual Meeting*, 2015, poster.
31. D. E. Alonzo, *et al.*, *Cryst. Growth Des.*, 2012, **12**, 1538–1547.
32. M. Manne Knopp, *et al.*, *Eur. J. Pharm. Biopharm.*, 2015, **105**, 106–114.
33. D. Dhaval, *et al.*, *Mol. Pharmaceutics*, 2014, **11**, 1489–1499.
34. P. Gao and W. Morozowich, *Expert Opin. Drug Delivery*, 2006, **3**(1), 97–110.
35. S. White, *et al.*, *Diabetes Technol. Ther.*, 2005, **7**, 896–906.
36. D. E. Geller, *et al.*, *J. Aerosol Med. Pulm. Drug Delivery*, 2011, **24**, 175–182.
37. C. Zhu, *et al.*, *Pharm. Res.*, 2014, **31**, 3006–3018.
38. M. M. Crowley, *et al.*, *Drug Dev. Ind. Pharm.*, 2007, **33**, 909–926.
39. M. Repka, *et al.*, *Drug Dev. Ind. Pharm.*, 2007, **33**, 1043–1057.
40. S. Radl, *et al.*, *Chem. Eng. Sci.*, 2010, **65**, 1876–1988.
41. O. M. Feeney, *et al.*, *Adv. Drug Delivery Rev.*, 2016, **101**, 167–194.
42. C. W. Pouton, *Eur. J. Pharm. Sci.*, 2000, 2(suppl. 11), S93–S98.
43. C. W. Pouton, *Eur. J. Pharm. Sci.*, 2006, **29**, 278–287.
44. D. C. Metz, *et al.*, *Aliment. Pharmacol. Ther.*, 2009, **29**, 928–937.
45. http://www.accessdata.fda.gov.
46. FDA, *Guidance for Industry Abuse-deterrent Opioids—Evaluation and Labeling*, April 2015, https://www.fda.gov/downloads/Drugs/GuidanceComplianceRegulatoryInformation/Guidances/UCM334743.pdf.
47. N. Babul, *J. Pain Symptom Manage.*, 2004, **28**, 59–71.
48. S. Stegemann, Non-gelatin based capsules, in *Pharmaceutical Dosage Forms Capsules*, ed. L. L. Augsburger and S. W. Hoag, CRC Press, Taylor & Francis Group, Boca Raton/London/New York, 2018.

CHAPTER 3

Soft Capsules

STEPHEN TINDAL

Catalent, 14 School House Road, Somerset, New Jersey, 08844, USA
*E-mail: stephen.tindal@catalent.com

3.1 Introduction

Soft capsule is a description that covers a range of one-piece capsule drug delivery systems, all of which share the common feature that a polymer based shell material is hermetically sealed around a non-aqueous liquid fill material without any gas headspace (Figure 3.1). The dosage form may also be variously described as soft elastic (gelatin) capsules, softgels or (one-piece) liquid-filled capsules. These descriptions are intended to differentiate the soft capsule from the liquid-filled hard capsule, which at first glance, appears very similar, but has some important differences.

Soft capsules have a range of different applications, but this chapter will focus on prescription drug delivery applications, where they are a unique drug delivery system that can provide distinct advantages over traditional dosage forms such as tablets, hard-shell capsules and liquids.

As a drug delivery technology, soft capsules almost always contain the active pharmaceutical ingredient (API) in the fill material. This is partly because the yield of API from the fill is much higher (typically greater than 95% compared with 65% from the shell due to the portion of the gel that goes to waste), partly because the shell weight in each capsule is less precisely delivered than the fill and partly because any API contained in the shell will need to be stable in the gel solution, which is stored at elevated temperatures. Small doses of API may

Drug Discovery Series No. 64
Pharmaceutical Formulation: The Science and Technology of Dosage Forms
Edited by Geoffrey D. Tovey
Published by the Royal Society of Chemistry, www.rsc.org

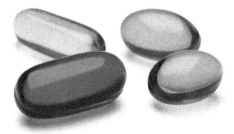

Figure 3.1 Opaque soft capsules. Image courtesy of Catalent.

be better delivered to the shell *via* spray coating to overcome these issues and the release profile may even be modified by careful selection of polymers.[13]

Irrespective of the location of the API, soft capsules are considered to be a solid dosage form.[9] When the API is contained within a solution in the fill, soft capsules provide many of the advantages of a liquid delivery system in a convenient solid dosage form.

Compared with tablet or powder-in-capsule solid dosage forms, soft capsules tend to utilize excipients that are mostly (non-water-based) liquid in nature or dissolved in liquid, and the materials tend to be different from the materials that are used to create solid dosage forms. Where the materials are used in several technologies, care may be needed to ensure that the specification is appropriate for use in soft capsules.

The soft capsule process is robust, with over 80 years of manufacturing experience and many commercial prescription products and over the counter products in the global market place. This makes soft capsules a good choice for new molecular entities, and the technology is also useful for reformulation of marketed drugs with the purpose of life-cycle extension.

When the liquid fill contains a drug in solution, and when other excipients are incorporated together to improve solubility in water or to enhance permeability, soft capsules may be able to increase systemic bioavailability or provide a different pharmacokinetic (PK) profile (such as faster uptake), which creates the main interest in the technology. Soft capsules can support modified release profiles by shell modification, coating or careful fill formulation. Soft capsules may also be used for topical or ophthalmic application (after radiation treatment to sterilize the product), for vaginal or rectal delivery and for oral delivery without systemic absorption.

This chapter reviews the key aspects of soft capsule formulation, manufacturing, and product development. A review of recent advances in this dosage form, such as non-gelatin-based soft capsules, modified-release/controlled-release soft capsules, and lipid formulation are also included.

3.2 Background

Encapsulation by hand was employed as a method of drug delivery as early as the 1830s. Initially, individual, empty capsules were created by hand dipping and later filled by hand. Early attempts to improve output involved making multiple capsules at once using plates or molds, adding the fill and sealing multiples together in a unit operation. R. P. Scherer effectively automated the process with a continuous rotary die encapsulation machine, filing a patent in 1931.[1] While the original and early patents have long since expired, there were relatively few producers or machine manufacturers in the world, due to economic and technological constraints. This has changed in the last two decades. Most encapsulation companies that exist today began with products for markets with simpler regulatory requirements, such as nutritional products, cosmetics and over the counter products. Catalent, who bought the RPScherer corporation in 1998, still dominate the global Rx (prescription) market.

3.3   Technology Strengths/Limitations

The major advantages of soft capsules for Rx application include the following:

- Improved oral bioavailability. Around 90% of small-molecule new chemical entities (NCEs) emerging from drug discovery teams will be limited by solubility, with around 20% having the further challenge of being poorly permeable.[2] By providing a drug in solution in a soft capsule fill, it may be possible to remove the rate-limiting step in a drug's uptake into the systemic circulation from the gastrointestinal fluid—namely, the need for crystalline drug substances to dissolve. In addition, some of the lipid excipients that may be employed in the soft capsule fill can also have an effect on permeability, either by permeation enhancement or by influencing drug efflux or transporter mechanisms.
- Pharmacokinetic profile changes. By formulating an NCE in a liquid formulation inside a soft capsule, and by selecting either a low or high viscosity solution, a low or high lipophilicity or by formulating a suspension, it may be possible to either advance or retard the release of the NCE.
- Excellent dose uniformity. When the API is delivered as a solution in the fill, the resulting soft capsules can benefit from very high dosing precision of the fill material (even when the dose is in the 1–5 µg range).
- Enhanced drug stability. Although there is a concern that drugs in solution are less stable than in the solid state, the normal degradation pathways for an aqueous solution may not exist inside a lipid formulation. It may be possible to avoid hydrolysis providing the API does not partition into the shell. In addition, since the shell is a superior barrier to oxygen,[17] certain drugs (such as high-potency omega 3 oils, retinoids and vitamin D analogues) that are prone to oxidative degeneration are more stable in a soft capsule. Light can easily be excluded by adding an opacifier to the shell.
- Superior patient compliance/consumer preference and pharmaceutical elegance. Results of studies done through the years show that consumers expressed their preference for soft capsules in terms of ease of

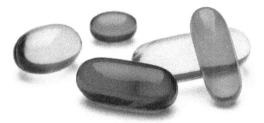

Figure 3.2 Clear soft capsules. Image courtesy of Catalent.

swallowing, perceived speed of delivery, lack of unpleasant odor or taste and modern appearance[3] (Figure 3.2).

- Better tamper evidence (tampering leads to puncturing and visible leakage).
- Safer handling of highly potent, hormonal or cytotoxic drug compounds.
- Product differentiation (through selection of novel shapes, colors, and sizes).
- Product life-cycle management. For example, product enhancement *via* faster onset of action.

The disadvantages of soft capsules include the following:

- Specialized manufacturing equipment and facilities with high start-up costs, fewer outsourcing options.
- Slightly higher manufacturing cost compared with tablets.
- Formulation complexity. Whilst the ability to modify a formulation provides opportunity, few companies maintain real expertise in lipid formulation, especially those that are synergistic with soft capsule encapsulation. In particular, it can be difficult to predict how a given formulation will perform *in vivo* and whether a given formulation will be commercially viable if the formulator has no experience with the soft capsule process.
- Raw material quality/stability/variability. Although there a wide range of excipients being sold for use in soft capsule formulations, some of them are unstable or are produced by suppliers who do not have a true pharmaceutical infrastructure. As a result, a lot of experience is required in the selection and storage of excipients to achieve pharmaceutical good manufacturing practice (GMP) requirements of the finished product and to ensure they are "fit for purpose"

3.4 Description

The soft capsule is a hermetically sealed, one-piece capsule shell with a liquid, gel or semisolid fill without a bubble of air or gas. The shape can be anything not too complex with a plane of symmetry but typically oval or rounded oblong. The fill material is typically a liquid (Figure 3.3).

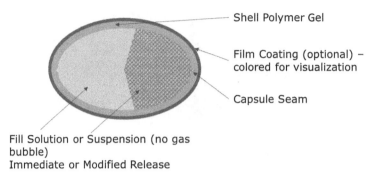

Shell Polymer Gel

Film Coating (optional) –
colored for visualization

Capsule Seam

Fill Solution or Suspension (no gas
bubble)
Immediate or Modified Release

Figure 3.3   Soft capsule schematic. Image courtesy of Catalent.

The shell is composed, at a minimum, of polymer, plasticizer and water. The content of water in the finished dried product is rather low and, being mostly bound to the shell polymer, inactive. Typically, the shell polymer consists entirely of natural gelatin as described in the original invention. Attempts to combine other natural and synthetic polymers with gelatin for specific applications have achieved limited success, particularly where the new combination does not sacrifice manufacturability, which means that gelatin, when present, is typically the main ingredient, and nearly always the only ingredient in Rx products. Non-gelatin soft capsules have been on the market in consumer products for several years, with the US FDA recently announcing the approval of an Rx product based on starch and carrageenan completely free of gelatin.[4] Manufacture of soft capsules without gelatin requires certain modifications and expertise but utilizes similar equipment to the standard gelatin-based process. Non-gelatin capsules may enable the use of different materials that are not compatible with gelatin and, depending on the process used to make them, a wider range of excipients in the fill.[15]

Though the soft capsules may be either clear or opaque, with or without added color, it is standard practice to only use a clear shell (clear colored or natural amber) when the fill is also a clear solution, and this results in an aesthetically pleasing and elegant dosage form. Opacifier is frequently added to either mask inconsistencies in the appearance of the fill material, or to protect the contents from light.

Soft capsules may be coated with a wide range of suitable coating agents, depending on the intended function of the coating. The practice is similar to coating tablets, usually in a perforated pan coater, but there are some important differences, such as a lower maximum bed temperature that is needed to avoid sticking.

The typical soft capsule shape for oral pharmaceutical products is oval, with a few oblong shapes and even fewer round shapes. Round shapes are not normally selected, due to difficulty during handling and transfer operations, where they are more prone to bounce upon hard surfaces and produce erratic flow. The size of the soft capsule made by the rotary die encapsulation process is most commonly represented by a numerical value, which

Figure 3.4   Soft capsules with twist off design feature. Image courtesy of Catalent.

represents its nominal capacity in minims (an old measurement of volume defined as 1 cc = 16.23 US minims or 15.59 UK minims). The minim scale is still used as it provides a convenient range of whole numbers (in the range 1–20 for oral applications). For example, a soft capsule made with 11 oblong tooling filled at 95% of maximum may have a capacity of 0.585 cc. It is important to realize that the nominal tooling capacity is not an absolute value, and actual fill capacity may vary between manufacturers. Since most rotary die encapsulation manufacturers make their own tooling, there can be several different geometries for a given size and shape, and each may be different to those of other manufacturers or available in several choices from one manufacturer.

Soft capsule tooling can be easily manufactured to provide any shape with a plane of symmetry and any size (to contain up to 25 ml) *via* appropriate die design (Figure 3.4). A recent survey has shown that smaller sized soft capsules are preferred within each shape category, with oval being the most popular shape.

Another way to encapsulate hermetically sealed soft capsules is by the concentric nozzle dropping technique to produce spherical, seamless capsules.[5] In this process, the liquid fill and molten gel are fed to two nozzles, one enclosed within the other such that as material drips out of each nozzle, the gel layer enrobes the fill. The resulting soft capsules are limited to smaller sizes and are always spherical. Manufacturing takes place using multiple nozzles, and soft capsules are typically formed in a stream of lubricant liquid. The technique must be difficult to practice, as the technology is relatively old, but there are far fewer manufacturers.

3.5   Equipment and Facilities

In addition to the cost of the unique manufacturing equipment required to manufacture soft capsules, there are several constraints that are not normally required for manufacture of other dosage forms. Several of the manufacturing steps require precise control of room conditions (both temperature and humidity), with the most significant infrastructure and energy needed for drying.

In the original process gelatin is dissolved in water as part of the manufacturing process. After encapsulation, the residual water needs to be dried out in order to provide a robust and stable finished dosage form. Applying some estimate figures of the water content of the shell formula when wet (35%) and dry (10%), it is relatively straightforward to calculate the amount of water that needs to be dried out from a typical batch size of, say, 1 million soft capsules of a typical intermediate size (such as 10 oval). If the shell weight is approximately 66% of the fill weight, and the fill weight is typically 500 mg, we arrive at a figure of $1 \times 10^6 \times 500 \times 0.35 \times 0.66 = 115.5 \times 10^6$ mg = 115.5 kg. Whilst this may not sound particularly difficult, the soft capsule ribbon is relatively thick, which is not conducive to rapid drying, and it is not easy to use heat to speed the process due to the risk of creating leakers or other undesirable physical changes in the shell. Historically, the only practical ways to achieve rapid drying were to spread the capsules into a thin layer on drying trays to increase surface area, to increase flow rate of air through the drying chamber, to control the incoming humidity and to try to ensure that air flow through the capsules does not create any dead spots in the drying chamber.

3.6 Shell Components

The invention of soft capsules relied on the use of gelatin to provide the basis for the shell.[1] This was due to the versatile properties of gelatin. Gelatin has very highly solubility in water, which not only enabled the formation of the shell for processing, but also ensured that the resulting product would disintegrate in the stomach. Gelatin is unusual in that it is a rapidly thermo-reversible polymer with a high elasticity, which enabled soft capsules to be filled up by injecting the fill material into an expandable shell and sealing the shells together by heat almost immediately after undergoing a sol to gel transition (from solution) to form the ribbons needed for encapsulation.

Gelatin is derived from collagen by extraction and polymer degradation to produce material with the appropriate molecular weight for the intended application.[6] Originally, soft capsule gelatin was derived from bovine or porcine sources, but considerable research has been undertaken to develop gelatin from other mammals and other species such as poultry and fish. Each source of collagen (country of origin, species, type of tissue used, manufacturing process employed) produces a gelatin with unique chemistry that influences cost and performance (appearance, processability and stability). As a natural material, gelatin is prone to variation, and manufacturers rely on a somewhat flexible manufacturing process to produce various sublots which are blended to ensure a final product that consistently meets pharmaceutical specifications. Switching gelatins from one type, site or manufacturer to another is not straightforward and requires expertise in order to ensure a satisfactory result, even when the gelatins appear to be the same grade.

For non-gelatin capsules made out of starch and carrageenan, a shell buffer is required (base) to ensure the proper chemistry.[14] Soft capsules made entirely from modified starch have begun to appear but so far have not been used in pharmaceutical products. Starch tends to change structure over time and this can lead to stability problems.

In addition to the base polymer, plasticizer and water are added to impart the desired properties for in-process manufacturing as well as the final product. Soft capsules made without any plasticizer would be too brittle. Soft capsules made without water would have too high a viscosity and processing would be extremely difficult.

Although common in the past, it is not necessary to include preservatives in the shell. Some products have ingredients added into the fill or shell to help mask any perception of gelatin odor that can build up in bottle packs over time.

If a product is needed that does not provide "immediate release" then it is possible to incorporate delayed or enteric properties into the soft capsule. This can be done by spray coating an appropriate polymer system onto the outside shell, or by incorporating delayed-release or enteric properties into the shell of the soft capsule prior to capsule formation.

Further details of the materials most often used to make soft capsules are provided in the excipient section.

3.7 Fill Formulation

A wide range of liquid and solid excipients have been employed to make soft capsule fill formulations, except those where water is present above approximately 10%. Care must be taken to ensure that any excipient is safe to consume based on the anticipated soft capsule intake, and will be approved by the relevant regulatory authority without need for toxicological data, which could delay an NDA submission or prevent approval.

Soft capsule fill excipients can be broadly classed into lipid excipients (*e.g.* corn oil), hydrophilic excipients (*e.g.* polyethylene glycol), and surfactants [which can be further divided into hydrophilic (polysorbate 80) and hydrophobic (capric acid monoglyceride) surfactants]. These terms are further described by Pouton in the lipid formulation classification system.[11] These range from simple oil formulations (type I) that do not disperse in water, and may or may not disperse in gastrointestinal fluid depending on whether they can be digested or not, to self emulsifying drug delivery systems (SEDDS) (type IV) that contain surfactants and co-solvents to help provide a spontaneous and stable dispersion in water.

Lipid formulations were the first to be used. The understanding of the mechanism involved has increased over the last 50 years.[7] These rely on the ability of the fill formulation to solubilize lipophilic molecules, to keep them in solution throughout the gastrointestinal tract, and even to stimulate secretion of bile to further increase the chance of solubilizing the molecule. Lipid formulations may deliver more patient variability, especially in those patients that have a compromised digestive system where digestion is important.

SEDDS were developed as an extension of lipid formulations. Originally to use the same high $\log P$ drug dissolved in lipid approach but using surfactants to speed the dispersion and to help reduce patient variability, as was the case for Cyclosporin. SEDDS were also found to help drugs that were less lipophilic, due to the presence of less lipophilic surfactants that can solubilize a broader range of drugs. However, it is still possible that bioavailability increase may be better from a Type 1 formulation, such as is the case for Cinnarizine.

In addition to lipid formulation, it is possible to encapsulate only a cosolvent (such as PEG 400) as long as it does not migrate to the shell and plasticize the polymer. This is typically used for biopharmaceutical classification scheme (BCS) class 1 molecules as the PEG 400 is merely acting as a solvent delivery system. Whilst soft capsules can contain co-solvents such as ethanol or water (commercial products are available), both are able to freely migrate through the polymer shell and are thus difficult to control in products with a shelf life of two years or more, and are better when restricted to less than 10% of the fill formulation, and also when their presence is not required to solubilize the API in the formulation. Other small molecules (for example propylene glycol, glycerin, triacetin and sorbitol) or any other effective plasticizers for gelatin should be similarly restricted or avoided in the fill formulation. Any excipient that is also volatile (such as ethanol) will need to be controlled by packaging to prevent loss over time. Other excipients that should be avoided are any that are incompatible with the polymer shell, such as strong acids and bases that can hydrolize the polymer, or excipients that contain or create functional groups that can cross-polymerize gelatin in the shell (such as peroxides or aldehydes).

It is also possible to promote active or passive transport. One example of passive transport is the enhancement of uptake *via* the lymphatic system. This is where, usually, a highly lipid soluble drug is preferentially taken up into the bodies lymphatic system and circulated systemically in that system rather than in plasma. Long-chain triglcyerides are typically associated with lymphatic absorption.

The wide range of formulation options available provides opportunity to target a wide range of BCS class molecules and to tailor a specific pharmacokinetic release profile, provided that the NCE dose can be dissolved or suspended in appropriate excipients and depending, amongst other things, on the partitioning behavior of the NCE.

The major problem for the formulation scientist interested in applying lipid formulation to a new molecular entity is how to compare the performance of a lipid-based formulation with that of a solid dosage form. In the case of the latter, the process of disintegration and dissolution is more straightforward. In the case of a lipid-based formulation, especially those that rely on digestion, lymphatic absorption or the presence of excipients known to interact with permeation or transport systems in the body, the correlation between dissolution and *in vivo* pharmacokinetics may be difficult to find. It is therefore more common to rely on animal pharmacokinetic data in order to compare formulation approaches. As an example, consider an NCE that is very lipid soluble ($\text{Log}\,P > 5$), not soluble in other types of excipients or water and is formulated into a digestible oil formulation. Upon ingestion, the NCE may never partition into gastrointestinal fluid at a level higher than its aqueous solubility, and thus may never be absorbed through the epithelium. However, the same formulation may, upon digestion (and stimulation of bile), increase the solubility in the (now more lipophillic) gastrointestinal fluid, even leading to supersaturation. Perhaps the molecule is absorbed along with the oil digestion process without ever needing to be in solution in the aqueous phase. It is difficult to predict from *in vitro* data which formulation would perform best *in vivo*.

In the case where API is likely to partition into the aqueous phase, release of API from the soft capsule may be faster than from a dosage form where the API is presented as a crystalline material (especially where the API crystals are relatively slow to dissolve and become the rate-limiting step for absorption in the gastrointestinal tract).

Whatever fill formulation is designed, the API must eventually be released from the formulation, even if systemic absorption is not required. Generally, the freedom to select a formulation is curtailed by solubility and stability of the API in the available excipients. Where it is possible to dissolve the API in a wide range of excipients, and thus to carry the required dose in a reasonable number of soft capsules, it may be possible to select from a range of formulation types with completely different properties achieved by simply altering the excipients and/or the ratios of those excipients. In this way, soft capsules offer a surprisingly complex range of characteristics that can be produced in the fill, whilst the overall manufacturing process may be based on experience from other products (providing that the shell can be selected from the library of formulae known to work well).

Drugs that cannot be dissolved at a sufficiently high loading may still be dosed as a suspension with bioavailability enhancement as the target, but, generally, there needs to be an advantage over traditional solid dosage forms to do so. For example, in the case where a drug is more soluble in intestinal fluid when a certain excipient that was part of the suspension is released into the intestinal fluid—perhaps mimicking a food effect. It may be necessary to employ particle size reduction to improve the rate of dissolution of the drug in this case.

Particle size reduction is important for physical stability of a suspension formulation. The API should not be soluble in the suspending agent in order to help prevent conversion to a different polymorphic form, or to agglomerate into larger particles during temperature fluctuations. Temperature cycling may be used to ensure this is not likely to happen in marketed product (International Committee on Harmonization (ICH) chambers may not discover the problem as temperature fluctuation is controlled).

Irrespective of the type of fill, there are a few other physical issues to be aware of when developing a soft capsule. Seal quality is important, and any material that interferes with seal quality should be avoided or minimized in the formulation, this can include fibrous particles with a length greater than 200 μm, highly viscous fill materials, or fill materials high in surfactants known to reduce soft capsule seal quality. Materials that produce color changes upon reaction with other ingredients should not be placed inside a clear-shelled soft capsule. White or very pale colored shells should only be used where the color of the fill will not show through the seams. It is not common for emulsion formulations to be encapsulated due to their physical stability issues. It is more common to encapsulate so called emulsion pre-concentrates, *i.e.* a mixture of lipids, surfactants and cosolvents that spontaneously produce a stable emulsion upon contact with water with minimal stirring (so called self emulsifying drug delivery systems, or SEDDS). These are called "micro emulsifying (SMEDDS)' or 'nano emulsifying (SNEDDS)' when the resulting droplet size is in the micron or nanometer range, respectively.

In general, while it is possible for any reasonably competent formulator to develop a fill formulation suitable for soft-capsule filling, there is considerable experience and expertise required to select materials that are of pharmaceutical quality, produce a stable product with a shelf life of 2 years and are compatible with all aspects of the rotary die encapsulation process.

An excellent summary of recent lipid formulations was provided in 2016 by Feeney *et al.*[7]

3.8   Product Development

There are a few key stages in the development of a soft capsule product. These are typically: fill formulation development, shell selection, prototype manufacture, prototype stability, process development, clinical supply, scale up and process validation). Analytical development that is required in parallel is discussed separately.

3.8.1   Fill Formulation Development Processes

The key stages of fill formulation development are: excipient solubility screen, excipient compatibility screen, fill formulation development, fill formulation characterization/dispersion test, fill formulation robustness and fill stability testing. Not all these steps are necessary for every program.

The target type of formulation may also be determined based on properties of the API. For most drugs, where insufficient solubility in intestinal fluid is a major cause of low bioavailability (DCS IIb), lipid-based solution formulations may be considered.[12] For drugs that are dissolution-rate-limited (DCS IIa), solution formulations based on polyethylene glycol may be sufficient, or a micronized or nano-milled suspension.

It is important to consider the target dose range for the API. It may not be straightforward to determine the appropriate dose at the time formulation development work is undertaken, without any clinical data. In addition, if the dose target was set based upon the limited bioavailability achieved with a different 'unsophisticated' formulation, it may be possible to reduce the target dose with, for example, a bioavailability-enhancing lipid-based formulation. Nonetheless, it is important for the soft capsule to carry the target dose in a number of soft capsules that a patient would be willing to take. One factor in favor of a bioavailability-enhancing lipid-based formulation, over a simpler approach, such as powder-in-capsule, is the ability to span a broader dose range and still achieve a linear pharmacokinetic profile during ascending dose studies that are typically performed in phase 1.

Once the type of formulation and dose target have been established, it is straightforward to perform a solubility and compatibility screen in excipients, and to determine whether the target dose range is achievable and stable in the types of formulations that are desired. Formulation development is usually performed to confirm that the API is soluble in a blend of the most favorable excipients. Fill formulation characterization/dispersion tests are performed to demonstrate that the API does not precipitate on contact with biorelevant media. Fill formulation robustness helps confirm that any target properties have been achieved (such as self-emulsification) and that the formulation is likely to maintain the drug in solution throughout the soft capsule manufacturing process, using suitably designed heat, water content and temperature challenges. For poorly water-soluble drugs, it is important that the fill material is able to solubilize the drug during the first 72 hours after encapsulation, when the fill material will typically experience a modest range of temperature fluctuations, and, more importantly, exposure to the shell, which has a high water content. Fill formulation stability can be performed, for example, if the API is known to be unstable or to help narrow down the number of available options if a lot of choices are available.

For solution formulations, solvents that provide adequate solubility of the drug can be selected, though it is typically necessary to limit some of them based on the upper safe limit that can be tolerated, especially where a number of soft capsules may be required to deliver the target dose range.

For suspension formulations, it may be important to avoid any excipient in which the API has high solubility (due to the risk of form or particle size change during normal temperature fluctuations), and the formulation will also need to include a suspending agent, to prevent sedimentation during manufacture and, preferably, during the shelf life as well.

3.8.2 Shell Compatibility

Soft capsules comprise two compartments, the fill and the outer shell. Depending on the constituents of each, there exists the opportunity for material to migrate between the two, either during manufacture, or during the shelf life. There is only ever a practically zero risk of migration when the fill is composed entirely of materials that are not soluble in the shell, and the fill consists of lipid materials in which the shell materials are not soluble. A thorough understanding of the most likely migration issues is typically only developed from years of manufacturing experience. In order to prevent issues later on, it is best to develop the fill and the shell together rather than to develop the fill and hope that a suitable shell will be available.

The risk of migration depends on the stage of the manufacturing process. During encapsulation, the shell has high water content and is more permeable to migration. At this time, there is a potential for migration of small molecules with high affinity for water, either from the shell to the fill or from the fill to the shell. It is possible for the fill water content of a PEG solution to increase above 10% in the first 24 hours after encapsulation, which can have an effect on the solubility of the API in the fill.

After drying is completed, there is still potential for migration over longer time periods. Since the shell water content is lower, the risk is of molecules migrating from the fill to the shell, particularly those that have a high affinity for the plasticizers in the shell.

Migration of plasticizers from the shell to the fill is likely when the fill consists of or contains hydrophilic co-solvents such as PEG. Depending on formulation, this can result in the spontaneous fracture of the capsule shells due to embrittlement of the shell and concomitant increase in volume of the fill when ambient humidity is low.

A procedure for evaluating long term migration is to handfill capsules that were previously manufactured and filled with air ("Airfills"), but these do not contain water. Although strips of wet gel can be used, the best way to evaluate migration during manufacture is by actually making capsules. This is typically more cost effective in facilities that have small-scale equipment in R&D laboratories.

3.8.3   Prototype Manufacture

It is important to be able to test formulations, to select between available options, without committing to the costs associated with full-scale manufacturing. Since the encapsulation process for soft capsules is continuous, and involves forming the wet shell ribbon, filling–forming–sealing and drying the resulting capsules, it is difficult to reproduce the technology any other way than by making capsules.

For a small organization, where the cost of a dedicated machine is prohibitive, prototypes are made on a commercial machine during a gap in the schedule. To provide more schedule flexibility, a machine may be dedicated to non-commercial work or entirely to non-GMP work in an area outside the main manufacturing area. The usual minimum batch size for a full scale machine is in the range 1.0–2.5 l, in order to limit the amount of API needed for prototype manufacture, it is preferable to utilize a machine with reduced capsule output, as can be achieved by modifying a commercial machine, or by investing in a lab-scale encapsulation unit,[16] where batches as low as

0.1 l could be made, but are typically >0.5 l to allow for yield loss and to provide sufficient capsules for a 3 month stability evaluation.

Whilst it can be beneficial to produce prototypes on a commercial machine, this is not usually essential for NCEs.

3.8.4 Process Development

For any new soft capsule product, there are a range of parameters for which a variety of information will need to be collected, in order to determine the appropriate risk (of failure) and to develop mitigation or control strategies. In order to minimize risk, it is better to select a fill and a shell formulation that are already used successfully in commercial products (with several year's data showing process capability). In the best case scenario, this leaves only the API as the novel ingredient, and the process development can focus on collecting data that will support a new drug application, paying particular attention to the issues that are caused by the new drug. In a second scenario, the fill and shell composition are both novel, and a new source of gelatin will be used. In this second scenario, even if the same work is performed as in the best case scenario, there is a very likely risk of failure (and delay), as the program cannot rely on a body of data that show that the process is capable. It would be possible to collect this data, but it can be very difficult to predict the cause of failure in a matrix, or very expensive to perform all the studies needed to reveal the risks. In addition, the best case scenario leaves no room for innovation of the fill formulation, which could be needed to ensure compatibility or performance with the new API. In the USA at least, most programs lie in between, even when the best case scenario could be appropriate to a new API, and this is largely due to the fact that fill formulation is not typically performed by formulators with access to process capability data, as this resides in only a few organizations. In general, it is a good idea to use established gel formulae and processes. It is not straightforward to develop a new gel process, or even to transfer a process from one manufacturer to another.

Process development may include selection of the appropriate equipment, and processing sequence. Fill mixing is typically a batch process where order of addition, temperature and mixing/shear speeds are most important. For a solution, it may be critical to prove that all of the API is in solution at the end of the process. The fill needs to be entirely free of air to ensure filling uniformity. The gel preparation is also typically a batch process where a homogenous solution free of air is required, but it may be difficult to identify any other process issues that are critical. There has been a lot of debate on the most important process parameters for soft capsule encapsulation, but casting drum temperature is typically on every short list as this controls the structure of the gel ribbon, which has a direct effect on encapsulation.

Drying is typically conducted in two stages, with a first tumble drier step performed at the machine, where energy and airflow is applied to rapidly reduce the water content to a point where the soft capsules will withstand tray drying without deformation. Tray drying is normally performed over several days at around room temperature with increased airflow. Some manufactures have developed longer tumble drier stages to reduce drying time, and save the labor and cost associated with stacking and de-stacking trays, which is typically done by hand. There are few examples where the speed or duration of the drying stage is important, so end-of-drying parameters are used to control this stage of the process. As such, end of drying is typically determined by the hardness of the soft capsules and by fill moisture testing in those products where there is a risk of case hardening (the shell is dry but the fill is not). It is good practice to ensure physical and chemical stability across the intended drying range (as would be the case for any parameter, but since oxygen permeation rate through the shell is directly related to shell moisture content,[17] drying could be directly related to API stability in the finished product, especially where oxidative degradation is anticipated).

During any program, there is usually a conflict between time pressure to progress the program to the next milestone and allowing experimentation to find the best design space for the process. Whether a program is successful depends on soft capsule formulation and manufacturing expertise and also on the ability to recognize and resolve issues as they occur. As such there is a need for a rigorous gating process to ensure that risks are assessed and necessary development work is performed. The most important gate is the one that occurs prior to process validation, where an assessment of the available data should be performed in order to ensure that the product is capable of commercial manufacture, but also capable of passing the increased scrutiny that is applied during process validation. Every process range should be challenged to establish whether they are supported by data, in order to help avoid deviations or failure during validation. It is not a good policy to expect the process validation protocol to collect data from process variation in a part of the design space that has not been previously studied.

A wider issue is the ability of the overall "Quality System" to be able to recognize abnormal trends and to define root causes, which becomes a critical success factor.

3.9   Clinical Supply

Supply of clinical batches for phase 1, 2 and 3 are usually provided from a single machine type with different sized batch processes for the fill material. Providing that the fill material is a solution and providing the fill mixing equipment is scalable, the soft capsule process is capable of scaling up with minor issues providing the product is close to the best case scenario as described above in the process development section. The same machines are typically used for commercial manufacturing, which also helps ensure robust scale up of the registration batch process to commercial scale.

3.10 Analytical Considerations

Like any dosage form, a range of analytical techniques are required to support the dosage form. There are a few issues to keep in mind when developing these methods to support a soft capsule product.

First, it is common practice when developing a tablet or powder filled capsule dosage form, to begin with the assay/related substance high-performance liquid chromatography (HPLC) methods developed by the API manufacturer. This approach may not be appropriate for a lipid-based soft capsule, due to the range of different peaks that may be derived from the excipients alone, all of which need to be separated from peaks derived from the active agent, and which are not similar to many of the water-soluble excipients that are used for solid dosage forms.

Second, the dissolution method for lipid-based formulations may be extremely challenging. It is not always possible to recreate the sink conditions of the human body, complete with bile salts and enzymes designed to digest several grams of fat each day, in a dissolution apparatus with a single medium.[19] As a result, many projects reach a compromise, developing a medium that provides acceptable results, but ignoring the *in vitro* to *in vivo* correlation problem, which may surface later in the project.

Any drug delivery system that is based on gelatin that does not fully resolve issues with aldehydes or other reactive groups in the fill, may at some point have to deal with the effects of cross linking, especially after storage at elevated temperature and humidity. Currently, United States Pharmacopoeia (USP) allows addition of enzyme into dissolution media to recreate conditions known to exist in the human body, though this is not allowed in other territories.[10] Considerable expertise is required in the interpretation of data to be able to estimate shelf life from short-term stability data, as the degradation mechanisms and kinetics are different to those of solid dosage forms.

3.11 Excipient Considerations

Excipients are a necessary inclusion in any pharmaceutical product but particularly in a technology that is enabling bioavailability enhancement, but their importance in ensuring the performance of the final product can be forgotten. Whether an excipient is important or not depends on the function in the finished soft capsule. Of all the preceding topics, the assumptions that scientists make about the capability of excipients to perform as expected are the most likely to result in unanticipated delay or failure in clinical programs, in the experience of the author. This is not intended as a criticism of the excipient suppliers but rather as a reminder that each excipient brings with it a range of associated chemistry, process variability and other issues that may have a different response in combination with a new API that were never observed before but might have been anticipated by a more thorough evaluation at the beginning of the program.

3.11.1 Gelatin

Gelatin typically accounts for 30–45% of the wet shell mass. As a polymer material of natural collagen origin, it is only due to the expertise of the manufacturers that the end result is able to meet the specifications needed to assure pharmaceutical quality. Gelatin is unlike other polymers in that there is not a clear linear relationship between molecular weight and physical–chemical properties (such as viscosity). This is because of the underlying chemistry of gelatin and its ability to form triple-helix structures (which changes the likelihood of a molecule to interact with others nearby). The actual structure of a given grade is derived from both the source and the extraction process used to break down the collagen (Type "A" acid or Type "B" base/alkali) as well as the final molecular weight.

The manufacturing process produces a series of extracts, each with a unique heterogeneous blend of low, medium and high molecular weight, ideally with no residual collagen and no very high molecular weight gelatin. The more aggressive, later extraction processes, employed to increase yield, produce the lowest molecular weight material. Manufacturers blend these extracts to achieve the final specification and manage the overall process and logistics to ensure consistent supply across all customer applications. It is not common to blend gelatins from various sources, perhaps due to complexities of bovine or transmissible spongiform encephalopathies regulations.

Soft capsules are typically made from bovine gelatin (bone or hide) with porcine bone or hide also being used. The proliferation of gelatins from other species (poultry, fish and other animals) and the evolution of synthetic gelatin have not yet found their way into substantial volumes of pharmaceutical production.

Gelatin is prone to cross linking in the finished product. This is a polymerization reaction that increases molecular weight, eventually resulting in material that will no longer pass the dissolution test. By paying very particular attention to gelatin source, to excipient and manufacturing quality, it is usually possible to manufacture a product with a 2 year shelf life, except in ICH Zone IV countries. The FDA allows enzymes to be added to the dissolution test, but this is not allowed in Europe or Japan. For more information see the analytical section.

3.11.2 Plasticizers

Plasticizers are used for two reasons: (1) to reduce the glass transition temperature (Tg) of the shell ribbon during manufacture to reduce process temperatures; (2) to make the finished product less brittle. Glycerine is commonly used as it edible, has the strongest influence on Tg and brittleness, and is always liquid, but it may migrate from the shell to the fill in some applications, or react to form esters with acid drugs. Other longer-chain polyols (such as sorbitol) are used where migration or reactivity is undesirable, but they are typically used as proprietary non-crystallizing blends, and blooming of sorbitol powder on the surface of a soft capsule is not unknown. Plasticizers based on sorbitol tend to produce a stiffer finished capsule texture,[18] but can be more resistant to high-humidity environments. Plasticizer usually accounts for 20–35% of the wet shell mass. The ratio of plasticizer to gelatin is important for performance of the finished product.

3.11.3 Water

Pharmaceutical grade water typically presents no special process issues for manufacturing the wet shell mass. Addition of water to the fill is limited to 10% due to the fact that water migrates out of the capsules during drying. Studies have shown that dried soft capsules do not require preservatives as the water activity is below the critical threshold. Water is a good plasticizer for gelatin, and gelatin capsules pick up water in high-humidity environments. Good packaging is recommended to ensure finished product stability.

3.11.4 Colors, Opacifiers *etc.*

Soft capsule shells are naturally a clear amber color, depending on the grade of gelatin that was used. If the contents need to be protected from light, it is normal practice to add titanium dioxide or iron oxide to the shell. In addition, a wide range of water-soluble dyes and pigments are available, with only a few being excluded due to stability issues. These can be combined with opacifiers to produce a range of clear or opaque colors.

It is also possible to add odor or flavor to the shell to offset the slight odor of the natural gelatin.

3.11.5 Ingredient Specifications

Soft capsules are a unique dosage form, and it is not unusual to find that there are important attributes of the material that were not considered important when the material specification or monograph was written. As such there are a few occasions where further testing is required to assure that an excipient is "fit for purpose"—for use in soft capsule manufacture.

An example is the presence of peroxides and/or aldehydes. These need to be controlled to prevent cross linking in the finished product. Typical specifications applied to excipients may not be sufficient to assure a 2 year shelf life in the finished product. Aldehydes can be present in a material even when not specified.

3.12 Packaging and Stability Considerations

The soft capsule shell is an excellent barrier to oxygen,[17] as long as the product is kept dry. This fact partly explains the early success of the dosage form in encapsulating materials like cod liver oil, vitamin A and vitamin

D, all of which would have a considerably shorter shelf life if provided in anything other than an unopened bottle. In order to maintain the shelf life, it is important to ensure that minimal water is allowed to migrate into the primary pack, as the soft capsules will rapidly take up the water and lose their oxygen barrier properties. The United States Pharmacopoeia describes how to evaluate moisture permeation in packaging.[8]

3.13   Manufacturing Process

The critical 'operation' in the manufacture of soft capsules, is the encapsulation step where the capsules are formed, filled and sealed. Unlike a two-piece hard capsule, the filling and sealing of soft capsules occurs in one single process on a machine that is typically derived from the rotary die encapsulation machine that was first patented by R. P. Scherer in 1932, rather than two steps or more that are needed to make hard capsules (where the manufacture of empty shells is separate from filling/sealing). All of the other steps in the soft capsule manufacturing process are either designed to ensure that the form–fill–seal process proceeds without failure, or to deliver a robust finished product that is easy to handle in downstream processes (such as packaging).

The full list of operations is: fill preparation; gel preparation; encapsulation; drying (tray and/or tumble); finishing; packaging (bulk and or primary).

Fill preparation and gel preparation are typically batch processes, with the fill preparation limited to one batch, and potentially several gel preparation batches will be needed to complete larger batch sizes. Encapsulation is a continuous process. Drying, finishing and packaging are typically also batch processes. Depending on the manufacturer, imprinting or other surface marking is done during encapsulation as a continuous process or during finishing as a batch process. Packaging is normally into a bulk container, but some manufacturers are also able to support primary packaging at the soft capsule manufacturing site, especially where drying is rapidly completed in a 100% tumble drier step.

Although there have been technological advancements along the way (materials of construction, speed, size and number of capsules produced per hour, electronic interfaces *etc.*) the encapsulation operation is essentially the same as the original process: A homogeneous mixture of shell polymer, plasticizer, water and sometimes colors/opacifiers *etc.* is provided to the machine as a liquid at above room temperature and free of air bubbles; A liquid solution or suspension of fill material is also fed to the machine, usually at room temperature and also free of air bubbles; The liquid gel is cast onto two chilled casting drums so as to form a wet, elastic ribbon wide enough to cover the working area of the encapsulation die, one on each side of the machine; the wet elastic ribbons are peeled from the drum, lubricated and conveyed between two die rolls that pinch the ribbons, and by their rotation, draw the ribbon through the sealing zone; The shape and size of capsules produced is in part derived from raised perimeter surfaces around each depression or

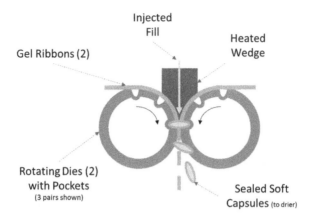

Figure 3.5 Schematic of soft capsule die and wedge. Image courtesy of Catalent.

'pocket'; the raised perimeters come together under pressure, aligned with the opposing die, and serve to cut out the shape and seal the outer rim of the soft capsules together; The fill material is injected between the two ribbons in the sealing zone by way of the encapsulation wedge, at appropriately spaced wedge holes; the encapsulation wedge rests on top of the ribbons, forming a hermetic seal on the upper part of the sealing zone; the injection of fill material through the wedge holes into the sealing zone causes the ribbons to expand into the pocket; the wedge is heated to raise the temperature of the wet ribbon and to help ensure that the two ribbons weld together to form a seal. The dies rotate at a constant speed, with multiple matched pairs of pockets cutting and sealing. The cold temperature of the die metal helps ensure the seal area is cold before exiting the sealing zone. The fill is timed to fill the pockets as they pass over the appropriate wedge holes (typically 4–12 depending on the size of the capsule), and alternate rows of pockets are filled in sequence until the die completes a whole revolution, and so the process continues (Figures 3.5 and 3.6).

It is common practice to run the machine with the same gel feeding both sides of the machine. It is possible to create two-colored capsules by feeding each side of the machine with a different gel melt, but any resulting imbalance in gel properties may lead to asymmetric capsule formation.

The thickness of the shell in the resulting capsules is typically 4–6 times greater than that for hard capsules.

As can be expected, any variation in fill, potency, viscosity/rheology, air bubbles or particles can affect sealing and fill weight or content uniformity. In addition, any variation in gel homogeneity, viscosity/rheology or air bubbles can affect seal quality.

Soft capsules must be dried to remove the process water. This is either done exclusively in a series of tumble drier baskets, or by a combination of tumble drying and tray drying, where the tumble drying is performed so that the capsules may withstand tray drying without deformation. Complete

Figure 3.6 Photograph of soft capsule dies and wedge. Image courtesy of Catalent.

drying is normally assured by monitoring hardness on daily basis until a specific target range is met, and where the finished product has an appropriate balance of physical (not brittle) and chemical stability (not too soft). Drying time depends on the fill and gel formulations used, on the size of the capsule, the conditions of the drying tunnel and the target range. Drying times in a commercial setting are typically 2–8 days (Figure 3.7).

After drying is complete, capsules are held for packaging. At this point, depending on the manufacturer, additional finishing operations can be performed, such as washing, lubrication, off-line printing, inspection, check-weighing, *etc.*

3.14   Key Process and Product Parameters

3.14.1   In-process Testing

As may be expected, it is impossible to confirm that every single soft capsule made on a rotary die encapsulation machine is exactly the same as all the others. In order to ensure that a lot of soft capsules conforms to the finished product specification, it is standard practice to perform a series of tests during or after manufacture, in order to give confidence that the machine was properly set at the beginning of the batch, and maintained the process within acceptable variation during the entire batch, right to the very end.

Figure 3.7 Photograph of soft capsule tumble drying. Image courtesy of Catalent.

As to which tests are performed, this may variably considerably, based on manufacturer, country of manufacture and regulatory agency, among other factors. Whether or not to apply all the tests below will depend on many factors, and these should be considered as recommendations rather than mandatory.

3.14.2   Set up (Pre-start)

At the start of a batch, before collecting soft capsules as part of the batch, it is standard practice to make a series of very short runs in order to verify that the proper fill and shell weight are produced, with adjustments made depending on the results, with the resulting capsules being rejected. This is done by carefully switching the wedge from a recirculation mode to a filling mode, as the pump is always running, whereupon the fill material is directed through to the tip of the wedge and starts to inflate capsules. At the same time the operator applies pressure to the dies to begin seal/cut out. After a short delay, samples are taken. It is typical to ensure that capsules from each wedge hole (*i.e.* across the whole die) are within range for fill weight, shell weight and seal quality. Capsules may be also checked for appearance (including lack of an air bubble) and print quality. Set up can be responsible for a large loss in yield of smaller batches if it takes a lot of iterations to create the right balance of process conditions.

3.14.3   During Encapsulation

After set up is completed, the wedge is switched to filling mode, pressure is applied to the dies and the run start time is recorded. Capsules are sent to the tumble basket. It is good practice to immediately take a fresh set of in-process samples as there has usually been a delay between setup and start (the time taken to perform all the tests). Fill weight, shell weight and seal quality are recorded based on batch size, confidence in the stability of the process and need to collect data. Other parameters are recorded in order to be able to correlate process conditions with output capsule quality, and to support trend analysis, typically machine speed, casting drum temperature, wedge temperature, gel temperature and others as considered appropriate based on the actual process employed. If there is high confidence that the process is under control, it may be appropriate to perform a reduced number of in-process tests (testing only two or three pockets across the wedge instead of all of them), provided that these are sufficient to perform a useful trend analysis, as part of a well-considered testing strategy.

As mentioned previously, it is common practice to make one fill batch lot and to attach multiple gel lots to the machine in order to complete the encapsulation. In addition to the in-process tests mentioned previously, it is good practice to recheck shell weight and seal quality after each significant gel change.

It is important to note that fill and shell weight testing are performed on wet capsules. This imparts a non-random error to the results (the capsules lose weight over time). There is a certain amount of skill required to obtain a consistent result from the typical procedure (solvent wash, weigh whole capsule, cut open capsule and squeeze out most of fill, solvent wash of empty shell, weigh empty shell). The final result may have a bias that is significant in terms of yield and potency, and there may be occasions where the standard approach needs to be modified to ensure the process is truly close to the mid-point of the range required.

3.14.4 During Drying

Soft capsules are sampled from throughout specific portions of the batch depending on where they are placed. For example from top middle and bottom of the first stack in the tunnel, where the stacks are placed so that the first stack receives fresh, dry air—there may be no need to test the second stack until the first meets the drying range.

3.14.5 Finished Product

Once the soft capsules have completed all steps in the process, the lot may be inspected and sampled for the final product release. The finished product specification for an active product will usually include appearance, identity of actives, assay, related substances, fill weight, dissolution or disintegration, content uniformity and microbial testing, but the actual test may depend on monograph or regulatory requirements.

3.15 Recent Technology Advances

3.15.1 Film Coating

Application of a film coating to soft capsule is not entirely new, but there have been few commercial Rx products until now. Due to the relatively large size, perforated pan coating is most often employed. Certain soft capsule lubricants may make it difficult to apply a coating, in which case the batch may be solvent washed before coating. Film coating becomes more difficult for larger capsules, especially oblong-shaped capsules, which do not fluidize as well in the pan. Film coating of soft capsules requires precise control of temperature and water content of the bed, in order to prevent capsules sticking together during coating. The finished product film coat needs to be flexible to compensate for the fact that the substrate material is somewhat flexible. All of these issues are sufficiently different to the standard practice of tablet coating to utilize different equipment, coating formulae and process conditions. Most coatings are either opaque or only semi translucent, meaning that printing has to be done after coating.

3.15.2 Non-gelatin Shell and Controlled Release Fill

In 2017, the FDA approved the first soft capsule based on a controlled release fill formulation.[4] This was achieved using high-temperature filling, made possible by the use of starch–carrageenan shell technology. This may be a useful approach to achieving a high AUC whilst limiting the C_{max}. The shell materials will appear on the FDA inactive ingredients database.

3.15.3 Vaginal Dosage Forms

Application of soft capsules as a vaginal drug delivery system has become established in the USA based on a few commercial OTC products.

3.16 Trends in Patent Activity

A search of worldwide patents with the word "soft capsule" in the title or abstract was performed, covering the period 1993–2017. Of the total of 397 patents, 83% were filed in China, 10% in the USA and eight other territories filed the remaining 7%. After a peak of activity between 2003 and 2008, there has been a steady year on year growth in patent filings ever since. Although not shown, there has been a steady growth in patent filings in the USA. Most applications contain inventions for specific traditional Chinese herbal medicinal treatments. There are also several inventions related to the machinery for soft capsule production, production of soft capsules based upon starch and for enteric coating.

3.17 Conclusions

Soft capsules provide a medium for encapsulation of a wide variety of fill formulations, for example those based upon lipid materials. These offer a different approach to solubilization of poorly soluble molecules compared with traditional solid dosage forms. Provided the formulation is designed with sufficient expertise to anticipate the several manufacturing issues that are known to occur, (*i.e.* within the scope of prior successful formulations), the soft capsule offers a commercially viable and rapidly scalable dosage form. For certain drugs, soft capsules can provide content uniformity, stability and safe handling that may be difficult in traditional solid dosage forms. Given that many early animal PK tests could be performed using solutions of API in lipids, soft capsules could have a head start on solid dosage forms in the race to market.

References

1. R. P. Scherer, Method and machine for making capsules, US Patent 1,970,396, 14 Aug 1934.
2. G. L. Amidon, *et al.*, A theoretical basis for a biopharmaceutical drug classification, the correlation of in vitro drug product dissolution and *in vivo* bioavailability, *Pharm. Res.*, 1995, **12**, 413–420.
3. W. J. Jones III, *et al.*, Softgels, consumer perceptions and market impact relative to other oral dosage forms, *Adv. Ther.*, 2000, **17**, 213–221.
4. FDA Summary Basis of Approval for Rayaldee, https://www.accessdata.fda.gov/drugsatfda_docs/nda/2016/208010orig1s000sumr.pdf.
5. R. P. Scherer, Fabrication of capsule shells and filled capsules, US Patent 2,331,572, 12 Oct 1943.
6. R. Schrieber and H. Gareis, *Gelatin Handbook, Theory and Industrial Practice*, Wiley-VCH, Germany, 2007.
7. O. Feeney, *et al.*, 50 years of oral lipid-based formulations: provenance, progress and future perspectives, *Adv. Drug Delivery Rev.*, 2016, **101**, 167–194.
8. United States Pharmacopeia, *General Chapter <671> Containers - Performance Testing*, USP 40, 2017.
9. United States Pharmacopeia, *General Chapter <1151> Pharmaceutical Dosage Forms*, USP 40, 2017.
10. United States Pharmacopeia, *General Chapter <711> Dissolution*, USP 40, 2017.
11. C. Pouton, *et al.*, *Eur. J. Pharm. Sci.*, 2006, **29**, 278–287.
12. X. Q. Chen, *et al.*, Application of lipid based formulation in drug discovery, *J. Med. Chem.*, 2012, **55**, 7945–7956.
13. S. A. E.-S. Fahmy, *et al.*, Development of novel spray coated soft elastic gelatin capsule sustained release formulations of nifedipine, *Drug Dev. Ind. Pharm.*, 2009, **35**(8), 1009–1021.
14. K. E. Tanner, *et al.*, Film Forming compositions comprising modified starches and iota-carrageenan and methods for manufacturing soft capsules using same, US Patent 6582727, 27 Jun 2002.

15. K. E. Tanner, *et al.*, Gelatin free softgels with improved stability for pharmaceutical fill formulations containing alkaline ingredients, *AAPS Annual Meeting*, Poster, 2004.
16. S. C. Tindal and F. Asgarzadeh, Laboratory scale softgel encapsulation (Minicap), *AAPS Annual Meeting*, Poster, 2006.
17. F. S. Hom, *et al.*, Soft gelatin capsules II. Oxygen permeability study of capsule shells, *J. Pharm. Sci.*, 1975, **64**, 851–857.
18. N. Cao, *et al.*, Effects of various plasticizers on mechanical and water vapor barrier properties of gelatin films, *Food Hydrocolloids*, 2009, **23**, 729–735.
19. H. Nishimura, *et al.*, Application of the correlation of *in vitro* dissolution behavior and *in vivo* plasma correlation profile (IVIVC) for soft gelatin capsules – a pointless pursuit? *Biol. Pharm. Bull.*, 2007, **30**, 2221–2225.

CHAPTER 4

Tablet Formulation

K. G. PITT

GlaxoSmithKline, Global Manufacturing and Supply, Centre of Excellence
for Oral Solid Dosage Forms, Priory Street, Ware, SG12 0DJ, UK
*E-mail: Kendal.5.pitt@gsk.com

4.1 Tablet Formulation

Tablets provide an accurate, stable dose of drug and when correctly formulated are capable of large scale economic production with a high degree of tablet uniformity both within and between batches. The drug is commonly referred to as the active pharmaceutical ingredient, (API). There can be more than one API in a tablet.

About 70% of all medication is administered as tablets. Tablets are manufactured by filling a die with powder and compressing using rigid punches, followed by ejection. During this process the loose powder in the die is transformed into a tablet of given shape and microstructure. Typical compaction pressures are in excess of 100 MPa.

Tablets must be strong enough to withstand subsequent operations, such as coating, packaging, transport and patient handling, but weak enough to disintegrate or dissolve in the body to achieve the desired bioavailability characteristics. Tablets have the advantage of ease of administration by the patient.

Key considerations in the design of a tablet formulation are both the dose and properties of the API. The majority of APIs alone will have a combination

Drug Discovery Series No. 64
Pharmaceutical Formulation: The Science and Technology of Dosage Forms
Edited by Geoffrey D. Tovey

of poor compression properties and poor disintegration or dissolution. Hence, requiring additional functional materials in the tablet formulation, which are referred to as excipients.

One of the simplest ways of formulating a pharmaceutical powder for compression is by mixing the active ingredients and excipients to form a uniform powder blend. However, a granulation step is often necessary to

- Maintain content uniformity
- Give acceptable flow properties
- Avoid particle segregation
- Assure physical and chemical stability
- Improve compactability.

The properties of the final tablet are a result of the formulation, the manufacturing process and the equipment.

The compression of powders is usefully described in terms of compactability, compressibility and tabletability plots. These are plotted in non-dimensional terms of pressure, solid fraction (density) and tensile strength which are independent of the shape and size of the tablet. Tensile strength is derived from the tablet breaking force (sometimes referred to as hardness) and the dimensions of the tablet.[1] Thereby allowing ready comparison of tablets of differing size and shapes. Compressibility is a plot of compaction pressure *versus* solid fraction and shows how increasing the compaction pressure results in an increase in the solid fraction of the tablet. A solid fraction of 1 is a completely solid material with no porosity (Figure 4.1). Compactability is a plot of solid fraction *versus* tensile strength and shows how the tensile strength of tablet increases with density (Figure 4.2). A tabletability plot

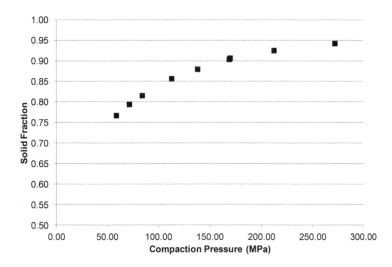

Figure 4.1   Compressibility plot.

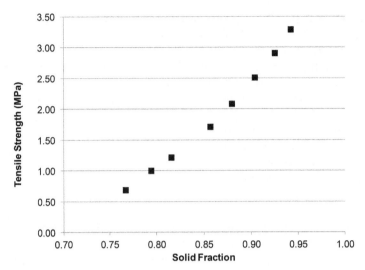

Figure 4.2 Compactability plot.

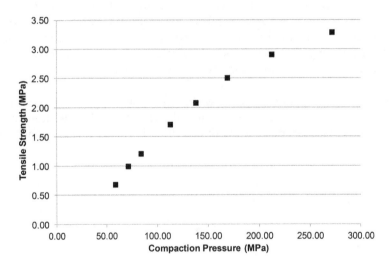

Figure 4.3 Tabletability plot.

is a plot of compaction pressure *versus* tensile strength and shows how the tensile strength of the tablet increases with increasing pressure. This plot normally shows that a limit is reached beyond which increasing the pressure does not increase the tensile strength of the tablet. This is the point at which cracking is induced in the tablet due to overstressing of the powders, which results in tablet damage. Typically this cracking occurs when the porosity of the tablet is below 10%. Maximum tensile strength of the tablet occurs at about 15% porosity. 15% porosity is equivalent to a solid fraction of 0.85 (Figure 4.3). Tabletability is sometimes simplified for a set tablet size and shape by plotting compaction force *versus* breaking force. This will still define the design space for compression of a given size and shape of tablet, but will not allow comparisons with other shapes and sizes of tablets to be made easily.

Typically a final tablet weight of above 50 mg is targeted to facilitate handling by the patient. Patient handling can be enhanced though to a certain extent by use of specially modified shapes. An example of these are "Tiltabs" which have an irregular shape which prevent them from lying flat.[2] An additional consideration is that the feed-back control loops on high-speed tablet presses tend to require tablet weights of 150 mg and above to be fully functional. Hence, if the dose is low then bulking agents will be needed to increase the weight of the tablet. It would be expected that the large amount of excipients in a low-dose tablet would result in an easily compactable and disintegrating formulations. The risk though of low-concentration formulations is that the API may segregate, leading to content uniformity issues. Low concentrations of finely sized API *e.g.* at levels of *circa* 2% and below can be stabilised though by means of ordered mixtures.[3] Ordered mixtures are formed when the fine drug particles adhere to the rough surface of an excipient following dry blending, thereby forming a physically stable free flowing mixture.

If the dose is high then although the API segregation risk is low, the compactability of the mixture or flow of the API can become an issue. There is a limited amount of excipients that can be added to the formulation before the tablet size becomes unacceptable. Tablet size acceptability tends to vary with the medical indication. So for example a tablet to treat human immunodeficiency virus (HIV) or an antibiotic can be up to 1.7 g in compression weight, whereas for less severe indications there tends to be a much lower size limit for patient acceptability.

The particle size and shape of the API are also key considerations for tablet formulation. The acceptable size of the API particle may be dictated by its solubility and dose. To increase the dissolution rate of a poorly soluble drug the particle size may be reduced by milling or micronising. An approximate guide for adequate dissolution rate is that the particle size in microns should be equal to or less than the API solubility in micrograms per millilitre. For example a 10 μg ml^{-1} soluble API would need a particle size of less than 10 microns.[4] However, the smaller the API particle size then the more prone the API is to electrostatic charging and to poor flowing and clumping, thus, leading to the need to granulate the API prior to compression for these small particle sizes. Below about 50 microns particle size is in the region in which electrostatic charging may become an issue and poor flow may be expected to occur. A different reason to reduce the mean API particle size is to improve the content uniformity of a tablet. This need to reduce API particle size is related to the dose. The smaller the dose the smaller the particle size and the tighter the particle size limit to achieve content uniformity.

The effect of the shape of the API also needs to be factored into formulation and processing design. So, for example, a needle-shaped API tends not to flow well. So if present at a high concentration this would drive the formulation towards the need for granulation. Rounded API crystals (aspect ratios of 1 to 2) though tend to have adequate flowability particularly after addition of excipients.

The critical quality attributes (CQA) of the finished tablet reflect the above requirements for efficacy and safety. The majority of oral tablets will have the CQAs of:

- Drug Assay
- Drug Content Uniformity
- Dissolution (or disintegration)
- Identification
- Impurities

The main requirements of the material for tablet compression are as described in the following sections.

4.1.1   Free Flowing

The compression mixture needs to be free flowing and hence should be as near spherical as possible with minimal surface roughness. The aim is to have rapid, reproducible powder flow so that compact weight variation is kept to a minimum even at high production rates. Flow can be readily assessed by means of a measuring cylinder to give a compressibility (or Carr) index[5] or by a shear cell to give a flow function (FFc) value.[6] For high speed tablet compression the target Carr index should be "fair", (less than 20%, Carr index), or have a FFc greater than 7.[7]

4.1.2   Good Compression Properties

The aim is to produce a tablet of sufficient mechanical strength to allow downstream operations, such as coating, packing, shipping and subsequent handling by a patient. Tablet mechanical strength is best expressed in terms of tensile strength, which is independent of the size of the tablet. Tensile strength is derived using the dimensions of the tablet and the crushing force measured from a tablet hardness tester.[8] The target tensile strength for commercial production is at least 1.7 MPa, preferably greater than 2 MPa.[1]

4.1.3   Low Ejection Shear Stress

The formulation should be designed so that the tablet will eject smoothly with a low ejection force, otherwise tablet cracking may occur. The ejection force that a tablet can withstand is dependent on its size. Hence it is useful to calculate the ejection shear stress of the tablet which is the ejection force divided by the central core ("belly band") which is in contact with the die wall during ejection. The target should be an ejection shear stress of less than 3 MPa.[9]

4.1.4 Good Content Uniformity and Low Segregation Potential

Particle size and distribution will dictate the ability to obtain good content uniformity which meets the pharmacopeial requirements. Nomograms have been produced to assist with this estimation of particle size effects on content uniformity.[10,11]

Segregation potential can be assessed by measuring the assay of API across a number of sieve fractions.[12] For low-dose drugs, segregation potentials of less than 10% should be targeted. Experience has shown that over 15% segregation potential will lead to segregation. Note that intermediate segregation potentials would not necessarily preclude commercialisation of the formulation. However care would have to be taken in designing the processing equipment to reduce the propensity to segregation. One formulation factor which can drive segregation is the density differences in input materials. For example between an organic excipient such as lactose monohydrate with a true density of 1.54 g cm^{-3} and an inorganic material such as dibasic calcium phosphate with a true density of 2.39 g cm^{-3}. If granulated together then they are unlikely to segregate. However for a direct compression they could separate. Similarly differences in particle size can drive segregation. Segregation can occur when the ratio of the mean particle size of the components exceeds 2.8.[13]

4.1.5 Rapid Disintegration and Dissolution

Disintegration is needed for instant-release tablets to ensure breakup of the tablet and the subsequent dissolution of the API so that it is available in the gastrointestinal tract for absorption. Pharmacopeial limits for disintegration tend to be in excess of 15 minutes. However if the disintegration time exceeds 10 minutes using standard pharamacopeial disintegration apparatus then this is an indication that the tablet is eroding rather than breaking apart rapidly into primary particles.

There is a risk, therefore, that if the dissolution specification is set at 15 minutes (or less) then the tablet will not reproducibly meet this dissolution specification with a 15 minutes disintegration time. In addition, if erosion is occurring rather than disintegration then there is the possibility that any solubilising agents will leach out, leaving behind an insoluble API core which does not break up to release the API. Hence disintegration times should be targeted at less than 10 minutes.[14] This becomes especially critical when dissolution specifications are set for time intervals of less than 15 minutes.

4.1.6 Low Friability

The tablets in friability testing are usually rotated in a defined drum at a set speed for a controlled number of revolutions. The amount of weight loss due from the compacts after the test is recorded as a percentage of their initial weight.

The pharmacopeial limit is normally less than 1% (United States Pharmacopeia (USP) reference) weight loss. However this 1% friability can produce tablets with marked surface damage, resulting in poor legibility of the identifying de-bossing. Hence friability specifications of less than 0.1% are more appropriate in these cases of de-bossed tablets. Generally if the tensile strength of the tablet is >2 MPa, then friability tends not to be an issue.

Identification is largely controlled by the engraving on the tooling punch face. Shape and colour of the tablet also contribute to patient identification and to the overall appearance of the tablet. Normally the tablet punch face is embossed so that it leaves a debossed tablet imprint. Occasionally direct printing using inks on to the tablet is employed. Printing has the disadvantage of an additional processing step which can become complicated for shaped tablets. It is used though when a thick coat has been applied to a tablet core when the coating would in fill and obliterate any debossing.

4.2   Impurity

Impurities can come from the API incorporated in the tablet or by degradation caused during the manufacturing process or in subsequent storage. Impurity control, particularly degradation in storage, can be affected both negatively and positively by excipients.

The main manufacturing indices to ensure tablet compression are summarized in Table 4.1.

4.3   Types and Uses of Tablets

The majority of tablets are required to break down (disintegrate) rapidly in the stomach to enable rapid release of drug. Other types of tablet formulation include:

4.3.1   Controlled-release Tablets

Suitable formulations can provide sustained-release characteristics whereby the dose is released at a controlled rate as the tablet progresses along the gastrointestinal tract.

Table 4.1   Manufacturing indices.

Criteria	Target
Powder flow (final blend)	Carr's index target <20
	Shear cell Fcc >7
Solid fraction	Tablet solid fraction <0.85
	Not greater than 0.9
Tablet ejection stress	<3 MPa
Segregation potential	<10%
	Not greater than 15%
Core tablet disintegration	Ideal <10 min
	Target <15 min
Core tablet friability	<0.1% (no tablet breakage, logo legible)

4.3.2 Enteric Coated Tablets

If the drug is inactivated at low pH or causes gastric irritation then an enteric coating is applied to ensure that disintegration does not take place in the stomach but readily occurs in the small intestine. Targeting further down the gastric intestinal tract, such as the colon, is also possible. This targeting is usually achieved by coating.

4.3.3 Buccal and Sublingual

Buccal tablets are placed in between the tongue and the cheek areas, whereas sublingual tablets are placed under the tongue. The advantage is that they avoid first-pass metabolism, for example glyceryl trinitrate. Alternatively they can be for local action in the mouth, for example, steroids or antiseptics. The tablets should not disintegrate in the mouth. They can be designed to slowly dissolve over 15–30 minutes.

4.3.4 Soluble Tablets

The drug may be formulated as a soluble tablet to be dissolved in water prior to administration. Typically they are formulated with a bicarbonate, for example, sodium or potassium bicarbonate, and with an organic acid, for example, citric acid. The resulting formulation effervesces when added to a tumbler of water.

4.3.5 Chewable Tablets

These tablets are formulated such that they can be readily chewed in the mouth and will also contain sweeteners or flavours if the drug has an unpleasant taste.

4.3.6 Vaginal Tablets

These are designed to be inserted into the vaginal cavity and to release medicament locally. The same design rules apply as for oral tablets. Shape and size of the tablet need to be considered and tensile strength needs to be adequate if an applicator is used.

4.4 Formulation Components

The active ingredient in pharmaceutical tablets is usually formulated with other ingredients (excipients) having specific functions. These include lubricants (to control friction between powder and tooling), glidants (to improve powder flow), binders (to improve strength) and disintegrants (polymers that swell in contact with water or allow water to be channelled in).

Normally a compression aid is also present to improve the compactability of the tablet formulation, as the API alone may not be particularly compactable. In addition, a filler (sometimes termed a diluent) is also used to make a tablet large enough to be subsequently handled. Although often described as inert, diluents may aid compaction or have other functional roles.

Initially, formulation activities usually commence by assessing the chemical stability of the API with any desired excipients by means of excipient compatibility studies. This is so as to exclude at an early stage in development any chemically incompatible excipients. The API is blended with selected excipients and then stored as a loose powder or compressed compact and placed under stability at stress conditions of elevated temperatures and humidities.

Mixing of the powders alone may be sufficient to be able to compress a tablet of the desired properties and is referred to as direct compression. Sometimes, though the powders need to be processed to form granules, for example, to improve flow or to prevent segregation of the active ingredient.

The ideal properties of a granule or direct compression mix for compaction are:

- Binding properties that confer physical strength and form to the tablet. If the tablet is subsequently designed to disintegrate in fluid then the porosity of the granules should be designed so that ingress of liquid can readily occur.
- Free flowing, hence, should be as near spherical as possible with minimal surface roughness. The aim is to have rapid, reproducible powder flow so that compact weight variation is kept to a minimum, even at high production rates.
- Relatively dust free to minimise any containment concerns.

4.5   Tabletting Material Preparation

The three common processes used to produce material for compression in the die are wet granulation, dry granulation (roller compaction or slugging) and direct compression.

4.5.1   Wet Granulation

The aim is to produce a granule which has bound in all the input materials so that they do not segregate but is still sufficiently porous to allow compression and disintegration. The density of the granule is one of the key properties that are manipulated during granulation and during subsequent compression.[15] The mixed powdered tablet constituents are converted to a wet mass by the incorporation of a granulating fluid, which is normally sprayed onto the powder. Typically the powder is in a moving powder bed in a bowl. Normally a high-shear granulator is used for the process, although planetary mixers were common in the past and are still occasionally used, particularly for older products. Granulation with planetary mixers is usually referred to as low-shear granulation. Alternative processes include fluidised bed granulation, where the powders are fluidised in a spray dryer and the solution sprayed in, and continuous twin-screw wet granulation. Twin-screw wet granulation is normally high-shear. Melt granulation is another variant. In this type of granulation process a polymer is melted and sprayed onto the powder bed and then forms granules with the rest of the powder as it cools down below its melting point.

Fluid bed granulation cannot densify the finished API–excipient agglomerate. So control of finished granule properties can be more challenging when the input API properties vary. Granule densification though can be altered by high-shear wet granulation, hence giving more scope to influence finished granule properties. Hot-melt granulation can extensively change the finished granule properties.

The granulating fluid in wet granulation is normally aqueous-based and may include the binder dispersed in it. Alternatively the binder, or a portion of it, may be in the dry mix. Binder level in the finished tablet is typically *circa* 5% for a polymeric excipient such as poly vinyl pyrrolidone (PVP) or hydroxypropyl cellulose (HPC). Mucilage binders, such as starches, tend to be higher, at 5–10%. Granulation conditions are highly dependent on the material properties of the powder blend and the intended properties of the granule. At a commercial scale, spray rates of between 10 and 30 g liquid min^{-1} kg^{-1} dry mix over 8–10 minutes are typically encountered, with total fluid addition in the range of 10–25% w/w. Higher fluid addition quantities can be used, but have the disadvantage of prolonged subsequent drying times.

Granules are formed by passing the wet mass through a screen which is then dried, rescreened (or milled) to break down agglomerates and blended with other tablet excipients such as lubricant and disintegrant. Drying tends to be by fluidised bed drying. Although the much slower and more manual tray drying is still occasionally used. Wet granulation has the disadvantage that it cannot be used for moisture- and heat-sensitive materials and that the granulation conditions may result in disproportionation of the API and its counterion.[16]

4.5.2 Roller Compaction (Dry Compression) or Slugging

Typically the API is blended with excipients and lubricants and then passed through a roller compactor to produce ribbons, which are milled to produce granules coarser than the original starting powders. For example, the starting materials may be sub 50 microns in size but the milled granules can be *circa* 500 microns. The tensile strength of the ribbons or slugs tends to be *circa* 1 MPa in order to facilitate onward milling. The advantage of roller compactors is that they are able to accept comparatively poorly flowing input materials compared with a tablet press. Dependent on roller compactor design, this input powder can have a Carr index of up to 40%. Slugging follows the same design principles as roller compaction, except that a tablet press is used to produce lightly compressed compacts of 25 mm diameter or greater which are then coarsely milled.

Originally dry granulation was used where the tablet ingredients needed to be granulated but degraded or chemically reacted in the presence of water or when heated. Roller compaction has the additional advantage of being amenable to Process Analytical Technologies (PAT), as it has a continuous particle forming step, so lending itself to Quality by Design (QbD). It is also comparatively insensitive to input material properties and variability when compared with direct compression. Increasingly, therefore roller compaction is preferred over wet granulation and direct compression. Roller compaction tends not to use surface binders, as used in wet granulation. Instead dry binders are utilised, (for example modified celluloses), typically at levels of 10% w/w or higher.

There is a limit to the compression pressure that can be applied in the roller compaction step as there is a finite amount of bonding energy within a powder system with which to form tablets. This energy has to be split between the first compaction and the second compaction so as to allow sufficient energy to allow a target tensile strength of 2 MPa to be achieved for the final tablet.[17] This reduction in compressibility is sometimes referred to as work hardening.

4.5.3   Direct Compression

This is the simplest of the three main preparation techniques but does not have the flexibility of any processing steps to deal with any variability in the input materials or to deal with poor flow. The material to be tabletted is screened if necessary, mixed and is then compressed. It does require input materials, both excipients and active pharmaceutical ingredient(s), to be more closely controlled than for the other methods, for example particles' size and shape, particularly to give adequate flow.

Direct compression is the most economical of the three granulation processes in terms of time, labour and equipment. It is particularly useful for water- or moisture-sensitive drugs or excipients. Against these advantages must be set the relatively high cost of some diluents used in direct compression formulae and the restricted range of excipients. A risk is that of segregation, as the API as it is not locked into a granule.

The granulation and processing choice may also be dictated by economic or by equipment availability or sourcing considerations as well as considerations such as the properties of the API. For example direct compression requires the least amount of processing equipment compared with other granulation processes. In addition direct compression is usually regarded as the most energy-efficient method of manufacture due to the reduced milling and drying steps compared with wet granulation.

4.6   Components of Tablet Formulations

A tablet formulation can contain in addition to the active pharmaceutical ingredient:

- Compression aids and fillers
- Binders
- Disintegrants
- Lubricants
- Glidants
- Anti-adherents
- Wetting agents
- Stability enhancers
- Colouring agents
- Flavouring agents

4.6.1   Compression Aids and Fillers

An ideal tablet formulation needs to have both brittle components and plastic components, thereby resulting in a tough and robust tablet. These components can be contributed by either the API if it is at high drug loading, or by the excipients. Plastic behaviour can be induced by granulation, for example by producing a porous deformable granule, or from the excipients. Typically a tablet formulation would therefore have an organic polymer such as a cellulose or a starch to add plasticity and a more brittle material such as a sugar or an inorganic phosphate or carbonate. A number of predictive approaches have been proposed.[18] But in practise formulation optimisation investigations tend to commence with polymer levels of *circa* 20%. The 20% level fitting with co-ordination number,[19] and percolation thresholds theory.[20]

A comparatively simple way of assessing the compression properties of an API (elastic, plastic, fragmenting and punch filming) was proposed by Wells.[21] An example of protocol is:

- Accurately weigh three 500 mg aliquots of drug and 5 mg (1%) magnesium stearate as lubricant.
- Blend two samples (A and B), with lubricant for 5 minutes and the third (C) for 30 minutes by tumble mixing.
- Load sample A into a flat face punch and die set and compress quickly to form a compact of 0.85 solid fraction (15% porosity), hold for 1 s and release. Eject the compact and store in a sealed container at room temperature overnight (to allow equilbration).
- Repeat with sample B, but hold the load for 30 s, before releasing the pressure.
- Compress sample C in precisely the same way as sample A.
- After storing each compact, crush diametrically to determine compact breaking load and diameteral tensile strength.

Interpretation:

	Plastic	Fragmenting
Compare A and B		
If the strength of	A < B	A = B
Compare A and C	A > C	A = C
	C < A < B	A = B = C

Elastic material will if made under conditions of:

A: Cap or laminate
B: Be very weak
C: Cap or laminate

Punch filming and sticking will be visible or can be quantified by dipping the punches in to a suitable solvent to dissolve the API and assaying.

Apparently contradictory results are occasionally seen with the above testing. A cause for this can be traces of amorphous material on the surface of the API caused, for example, by milling or micronising.

One of the most commonly used fillers and compression aids is lactose. A number of grades and forms of lactose are available and are selected both on the API properties and the processing method. Two anomeric forms of lactose, termed alpha and beta are known. Alpha lactose can exist as a monohydrate (sometimes termed hydrous lactose) and is primarily used for wet granulation. It is available in a range of particle sizes, typically below 100 microns. Alpha lactose can also exist as an anhydrous form, as can beta lactose. Direct compression grades of lactose tend to be modified lactose, such as anhydrous lactose or spray dried lactose, which consists of rounded agglomerates with a particle size above 100 microns. Particle size of lactose is not specified in the pharmacopeias. Spray dried lactose is a mixture of alpha lactose monohydrate and amorphous lactose. The amorphous component enhances the tabletability of the lactose. Note that different sources of spray dried lactose can have different levels of amorphous material. Also that on storage the amorphous component will crystallize back to alpha lactose. Consequently source and age of spray dried lactose need to be controlled. Lactose can enhance the dissolution of poorly soluble API. One disadvantage of lactose is that it can initiate a Maillard reaction with active agents containing a primary amine (or anything which degrades to a primary amine) resulting in a brown-mottled appearance on the tablet.

Microcrystalline cellulose is frequently used as a compression aid. A number of particle sizes, shapes, bulk densities and moisture levels are commercially available. Common sizes are 50 microns, used for wet granulation, and 100 microns, for direct compression. Roller compaction tends to use wet granulation grades of excipients intragranularly, when flow is not critical, and direct compression grades extragranularly, when flow is potentially an issue for good tablet uniformity. There are also elongated-shaped microcrystalline celluloses (aspect ratio up to 10:1, compared with 1–2:1 for standard grades). These are used in high-dose-drug direct compression when only a limited amount of polymeric material can be added to the formula. The elongated shape increases the area for bond formation for a given weight of cellulose. The disadvantage is that they have poor flow, meaning that poor tablet weight uniformity can limit their use.

Mannitol is used when, for example, chemical incompatibilities are found with lactose. It also has a cooling effect in the mouth so is in some chewable or buccal tablets. It comes in a variety of particle sizes and polymorphs and spray dried forms to enable grade selection for wet granulation, roller compaction or direct compression. It needs high levels of lubricant (for example up to 2–3% magnesium stearate) compared with other excipients to offset its high-friction properties.

Alternatives to microcrystalline celluloses include modified starches, such as pregelatinized maize starch, or modified cellulosics, such low substituted hydroxyl propyl cellulose. These can have disintegrant properties as well.

Examples of inorganic compression aids and fillers are dicalcium phosphate, calcium carbonate and sodium carbonate and are brittle materials. These too come in fine particle grades for wet granulation and coarser ones for dry compression. Dicalcium phosphate dehydrate is frequently used as the intragranular component of a roller-compacted granule, together with a cellulosic such as microcrystalline cellulose, as an alternative to lactose or mannitol. Calcium phosphate for multivitamins has the advantage that it is an active ingredient as it is a source of minerals. Similarly, the carbonates in large doses have antacid properties. These inorganics may well need higher levels of lubrication compared with organic formulations when used on their own. These increased lubricant levels may be partially due to the increased compaction pressures (up to 500 MPa) compared with organics (typically up to 300 MPa). From a process design perspective these higher pressures would also drive the need to compress with high mechanical strength steel, as standard tooling steels are typically rated to 300 MPa so can deform at these higher pressures, leading to tablet issues.

Other sugars, such as dextrose, sucrose and sorbitol, are sometimes used. Sorbitol, in particular, is used for its sweetening action. All these sugars though are subject to moisture uptake, resulting in tablet softening, so the finished products require suitable moisture-protective packaging.

Compression aids for granulated materials, can be divided up both intragranularly and extragranularly thereby promoting both formation of the granule and compression of the finished tablet. The grades used extragranularly tend to be the coarser and more free flowing direct compression ones so as to maintain good flow of the final tabletting mixture.

4.6.2 Binders

Binders in wet granulation tend to be polymeric. Synthetic examples are poly vinyl pyrrolidone (PVP) and hypromellose, sometimes called hydroxypropyl methylcellulose (HPMC). They come in a range of molecular weights and resulting viscosities. Lower molecular weight ones are used for instant-release tablets. Higher molecular weights are used for controlled-release or coating applications. The degree of polymerisation and substitution of methoxy and hydroxypropyl groups are key to HPMC performance. The polymers can be added in solution or dry. Typically levels of 4% or above are needed to ensure adequate surface coverage and binder activity in large-scale wet-granulation manufacture. Modified starches, such as pregelatinized maize starch, can also be added to the dry mixture prior to addition of liquid.

Starch as a binder has to be used as mucilage in water prior to application as a binder. This mucilage requires the starch to be dispersed in cold water and then heated to 64–68 °C. The temperature is then maintained for 5–10 minutes until it has gelled. Starch, as a natural product, can be subject to source and seasonal variation in properties such as lipid content, leading to product performance issues. Historically, starch-containing products have been shown to have strain-rate sensitivity on tabletting. For example, the tablets have varying tablet properties, such as hardness, dependent on the tablet press speed.

One practical consideration for when the binder is dissolved in the granulation fluid is the viscosity of the resulting liquid for spraying. A liquid of too high a viscosity will not be able to be pumped and sprayed.

Poly ethylene glycol (PEG) at molecular weight in excess of 5000 can be used as binder in hot-melt granulation. In this type of granulation process the PEG is melted, sprayed onto the powder bed and then forms granules with the rest of the powder as the PEG cools down below its melting point. The granulation is performed using either a high-shear granulator or twin screw. PEG binder levels are in the range of 2–15%.

4.6.3 Disintegrants

These assist the disintegration of the tablet and can act by two main mechanisms. One is to act as a water-soluble path for the water to penetrate into the compact. The other is by swelling up and applying pressure, which breaks apart the tablet.

The disintegrant can be divided up both intragranularly and extragranularly for granulated materials. Thereby promoting disintegration of the tablet into granules and then subsequent disintegration of the granules into smaller particles. Disintegrants are used at distinct ranges dependent on their mode of action (Table 4.2). Examples of "superdisintegrants" are cross linked poly vinyl pyrrolidone (1–3% w/w), sodium starch glycolate (2–8% w/w) and Croscarmellose sodium (0.5–5% w/w).

If the disintegrants are used at too high a level then instead of helping to disintegrate the tablet they can produce a gel matrix, which inhibits tablet break up. Most of the disintegrants come in a range of particle sizes and degrees of substitution, thereby allowing optimisation of gelling *versus* disintegration to be achieved for a given formulation composition and granule particle size distribution.[22] These differences in grade are

Table 4.2 Typical disintegrant ranges.

Material	Range commonly used as disintegrant (%)
Cellulose	1–10
Cross linked polyvinyl pyrrolidone	2–5
Croscarmellose sodium	0.5–5
Sodium starch glycolate	2–8
Starch	2–10

not always fully described by pharmacopeial monographs. Hence care must be taken when interchanging between different sources of a disintegrant. Starches and celluloses (such as microcrystalline cellulose) can also enhance disintegration.

An effervescent base can be used to promote disintegration by the liberation of gas. This formulation approach consists of citric or tartaric acid together with sodium bicarbonate, potassium bicarbonate or calcium carbonate. These react in contact with water to liberate carbon dioxide which disrupts the tablet.

4.6.4 Lubricants

Lubricants act by forming a layer between the tablet and the die wall and punch faces to reduce shear stresses. The smaller the amount of shear stress then the less likely that cracking of the tablet on ejection will occur. Lubricants are generally incorporated immediately before compression and so are extragranular and act by coating the granule surface. Their disadvantage is they tend to be hydrophobic so can slow water penetration, thereby slowing disintegration and dissolution. In addition, they can interfere with the bonding action of the excipients or API, resulting in unacceptable low tensile strength. These unwanted effects are related to the amount of lubricant and the intensity or shear force of blending. Hence, both lubricant level and amount of shear need to be carefully optimised. Incorporation into a tablet formulation of brittle materials, which shatter on compaction to expose clean surfaces, can reduce these negative effects of over lubrication. The most common lubricant is magnesium stearate which is typically present at 1%. In roller compaction and slugging lubricant is equally split between the initial blend for roller compaction and the final blend for compression, for example, 0.5% and 0.5%. Exact lubricant levels are formulation-dependent. Other lubricants are available but all have a balance between lubrication effect and the detrimental effects on tablet strength or water penetration, which can vary from formulation to formulation. Examples of other common lubricants are sodium stearyl fumarate, stearic acid and PEG, but at higher levels than used with magnesium stearate, for example up to 8%. Magnesium stearate is insoluble, so it can form a scum with

Table 4.3   Typical levels of lubricants.

Material	Range commonly used (%)
Magnesium stearate	0.1–1
Polyethylene glycol	2–15
Sodium stearyl fumurate	0.5–2
Stearic acid	0.1–2

soluble tablets after addition to a tumbler of water. Some of the other lubricants, such as PEG, will not form a scum so are used to give clear dispersible tablet solutions (Table 4.3).

Magnesium stearate surface area (and hence particle size) is a key property that will influence its lubrication efficiency and mechanical strength and water penetration. Hence the surface area needs to be understood and controlled as changing from for example a surface area of 5 m^2 g^{-1} to 10 m^2 g^{-1} samples would need a reduction in the concentration or in lubricating time or energy. Similarly magnesium stearate can exist in a number of hydrate forms whose lubrication and hydrophobicity properties will vary due to the shape of the crystal. Magnesium stearate dihydrate is plate-like and will shear more readily, whereas the monohydrate and trihydrate are needle-shaped. A further variable that will affect the performance of different sources of magnesium stearate is that magnesium stearate is a mixture of palmitic and stearic acids with the exact ratios not defined in the pharmacopeias.

Consequently the properties of the magnesium stearate used in both formulation and production need to be well characterised and understood. The information typically is available from certificates of analysis from the supplier.

A subset of lubricants are anti-adherents, which can reduce sticking to the punch face. The most common one is talc. Levels are formulation dependent but are in the range 1–5%.

4.6.5   Glidants

Glidants are added to increase the flowability of the final tablet mixture. They are added in particular to direct compression mixtures. Glidants should not be needed to be added to wet-granulated and roller-compacted formulations as, in theory, one of the outputs from these processes is to produce free-flowing powders. Colloidal silica is typically used at levels between 0.1 and 0.5% and added extragranularly. However the silica needs to be blended in first before addition of a lubricant, such as magnesium stearate. Otherwise if blended together at the same time then the lubricant and glidant preferentially coat each other (and not the rest of the tablet mixture) and so have reduced glidant and lubricant effects on the tabletting mixture.

4.6.6 Wetting Agents

Surfactants are incorporated into the tabletting mixtures to assist with the wetting and hence dissolution of hydrophobic drugs. Typical examples are sodium lauryl sulphate and sodium docusate. Sodium docusate is a gel so tends to be dissolved in the granulating fluid. Sodium lauryl sulphate is a powder and so can be added dry. Levels are API-dependent but are *circa* 0.25–0.5%.

4.6.7 Flavours

Flavours are added to give an acceptable taste, for example, to mask a bitter API. They tend to be based around aromatic volatile oils. Hence, if used in a wet granulation process they should only be added after any drying step.

4.6.8 Colouring Agents

Colours are mainly added to assist with product identity or for branding considerations. They can be added in at any stage of the process, but if added extragranularly can lead to mottling. A key concern regarding colours is the regulatory acceptability of the selected colour for the intended market. Frequently only iron oxide colours are permitted, which will severely restrict the colour palette available.

4.6.9 Stability Enhancers

This is a complex area and depends initially on understanding the degradation mechanism of the API which is to be controlled and the triggers for it. For example, if degradation is free-radical-mediated then differing combinations of antioxidants, such as butylated hydroxy anisole (BHA), butyl hydroxy toluene (BHT) and propylgallate, subject to current regulatory acceptability in the intended marketing region, may need to be examined. If it is a transition-metal-mediated reaction then a chelating agent, such as citric acid, may be required. Similarly, if acid–base reactions are the cause of the degradation then acidifying agents, such as tartaric acid, may be needed or alkalizing ones, such as sodium bicarbonate.[23]

4.7 Tabletting Problems and Solutions

Tablets may exhibit a number of defects which may be immediately apparent or appear only after storage. In the following sections a number of common compaction problems are identified and potential solutions discussed both from a formulation and from the interactions of the formulation with processing and equipment. The aim is to formulate a robust product which is insensitive to the natural variance in excipients and machine running.

So that there is a robust broad design space to operate in rather than a narrow design spot.

4.7.1   Cracking

A major problem that can occur during or after tablet manufacture is cracking. This can manifest itself in a number of ways, depending upon the material properties of the formulation. It can range from surface cracking through to capping and lamination. Capping is when the upper or lower cap of the tablet separates horizontally, either partially or completely, from the body of the tablet. This can occur during ejection from the tablet die or during subsequent operations, such as coating, packing or shipment. Lamination is when cracks form within the body of the compact, resulting in the tablet splitting apart into layers.

Cracking can be caused by a number of factors that may contribute to these problems. One potential cause is the inadequate removal of air from the granules in the die-cavity before and during compression. The initial volume of granules may be several times that of the compact into which they are compressed, particularly for high-porosity granules. During compression both particles and air will be compressed. The reduction in volume is due to removal of air. This air will need to escape from the compact otherwise there is the potential for this entrapped air pressure to blow apart the compact on ejection. The entrapped air interferes with granule bonding while its subsequent expansion at the ejection stage detaches the cap or laminates the tablet. Air removal can be facilitated by using dies which are tapered outwards toward the top of the die to allow the air to escape. Large numbers of fine particles or too small a top punch/die-bore clearance all hinder escape of air from the die cavity and may be another cause. Fine material can seep downwards through the clearance and compact to form a tough film which hinders free movement of the punch.

For example, if on the punch and die drawings the tooling clearance is shown to be 25 microns and the compression mixture has a large 25 micron component there is the potential for powder to get lodged between the punch and die clearance. Decreasing the tabletting speed will also increase the time available for the air trapped between the granules to escape, thereby leading to the potential for decreased air pressure in the die, particularly for high-porosity beds. High-speed compaction using tooling with deep curvatures, such as are used for coated tablets, contributes to air entrapment at the top punch. Hence, using shallower tooling can alleviate this cause of cracking.

A second cause of cracking is associated with undue elastic compression of the tablet due to the use of too high a pressure at the compaction stage. During ejection, elastic recovery of that part of the tablet protruding from the die gives rise to lateral forces, which rupture the intergranule bonds and the tablet then caps or laminates. One source of over compression will occur when the porosity of the formulation drops below 10%. The maximum tensile strength of a formulation is at 15% porosity (which is equivalent to a solid fraction of 0.85). Hence, if the desired tensile strength is not being achieved than the main formulation approach would be to increase the binder or compression aid level. In addition, if porosity drops below 15% then water penetration normally becomes harder and so disintegration times become prolonged.

The viscoplastic–elastic behaviour of the formulation components may also be a contributor to cracking due to the elastic nature of some of the materials causing the tablet to spring apart. The response is often time-dependent. As speed is increased, the relative elastic component of a given material also increases, giving rise to a higher incidence of cracking. Hence as compression speeds are increased, the occurrence of cracking and lamination of compacts tend to become more prevalent. Hence reducing compaction speed may help. An alternative processing method of overcoming this and effectively increasing the relaxation time is to use pre-compression prior to the main compression. Most rotary presses have two rollers in series after the die filling step, which apply the pressure to the tablet prior to compressing it. The first roller is a pre-compression roller and the second is the main compression roller. Typically the pre-compression is at 10% of the main compression pressure. However this is very much formulation-dependent. The optimal tensile strength of some formulation being produced from having a pre-compression larger than the main one.[24]

Elastic recovery itself will not necessarily result in lamination. Lamination will only occur if the inter-particle bonding cannot accommodate this elastic recovery. Hence the formulation options are to either increase the binder level, or change the type of binder in the granule. An alternative approach is to incorporate polymeric materials as compression aids, such as celluloses, which undergo less elastic recovery or which can absorb the stresses. For organic material the moisture content can also be important as the level of residual moisture in a polymer can affect its deformation properties. Typically the drier the material the more brittle it becomes as water tends to have a plasticizing effect.

Unwanted viscoelastic behaviour can sometimes be shown by high residual die wall stress. Die wall stress is the force remaining in the tablet in the die after compression and prior to ejection. It is measured using an instrumented die which records the stress exerted by the area of tablet in contact with the die wall (the "belly band"). Residual dies stress values below 20 MPa should be targeted. If high values are obtained then they can be reduced by including auxetic materials, such as low-substituted hydroxypropyl celluloses or modified starches.[25] Tapered dies can also be used as they have the advantage of increasing the volume available for the tablet to expand into radially, hence reducing residual die wall pressure.

Sticking of the compact to the die wall or punch components can also induce stresses resulting in cracking and can be controlled by adjusting the lubricant levels. Lubricants will minimise die wall friction and prevent the adhesion of the granules to the punch faces and, hence, can be manipulated to overcome cracking and lamination. Varying the level of lubricant or the ratio of external to internal lubricant can both help. An alternative approach is to spray the lubricant into the punch and die cavity immediately before die filling and hence directly coat the surfaces of the tooling. This latter approach requires modification of the tablet press. One approach is to spray the lubricant dry into the punch and die cavity. The other approach is to disperse the lubricant into a volatile solvent and to spray the tooling surfaces *via* a liquid nozzle. Magnesium stearate, as well as other lubricants, has been applied by both methods.

Wearing of the dies bores, particularly by hard inorganic fillers, can also lead to tablet damage. Wear in the dies takes place usually about the point of compression and results in a circular depression within the die. A compact compressed in this cavity has therefore to be forced out through the smaller aperture in the top of the die resulting in shear and lamination. This can be resolved by using wear-resistant steel for the dies or specially hardened die inserts.

4.7.2   Low Tensile Strength

In general, the higher the compaction pressure then the denser the compact will be and hence the higher the resulting tensile strength of the compact. Consequently too low a compaction pressure will lead to low tensile strength or "soft" and crumbly compacts. Alternative reasons are excessive coverage of the granulation by a lubricant, such as a stearate, reducing the potential to form strong interparticle bonds. This over lubrication can be caused by:

- Too high an initial level of the lubricant,
- Excessive shear during the lubrication stage,
- Excessive lubrication time.

Over lubrication, particularly during formulation development assessment, can also occur if incomplete sets (for example half sets or singles sets) of tooling are used on rotary presses to conserve granule usage. This will be because of extended residence time in the feeder resulting in overworking of the granule particularly feeders with paddles.

An additional cause can be the weakening of the intergranular bonds by air entrapment, even when this is not sufficient to cause capping.

4.7.3   Picking and Sticking

In some instances, a small amount of the compact material may stick to the tooling surfaces' faces and is referred to as sticking. As compacts are repeatedly made in this station of tooling, the problem gets worse as more and more material gets added to that already stuck to the punch face. The problem tends to be more prevalent on upper punches. The root cause is usually insufficient or a limited extent of lubrication, although surface roughness of the tooling can also play a part. An alternative approach is to use coated tooling. The coating though has a finite life and so can only be used for a limited number of batches before replacement or re-coating. Picking is compressed granule adhering to embossed detail on the punch face and is more often observed for compacts with fine embossing, where the design of such geometric details becomes important. Hence, it can be alleviated by careful lettering design.

4.7.4   Pitted or Fissured Surface

The most likely cause of a fissured surface, if it is not due to picking or sticking, is the presence of granules that are uniform in size and lack the smaller particles to fill the voids. Generally the problem can be resolved by broadening the particle size distribution of the granules, for example, by changing the granulation or milling conditions or for direct compression by changing the input materials. Provided that this does not lead to other problems such as cracking or segregation.

4.7.5   Chipping

Sometimes compacts after leaving the press, or during subsequent handling and coating operations, are found to have small chips missing from their edges. This fault is described as "chipping" and, in addition to the obvious formulation deficiencies, may be caused by compaction conditions which make too soft (low mechanical strength) or too brittle tablets. Incorrect machine settings, especially the ejection take-off plate being set too high, and excessively harsh handling of compacts after they leave the press, may be additional factors. Friability testing is employed as an indicator of an inherent tendency for a given batch of product to chip.

4.7.6   Binding in the Die

This is characterised by excessive side scraping of the die with the compact ejection forces being high, with the resulting compact edges being rough and scored. The root cause results from high die wall friction. This in turn could be caused by poor lubrication or blemished and worn dies or tooling. An alternative cause is too large a clearance between the lower punch and die bore, resulting in trapping of powder which is compacted to form a hard film that hinders free movement of the lower punch.

4.7.7   Uneven Weight Control

Poor weight uniformity is usually due to poor die filling. This can be due to either poor flow characteristics of the granule, or due to inadequate filling mechanisms on the compression machine. Granules or powders that are too large, too fine or contain a large proportion of fine material, or are incorrectly lubricated or have components with widely differing densities or sizes, may all contribute to weight variation.

If it is due to poor granule flow then the addition of glidants, such as silica or talc, can be employed. Some particles may acquire a frictional electrostatic charge when handled and this mutual repulsion of the particles and may be sufficient to impede die filling. Talc (at up to 1%) or sodium lauryl sulphate (at up to 2%) are substances which have been used to reduce this charging[26] and which can also have lubricant and anti-adherent properties. Lubricants, such as magnesium stearate, may or may not promote granule flow, depending on the level at which they are used. Higher levels tending to impede flow. Occasionally, with high-weight tablets, more uniform weight and improved appearance can be obtained by slowing the machine speed so allowing more time for die-cavity filling.

4.7.8   Disintegration and Dissolution

Disintegration is the time taken for the tablet to break apart into its primary particles in a fluid, normally aqueous. Dissolution is a measure of the release of the active ingredient from the compact into solution.

Protracted disintegration or dissolution can be because the tablet either rapidly breaks down to form large particles that persist for a long period, or fine particles are produced, but the overall disintegration time is excessive. This can be caused by:

- Conditions that inhibit the penetration of water, such as a high degree of compaction. Water can generally only gain access to the inside of a compact *via* pores. Hence if the compact is compressed at high pressure then its porosity is likely to be too low to allow water ingression. Below 15% porosity is where water ingression may become limited.
- Hydrophobic tablet ingredients, excessive quantities of fatty lubricant or high shearing during lubrication can all lead to disintegration and dissolution issues. Water will not readily penetrate hydrophobic powders. A potential issue therefore is the use of hydrophobic lubricants, such as the stearates, which, if in high concentrations, can prevent penetration of water and can decrease dissolution and disintegration. Addition of a wetting agent in the granule formulation can assist in the penetration of water into the compact. Alternatively, a hard polyethylene glycol may be employed to provide soluble entry points into the tablet structure.
- Inefficient bursting action resulting from the use of the wrong type of disintegrant or insufficient levels of it. Occasionally too low a degree of compression may lead to a long disintegration time as the disintegrants may not have a dense enough structure to deform against and break open. Sometimes more efficient bursting action is obtained if part of the disintegrant is split intragranularly as well as extragranularly. Too much binder or too much disintegrant can also lead to gelling and dissolution issues.

4.7.9 Mottled Appearance

This is typically seen with coloured granules. This can be due to dye migration to either the small or large granules during the granulation process. Alternatively it can be an optical phenomena due to the smaller particles providing a background of a slightly different hue which shows up the larger granules on the compact surface.

4.8 Formulation Considerations for Specific Tablet Dosage Forms and Processes

4.8.1 Multilayer Tablet Formulation

Multilayer tablets are formed by compressing two or more layers together to form a tablet. The sequence is that the first, bottom layer is filled into a die using the normal filling mechanisms, such as feeder paddles and punch pull down of a tablet press. This first layer is then normally tamped lightly, typically less than 10 MPa. The next later is then filled, but usually without punch pull down to assist die fill. This is then lightly tamped (or pre-compressed) before the main compression occurs. The tablet is then ejected as normal.

Formulation of multilayer tablets, such as bilayers and trilayers, is inherently the same as for monolayers. However there are some additional considerations which need to be factored in for robust manufacture. In particular, the powder for the second layer needs to be able to flow very well, for example having less than 10% Carr index, as punch pull down cannot be applied to assist die filling. Particles size of the powders for compression cannot be too fine as otherwise there will be cross contamination between the two layers due to carry-over on the die table. Lubrication also needs to be carefully controlled so that the two layers adhere to each other. High levels in particular should be avoided. In addition, highly elastic materials should be avoided otherwise the two layers can spring apart.

4.8.2 Minitablets

Minitablets are usually defined as tablets with a diameter of less than 3 mm.[27] They are produced with multiple tip tooling. Minitablets are used particularly when a wide dosage range is desired and when swallowing or administration of normal-size tablets could be an issue. For examples paediatric dosing or for administration to dogs or cats. Minitablet formulation is the same as for larger tablets. However the tabletting mix requires excellent powder flow due to the small diameter of the dies. The force weight control consideration for modern presses mean that a combined tablet weight per tool of at least 150 mg should be targeted. So, for example, if the mini tablet weighs 15 mg then there should be at least 10 multitips per tablet punch. Even with excellent flow and precise control of tablet forces there can still be high weight or content uniformity variation. The impact of this high variation can be minimised by having doses of at least three to five minitablets per administration.

4.8.3 Continuous Processing Formulation Considerations

Continuous processing can be applied to all three main types of granulation. One additional demand of both the excipients and API is that they have adequate flow for weight metering purposes for accurate continuous dispensing of materials. This could well be a challenge, in particular for transferring batch wet-granulation processes to continuous processes where the prime reason for wet granulation was the poor flow of the API. Similar complications can also arise with transfer of direct compression and roller compaction to continuous processing where a well-flowing tabletting mixture was as the result of a poor-flowing API being diluted out with a well-flowing excipient. Binder addition and wetting times also need to be assessed when transferring batch wet-granulated processes to twin screw.[28]

An additional consideration for continuous direct compression is the effect of over lubrication during the lubrication stage.[29] In batch direct compression the premixed components are compressed into tablets with the lubricant mixed only for a short time prior to compression in order to minimise over lubrication. In contrast, with continuous direct compression, all the API and excipients, including the lubricant, are mixed together under the same conditions. Selection of brittle inorganic excipients, which are not sensitive to over lubrication, may need to be performed to prevent this effect.

Another difference between batch and continuous processing occurs if the tablets are to be film coated. This is with regards to the time for tablet relaxation, particularly the time taken for the tablet strengths to increase or decrease. In a batch commercial manufacturing process the compressed tablets will have at least an hour of relaxation time from the end of compression to initiation of coating because, for example, the tablets need to be reconciled and moved to a different processing room for coating. For a continuous line the time from the end of compression to start of coating might only be a few minutes. The original work in this area of compact ageing[30] showed the potential for mechanical strength changes in comparatively short time periods. In particular sodium chloride compacts doubled in mechanical strength in a 60 minutes period post compaction. This work is also of interest as it indicates the effects of temperature on the rate of tablet strength increase. Hence, material relaxation post compaction becomes an additional area of assessment for continuous processes, which is not necessarily the case for batch processes.

Hence, the need for enhanced flowability and for measurement of material relaxation for both API and excipients, plus wetting and penetration times for wet binders, will need to be assessed to allow processes to be successfully run continuously.

References

1. K. G. Pitt and M. G. Heasley, *Powder Technol.*, 2013, **238**, 169.
2. G. D. Tovey, *Pharm. J.*, 1987, **239**, 363.
3. J. Hersey, *Powder Technol.*, 1975, **11**, 41–44.
4. R. J. Hintz and K. C. Johnson, *Int. J. Pharm.*, 1989, **51**, 9.
5. R. L. Carr, *Br. Chem. Eng.*, 1970, **15**, 1541.
6. A. W. Jenike, *Utah Eng. Exp. Stn., Bull.*, 1964, **123**, 1.
7. C. C. Sun, *Powder Technol.*, 2010, **201**, 106.
8. *United States Pharmacopeia and National Formulary*, USP 41-NF 36, General Chapter 1217 United States Pharmacopeia Convention, Rockville, MD, 2018.
9. K. G. Pitt, R. J. Webber, K. A. Hill, D. Dey and M. J. Gamlen, *Powder Technol.*, 2015, **270**, 490.
10. B. R. Rohrs, G. E. Amidon, R. H. Meury, P. J. Secreast, H. M. King and C. J. Skoug, *J. Pharm. Sci.*, 2006, **95**, 1049.
11. S. H. Yalkowsky and S. Bolton, *Pharm. Res.*, 1990, **7**, 962.
12. W. J. Thiel and L. T. Nguyen, *J. Pharm. Pharmacol.*, 1982, **34**, 692.
13. R. Jullien, P. Meakin and A. Pavlovitch, *Phys. Rev. Lett.*, 1992, **69**, 640.
14. M. Leane, K. Pitt and G. Reynolds, *Pharm. Dev. Technol.*, 2015, **20**, 12.
15. S. van den Ban and D. J. Goodwin, *Pharm. Res.*, 2017, **34**, 1002.
16. J. M. Merritt, S. K. Viswanath and G. A. Stephenson, *Pharm. Res.*, 2013, **30**, 203.
17. S. Malkowska and K. A. Khan, *Drug Dev. Ind. Pharm.*, 1983, **9**, 331.
18. R. J. Roberts and R. C. Rowe, *Chem. Eng. Sci.*, 1987, **42**, 903.
19. M. Suzuki and T. Oshima, *Powder Technol.*, 1985, **44**, 213.
20. C. Kurrer and K. Schulten, *Phys. Rev. E*, 1993, **48**, 614.
21. J. I. Wells, *Pharmaceutical Preformulation- the Physicochemical Properties of Drug Substances*, Taylor and Francis, London, 1988.
22. P. Zarmpi, T. Flanagan, E. Meehan, J. Mannb and N. Fotaki, *Eur. J. Pharm. Biopharm.*, 2017, **111**, 1.
23. H. Nie, W. Xu, L. S. Taylor, P. J. Marsac and S. R. Byrn, *Int. J. Pharm.*, 2004, **517**, 203.
24. O. F. Akande, M. H. Rubinstein, P. H. Rowe and J. L. Ford, *Int. J. Pharm.*, 1997, **157**, 127.
25. H. Takeuchi, S. Nagira, H. Yamamoto and Y. Kawashima, *Int. J. Pharm.*, 2004, **274**, 131.
26. R. Chopra, F. Podczek, J. M. Newton and G. Alderboran, *Eur. J. Pharm. Biopharm.*, 2002, **53**, 327.

27. A. Aleksovski, R. Dreu, M. Gašperlin and O. Planinšek, *Expert Opin. Drug Delivery*, 2015, **12**, 65.
28. P. Beer, D. Wilson, Z. Huang and M. De Matas, *J. Pharm. Sci.*, 2014, **103**, 3075.
29. M. A. Järvinen, J. Paaso, M. Paavola, K. Leiviskä, M. Juuti, F. Muzzio and K. Järvinen, *Drug Dev. Ind. Pharm.*, 2013, **39**, 1802.
30. J. E. Rees and E. Shotton, *J. Pharm. Pharmacol.*, 1970, **22**, 17S.

CHAPTER 5

Suspension Quality by Design

BRIAN A. C. CARLIN

Lawrenceville, NJ, USA
*E-mail: brianac.carlin@gmail.com

5.1 Introduction

Pharmaceutical suspensions are dispersions of particulate drug(s) in a liquid vehicle, usually aqueous.[1] The particle size can range from the colloidal (nanometre) to millimetres. Depending on their density, finer particles may be self-suspending if the energy of Brownian motion exceeds the gravitational force on the particle. The critical particle size for polystyrene latex particles of relative density 1.05, suspended in water, is 650 nm.[2] Nanoparticles have greater surface-to-volume ratios, which makes them more susceptible to Ostwald ripening[3] where smaller particles preferentially dissolve and recrystallize on larger particles, giving a progressively coarser particle size distribution.

Although there is no absolute cut-off from colloidal dispersions, in practice the lower end of drug particle sizes for suspension development is usually around one micron, often as a result of air-jet milling ("micronisation") to ensure that dissolution of poorly soluble biopharmaceutical classification system (BCS) Class II/IV drugs is not limiting on oral bioavailability.

A good suspension should be capable of suspending millimetre-sized particles but, if not already constrained by dissolution, there may be an upper limit on particle size due to gritty mouthfeel, or delivery (*e.g. via* a nozzle for

Drug Discovery Series No. 64
Pharmaceutical Formulation: The Science and Technology of Dosage Forms
Edited by Geoffrey D. Tovey
© The Royal Society of Chemistry 2018
Published by the Royal Society of Chemistry, www.rsc.org

nasal sprays). This chapter focuses on aqueous suspensions of drug particles in the 1–100 micron range for oral, topical or nasal delivery.

5.2 Suspension Quality by Design (QbD)

The quality of a design depends on how well the product performs against predictions based on the design inputs. In QbD product critical quality attributes (CQAs) are selected to meet a quality target product profile (QTPP), the "prospective summary of the quality characteristics of a drug product that ideally will be achieved to ensure the desired quality, taking into account safety and efficacy of the drug product".[4] Critical material attributes (CMAs) and critical process parameters (CPPs) can then be identified which need to be controlled to guarantee the finished product CQAs. The desired quality may also include manufacturing robustness (right first time) as stated in the proposed FDA quality metrics initiative.[5]

Suspensions are designed to prevent sedimentation and aggregation of the dispersed drug particles, therefore the number one CQA is dose uniformity. Dose uniformity is controlled by the suspension rheology, more specifically the elastic or solid properties of the vehicle. Apparent viscosity should never be a CQA for reasons of dose uniformity, because viscosity is a liquid property, not predictive of the structure or suspending power of the vehicle. Viscosity controls the rate of particle sedimentation but does not prevent sedimentation. Increasing viscosity to slow sedimentation is counterproductive, as the ability to pour or spray the suspension will eventually be affected. Drug particles which do sediment in a high-viscosity vehicle may not be redispersible by shaking. Viscosity only becomes a suspension CQA if it needs to be controlled to ensure pourability or sprayability. High viscosity is less of a constraint for semisolid suspensions, such as topicals, but even then it is not the high viscosity that suspends the drug, but the rheology.

The distinction between viscosity and rheology is eloquently illustrated by Faith Morrison in her discussion of why the surface of thinner mayonnaise retains the disturbance from whoever made the last sandwich whereas the disturbed surface of thicker honey is self-levelling within seconds.[6] Viscosity is not structure and should not be used as a design criterion for suspensions.

5.3 Suspension Types

Suspensions can be divided into three types, dependent on the degree of aggregation of the suspended particles; permanent, flocculated or coagulated (caked) as illustrated in Figure 5.1.

In a permanent suspension the original dispersion at time of manufacture is maintained throughout the product lifecycle. There is no

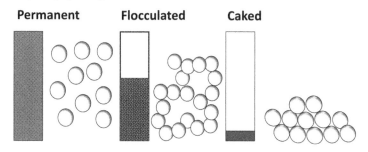

Figure 5.1 Types of suspension: permanent, flocculated or coagulated (caked).

sedimentation or aggregation of the particles and drug concentration is uniform throughout the vehicle. Shaking the bottle is not required for dose uniformity, but may be required for ease and smoothness of pouring. Structure is imparted throughout the vehicle by use of a structured vehicle former (SVF) as suspending agent. Structuring of the vehicle is independent of the suspended drug. Although cited by Remington[1] the concept of the structured vehicle suspension is relatively unaddressed in the literature. Googling "suspension" and "structured vehicle" (excluding automotive and investment usage) gives 1270 hits (0.3%) *versus* 416 000 for "suspension" and "flocculation".

A floc is a weak, open aggregate which forms on sedimentation but is easily redispersible on shaking the bottle. Unless the floc volume approaches 100% of the vehicle such suspensions usually have a characteristic clear supernatant layer above the floc. As the drug is suspended in the floc the drug concentration is not uniform throughout the vehicle. Shaking the bottle is required to redisperse the floc and restore content uniformity prior to dosing. Structure in the floc can be imparted by use of a floc former as a suspending agent, independent of the drug, or by using a flocculating agent to flocculate the drug itself. As low floc volumes are inelegant, and more difficult to redisperse, flocculating the drug itself is more suited to high drug loadings.

A coagulate is a strong close-packed aggregate, which is not redispersible on shaking the bottle, often referred to as caking. As such it represents the failure of the suspension to resist both sedimentation and aggregation. With a clear supernatant a coagulated dispersion is similar in appearance to a low floc volume suspension, the difference being that the floc is redispersible on shaking, whereas the coagulated dispersion (cake) is not. The term "clay" has also been used for a cake[7] but this usage is avoided in this chapter to avoid confusion with (smectite) clay suspending agents.

The foregoing discussion assumes that the supernatant layer above the sedimenting suspension is clear. In a suspension with multiple insoluble components the possibility of suspended material obscuring drug

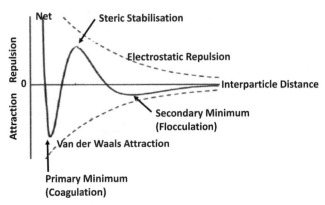

Figure 5.2   Interaction forces on close approach of two particles.

sedimentation should be ruled out during development by analytical confirmation of content uniformity.

The terms "coagulation" and "flocculation" are used in this chapter in accordance with the Derjaguin–Landau–Verwey–Overbeek (DLVO)[8,9] primary and secondary minima respectively. DLVO theory looks at the net interaction of electrostatic repulsion and van der Waals attraction as two particles come together in a liquid medium. As shown in Figure 5.2 attraction (negative in DLVO theory) increases with decreasing interparticle distance until overlap of the electrical double layers results in a much higher (positive) repulsion. This gives the primary minimum corresponding to irreversible coagulation (at least in terms of shaking the bottle). Such a suspension would rapidly cake. Sedimentation promotes aggregation and larger aggregates sediment faster.

Figure 5.2 also shows the imposition of a barrier to close approach, a shallow secondary minimum associated with flocculation. The particles aggregate at a greater interparticle distance due to interaction of adsorbed polymer. The shallowness of the secondary flocculation minimum reflects the ease of redispersibility after sedimentation. For stability there must also be an energy barrier between the primary and secondary minima otherwise supplying energy by shaking the bottle would promote coagulation rather than redisperse the particles. This maximum corresponds to steric stabilisation by the adsorbed polymers or surfactants on the particles, which can involve multiple mechanisms. The complexity of steric stabilisation and flocculation has been reviewed by Shi.[10]

A structured vehicle keeps the suspended particles at an interparticle distance in excess of the closer interaction distances. DLVO is a convenient framework for discussion of flocculation and coagulation but is not a practical way to design suspensions. There are also other non-DLVO forces, such as hydration force and hydrophobic interactions.[11]

5.4 Suspension Formulation Strategies

Three formulation strategies for suspensions are normally listed:[1]

1. Structured vehicle
2. Flocculated suspension
3. Flocculated drug in structured vehicle

The latter is not a logical design approach. If the structured vehicle is functional the drug particles are prevented from sedimenting or coming close enough to form a floc. In addition to the added complexity of incorporating redundant flocculating agent(s) there is the risk of flocculating the structured vehicle.

Flocculation by design ("Controlled Flocculation") can be achieved by use of a flocculating agent to flocculate the drug and other components. In flocculating the drug itself, not only do you have to tailor the flocculating agents to the specific drug but there may be additional CMAs imposed on the drug substance. From a design point of view it is better to uncouple the structuring from the drug substance, using the structured vehicle approach.

The term "structured vehicle" implies that solid properties, such as gel strength, are imparted to the liquid vehicle, without which sedimentation will occur. Unfortunately, there is no official definition of this term and it is often inappropriately used in conjunction with the terms "viscosity" and "sedimentation". Viscosity is a liquid property and sedimentation implies absence of structure. Structured vehicles will exhibit a wide range of apparent viscosities but at rest, under the very low shear exerted by a suspended particle, the apparent viscosity will tend towards infinity. An apparent viscosity is never a design criterion for structuring a vehicle, although it may have to be controlled for reasons other than suspension, such as pourability or sprayability.

5.5 Suspending Agents

Many excipients have been listed as suspending agents but it would be more accurate to describe the majority as excipients which have been used in suspensions. For example, surfactants are often used in suspensions as stabilisers or dispersants, and would not function as stand-alone suspending agents. The list of suspending agents in the latest edition of Remington[1] has been considerably reduced to essentially just two categories, clays and water-soluble polymers. Further information on the suspending agents cited in this chapter may be found in the Handbook of Pharmaceutical Excipients.[12]

5.5.1 Clays

Smectite is the mineralogical term for a group of trilayer clays which includes Bentonite (montmorillonite, Vanatural®) or magnesium aluminium silicate (Veegum®).[13] Once the clay is hydrated, liberating the plate-like or flaky crystals, the weakly positive platelet edges are attracted to the negatively charged platelet faces. A three-dimensional colloidal structure forms, commonly called the "house of cards", imparting structure to the aqueous dispersion.[13] Such electrostatic stabilisation is sensitive to ionic strength and pH. Increasing ionic strength shrinks the repulsive electrical double layer and allows closer approach between edges and faces. Initially, this can strengthen structure but too close an approach risks van der Waals interaction between the faces, collapsing the house of cards and losing structure.[13] Changing pH can also reduce the charge. The mobility of charged particles under the influence of an external electrical field can be used to calculate the electrokinetic (zeta) potential. Zeta potentials below 10 mV are associated with rapid agglomeration whereas values above 40 mV are typical of stable systems.[14] Electrostatically stabilised systems may also be sensitive to high shear but are relatively insensitive to temperature changes compared with polymer solutions. Clays may be combined with soluble polymers as stabilisers. The term "clay" has also been used to refer to compact sediments,[7] referred to as "cake" in this chapter.

5.5.2   Water-soluble Polymers

Dividing polymer-suspending agents into categories such as natural, semi-synthetic (cellulosic), or synthetic, or anionic *versus* non-ionic,[1] does not correlate with utility as suspending agents, except for some anionic polymers, which may exhibit counter-ion-dependent rheologies. Kappa-carrageenan gives strong brittle gels with potassium, iota carrageenan gives thixotropic gels with calcium and lambda carrageenan is non-gelling. Alginate (a copolymer of mannuronic and guluronic acids) will also give strong brittle gels with calcium dependent on the number of multiple guluronic acid sequences (blockiness). Brittle gels are of limited utility for suspensions as they break or crumble under stress, rather than flow.

At high enough concentrations, all water-soluble polymers will provide structure to suspend drug but the key application-dependent design consideration is whether useful structure will be provided before viscosity becomes limiting. For semi-solid topical applications or toothpastes, which are extruded by exerting pressure on a tube this is less of a restriction. However, if the liquid product has to be poured or sprayed most water-soluble polymers will not provide structure at concentrations of acceptable viscosity. Viscosity slows but not does stop sedimentation. Viscosity also hinders redispersion on shaking the bottle.

Polymers are often characterised and specified by dilute solution viscosities, which may be dependent on average molecular weights. Unfortunately, such apparent viscosities are not predictive of rheological performance at the higher concentrations necessary for suspension. Above a certain critical concentration C^*, corresponding to onset of polymer coil overlap, a gel network may form, giving structure which can prevent sedimentation.[7,15] Fu *et al.*[16] found that dilute solution viscosities of several grades of sodium alginate were not predictive of viscoelastic properties at higher concentrations. In contrast to clays, polymers are less sensitive to electrolytes, pH and high shear but exhibit temperature-dependent thinning.

A better measure of the utility of polymers as potential suspending agents is the rheology *vs.* concentration profile. By this measure most polymers listed by Remington[1] are thickening agents, where excessive viscosity limits their use in liquid suspension applications to concentrations below the onset of structure. Only xanthan, carboxymethylcellulose and carrageenan (iota) would be regarded as polymer SVFs for liquid suspension applications.

- Xanthan has an unusually low critical concentration for coil overlap (0.01%), related to its conformation as a helical structure with a large axial ratio.[7] As xanthan is described as more pseudoplastic than thixotropic[17] there is potential for a conflict between suspension and viscosity if a mobile liquid suspension is required at the normal use level of 0.2–0.3% w/v.
- Thixotropic grades of carboxymethylcellulose are restricted to a low degree of substitution (0.7) and non-uniformity of substitution. The unsubstituted cellulose regions can hydrogen bond to a similar region on an adjacent molecule, leading to the build-up of a loose gel network.[18] The non-uniformity of substitution, or blockiness, of the carboxymethylcellulose is therefore a CMA for suspension.
- Iota-carrageenan molecules cross-link *via* divalent ion (calcium) chelation.

5.5.3   Dispersible Cellulose

"Microcrystalline cellulose and carboxymethylcellulose sodium" NF,[19] is a composite or coprocessed excipient, defined as "a colloid-forming, attrited mixture" rather than a blend. Zhao *et al.*[20] showed that blends of the components, even when co-spray-dried, could not match the rheological performance of commercial co-attrited co-spray-dried material, Avicel® RC591.[21] This SVF is also known as dispersible cellulose BP.[22] The key advantage of dispersible cellulose as an SVF is that structure is provided without the higher viscosities associated with the use of soluble polymers. In quality by design the ability to uncouple structure from viscosity avoids the criticality or product weakness inherent in balancing structure *versus* viscosity.

5.6   Suspension Rheology and Rheometry

Suspensions cannot be designed or controlled by viscosity. Some understanding of other rheological parameters more relevant to structure is required.

5.6.1 Suspension Rheology

Rheology is the study of flow and deformation. A liquid will flow under stress or pressure and a solid will deform and possibly break. Both stress and pressure are force per unit area, the difference being that the force is tangential (shear) for stress and normal for pressure. Liquids cannot suspend particles unless the density of the particle matches that of the liquid (neutral buoyancy) or the particles are so small that Brownian motion outweighs the effect of gravity. Increasing the relative density of an aqueous vehicle beyond 1.2 (typical of high-solids syrups) is not feasible. Many organics have a relative density of approximately 1.5, with inorganics even higher (*e.g.* titanium dioxide, 4.2).

Viscosity (η) is a measure of the resistance of a fluid to an applied stress (τ). The response of a liquid (shear gradient, γ) to a disturbance is damped as the viscosity increases. The higher the viscosity the lower the fluidity.

$$\eta = \frac{\tau}{\gamma} \qquad \text{Pa s} = \frac{\text{Pa}}{1/s}$$

Newtonian liquid has a viscosity that is independent of the shear rate as shown in Figure 5.3. Water is an example with a single viscosity of 1 mPa s.

Newtonian liquids have no structure so cannot be used to suspend particles. Increasing the viscosity will slow but not stop sedimentation, and render redispersion by shaking impossible. Most polymer solutions and dispersions of solids are non-Newtonian and exhibit a range of viscosities dependent on shear rate (pseudoplastic) or dependent on both shear rate and shear history (thixotropic), as shown in Figure 5.3.

Pseudoplastics show no time dependency because structure rebuild is fast enough to recover from the destructuring caused by measurement. They are described as shear thinning because the viscosity decreases with increasing

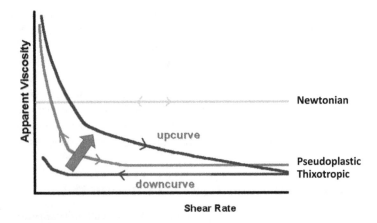

Figure 5.3 Viscosity profiles *vs.* shear for Newtonian, pseudoplastic and thixotropic liquids.

shear and *vice versa*. The same viscosities are measured when decreasing the shear rate of measurement as were measured when increasing the shear rate. Shear rate dependence may break down at extremely low or extremely high shear rates, the lower and upper Newtonian viscosities, which are experimentally inaccessible to simple viscometers. The upper Newtonian viscosity represents total destructuring, where polymer entanglements and/or particle interactions have been streamlined out under very high shear rates. The upper Newtonian viscosity is relevant to high-shear applications such as spraying. The lower Newtonian viscosity, sometimes referred to as zero shear viscosity, reflects shear rates too low to have a net destructuring effect, and is relevant to suspensions.

Pseudoplastics are unsuitable for liquid suspensions. A bottle of ketchup liquefies when shaken in the bottle but is immediately semi-solid and not easily pourable once the shaking stops (back to zero shear). However, they are ideal for semi-solid applications such as toothpaste, which flows under extrusion from the tube but immediately regains its semisolid structure on the brush, without flowing off.

Thixotropes are pseudoplastics where the rate of recovery is slower than the rate of destructuring due to the initial measurement. Viscosity decreases with increasing shear but does not recover to the same extent when the shear is decreased. Over time (as indicated by the arrow on Figure 5.3) structure rebuilds and the original shear thinning profile is regained. Apparent viscosity depends on both shear rate and shear history. Applying a specific shear rate to a thixotrope at rest will give a time-dependent decrease in viscosity and a subsequent time-dependent recovery of viscosity when the shear is discontinued. This is the ideal rheological profile for a liquid suspension. Structure and no sedimentation on standing in the bottle, destructuring on shaking the bottle (applying shear), remaining destructured and mobile for pouring, and finally regaining structure on standing.

If shear stress is plotted against shear rate as in Figure 5.4 Newtonian systems will give a straight line with zero intercept where the gradient is

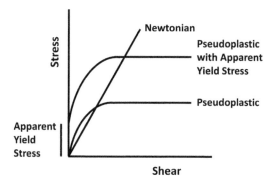

Figure 5.4 Stress *vs.* shear rheograms for Newtonian and pseudoplastic liquids, with or without an apparent yield stress.

the viscosity. Pseudoplastics have high attack with decreasing rates of stress increase as the system progressively destructures at higher shear rates. A positive intercept on the *y* axis may be observed in some non-Newtonian systems, commonly referred to as a "yield stress". Qualitatively this is a good indicator of structure and implies that a minimum force must be applied to initiate flow (hopefully higher than that exerted by a suspended particle). However, it is more correct to describe such intercepts as a method-dependent apparent yield stress. If the time of holding at each stress is increased then a lower apparent yield stress will be measured.

Whether pseudoplastic or thixotropic, a non-Newtonian system cannot be characterised by a single apparent viscosity, especially if the shear rate of measurement is unknown or unrelated to the application shear rates. As a minimum, a two-point viscosity test should show higher values for a sample tested after recovery than the value for a freshly destructured sample. Viscosity is generally an inappropriate attribute to design or control the quality of pharmaceutical suspensions, especially in the absence of meaningful product rheological profiles. More appropriate measurements of structure, such as gel strength or thixotropy, are discussed under rheometry.

5.6.2   Suspension Rheometry

Rheological testing of suspensions can be divided into rotational (destructive of structure) and oscillatory (non-destructive at low amplitude). Rotational rheometers may be controlled stress, controlled shear or both. In controlled stress, stress is the independent variable and the shear rate induced in the sample is the response or dependent variable. In controlled shear, sample stress is measured in response to changes in shear rate. The sample is sheared in a narrow gap between a stationary cup and rotating bob, or between a stationary plate and either a rotating cone or parallel plate as shown in Figure 5.5. Oscillatory rheometry utilises the same sample geometries. The cup and bob is used for more liquid samples. The angle of the cone

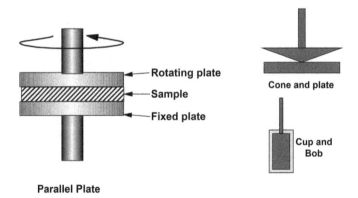

Figure 5.5   Rotational and oscillatory rheometer geometries.

is designed to ensure a constant shear rate across the radius but cannot be used if there are particles larger than the minimum clearance near the centre of the cone. The parallel plate gives an average shear rate but can be set at a gap large enough to accommodate large particles. Shear rate is the speed of the rotational device (ms^{-1}) divided by the gap (m) which gives shear in reciprocal seconds (s^{-1}).

Simple viscometers employ rotating spindles, bobs or plates immersed in the sample. These "infinite gap" viscometers provide no information on the shear rate associated with the apparent viscosity calculated from the torque on the rotating spindle.

5.6.2.1 Rotational Rheometry

Rotational rheometry imposes an infinite strain on the sample and is therefore destructive of structure. Sample handling and loading onto the rheometer is also destructive and needs to be controlled, including equilibration time after loading and avoidance of evaporation. The hysteresis technique[23] consists of increasing and decreasing the shear rate between zero and a maximum value, either as a continuous ramp or a series of small steps. When shear stress is plotted *versus* shear rate, a thixotropic sample will describe a hysteresis loop. An example rheogram is shown in Figure 5.6 using a controlled shear ramp to a maximum shear rate (upcurve) and back to zero again (downcurve).

The spike on the upcurve is an artefact of the controlled shear mode, as the instrument does not know how much stress to apply to the sample to achieve the set shear rate. Such spikes are not seen in controlled stress mode. Such spikes in controlled shear, and apparent yield stresses in controlled stress, are method-dependent in that they will be decreased with longer ramp times, allowing more time for flow to occur. However, they are good qualitative indicators of structure.

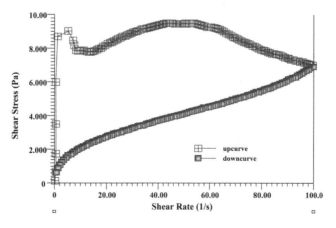

Figure 5.6 Thixotropy of 1.2% w/v dispersible cellulose dispersion.

The downcurve reflects a partially destructured sample being measured on decreasing the shear relative to the initial degree of structure encountered during the upcurve. The area between the curves, or hysteresis, is often regarded as a measure of thixotropy but in practice it merely reflects the degree of destructuring due to the measurement. With repeated measurements, the cumulative destructuring will eventually lead to a convergence of the downcurves, where measurement destructuring is balanced by rebuild. Applying higher shear will lead to more destructuring, so such a hysteresis loop is better described as a relative thixotropy.[24] Dapčević *et al.*[25] recommend holding the sample at constant high shear between the upcurve and downcurve so that the downcurve represents the fully destructured sample.

An alternative method of characterising thixotropy[24] is to shear the sample at a given shear rate until a steady-state viscosity is obtained, and then stepping up the shear rate and holding until a new lower steady state viscosity is achieved. Shear rate can then be stepped down to check reversibility. The growth of viscosity after a sudden decrease in shear rate provides the clearest indication of thixotropy.

5.6.2.2   Oscillatory Rheometry

Instead of rotating, the bob, cone, or plate, can be oscillated back and forth, at an amplitude low enough not to exceed the elastic limit of the sample. The measurement does not destroy the structure. A sinusoidal strain or stress can be applied in several modes, and from the strain, the resultant stress, frequency and phase shift the storage modulus G' (elastic, solid component) and the loss modulus G'' (viscous, liquid component) can be calculated.[7] G'' is a measure of gel strength.

A preliminary strain sweep at a fixed frequency (Figure 5.7) with increasing amplitude is used to determine the linear viscoelastic region, beyond which G' decreases as the increasing amplitude (strain) starts to destroy the

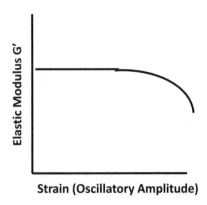

Figure 5.7   Strain sweep: elastic modulus *vs.* amplitude.

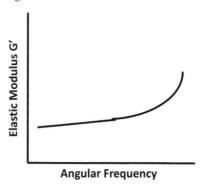

Figure 5.8   Frequency sweep: elastic modulus *vs.* frequency.

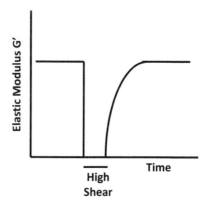

Figure 5.9   Elastic modulus *vs.* time, before and after high-shear destructuring.

structure. Oscillatory testing is otherwise conducted in the linear viscoelastic region.

A frequency sweep (Figure 5.8) applies an oscillatory strain within the viscoelastic region and observes the effect of increasing the frequency. Elastic effects predominate at higher frequencies so G' will increase.

Having selected a fixed amplitude (within the linear viscoelastic region) and frequency, a time sweep can be conducted where the time course of G' before and after shear is measured (Figure 5.9). Rotational shear between the before and after time sweeps is preferred to merely increasing the amplitude beyond the elastic limit. This is a useful method for measuring structure (re)build.

5.7   Suspension Formulation

5.7.1   Chemical Stability

The chemical stability of a suspension is dependent on the specific drug but there are some general considerations. Gross instability in the presence of water is a contraindication to formulation as an aqueous suspension.

Because the drug is in suspension the stability will be generally be greater than if it were fully in solution, but less than if presented as a solid dose form. As some of the drug will be in solution, buffer and/or sequestrant may be required if the drug exhibits pH-dependent stability or metal ion sensitivity.

Chemical stability may be worse in proportion to the water content of the suspension and dependent on how soluble the drug is in water. If the drug has a pH-dependent solubility profile, buffer may be required to minimise the amount of drug in solution.

If used, the stability of any preservatives must also be monitored, including both chemical degradation, and partition into plastic packaging.

5.7.2 Physical Stability

Before developing a formulation, the physical stability of the drug in water should be checked, including temperature cycling if the drug has a temperature-dependent profile. No matter how poorly soluble the drug there is always the possibility of phase-mediated polymorphic conversion, hydrate formation or crystal growth (coarsening) by Ostwald ripening.[3] More stable forms are less soluble so dissolution might be affected, in addition to other effects such as grittiness. *In situ* crystal growth can be exacerbated by water content, drug solubility and temperature fluctuations.

A functional structured vehicle should prevent sedimentation and flocculation but these are characteristic attributes of a suspension utilising controlled flocculation. Floc volume and redispersibility should be included in the stability program for a controlled flocculation suspension. Rheological characterisation is more difficult for controlled flocculation suspensions as the structure resides within the floc rather than the vehicle. If samples on the rheometer settle during testing the rheology will reflect the supernatant rather than the floc. Drug content uniformity should be monitored by top, middle and bottom sampling of the vehicle and floc for structured vehicle and controlled flocculation suspensions respectively.

5.7.3 Microbiological Considerations

Preservatives are required for a non-sterile multidose presentation. In addition to monitoring for potential degradation and partitioning into packaging the efficacy of the preservatives in the finished product must be demonstrated throughout the product shelf-life. Preservative efficacy testing should not be confused with microbial quality control. Absence of organisms is not predictive of ability to resist microbial contamination in the hands of the patient. Preservatives should also not be used to cover poor microbiological good manufacturing practice (GMP). The initial bioburden should low, consistent with GMP.

Preservative efficacy is a function of the formulation. In addition to individual preservatives other factors may influence efficacy. Preservative efficacy is greatly enhanced by minimising the water content, by including high levels of sorbitol, sugar or glycerol. Multiple preservatives may be synergistic, especially in combination with propylene glycol. Efficacy decreases with increasing pH, limiting preservative options.

Sodium benzoates and sorbic acid are effective below pH 5 at a level of 0.1% w/v. Combinations of methyl and propyl parabens (0.2/0.02% w/v) are effective at lower pH but propyl–butyl combinations are required above pH 7 due to hydrolysis of the methyl ester at higher pH (see formulation example in Section 5.7.4). Benzalkonium chloride and phenyl alcohol are often used in suspensions for nasal use.

Standardised challenge tests are described in the Pharmacopoeiae (*e.g.* Chapter 51 Antimicrobial Effectiveness Testing in the USP[19]) but the officially specified organisms should be supplemented with potential organisms likely to be encountered in a specific manufacturing plant, or in the hands of the patient. An inoculum of approximately 10^6 organisms must either not grow, or decrease by specified orders of magnitude, depending on the type of organism and product type.

5.7.4 Formulation

Drug, suspending agent and water are the minimum number of components for a suspension. High water contents are more tolerable in nasal spray suspensions where a low-viscosity vehicle is required to sufficiently reduce droplet size on atomisation to ensure delivery beyond the nasal valve in the posterior two thirds of the nasal cavity.[26] As a suspending agent dispersible cellulose has the best combination of low viscosity for atomisation with high structure to reduce clearance on deposition in the posterior nasal cavity.

It is desirable to limit the amount of water in oral suspensions to minimise those physical and chemical problems exacerbated by water content. Preservation of low water systems is also easier.

As discussed under suspension strategy the best design approach is to structure the vehicle, which can be independent of a specific drug. The SVF can be a water-soluble polymer, clay or dispersible cellulose. Soluble polymers will impart much more viscosity to the system than clays or dispersible cellulose. Although viscosity is not a critical attribute for structure it may need to be controlled for other reasons, such as pourability or sprayability. If the optimal level for structure exceeds the maximum tolerable level in terms of finished product viscosity, then there are two competing technological objectives, which have to be balanced, suspension efficacy *versus* pourability. This results in a criticality in the suspension which will cause problems on scale-up and predispose the suspension to effects from raw material variability.

There are several reasons why this is a design weakness:

- Balancing two conflicting objectives results in a very narrow operating window. Any raw material, process, or finished product drift can lead to sub-optimal product.

- The structure imparted by the suspending agent depends on a percolation threshold where a contiguous network is formed. Any drift straddling the threshold will lead to non-linear disproportionate effects.
- Viscosity may also increase on storage, posing stability problems.

If finished product viscosity is of concern then it is better to uncouple viscosity from structure by using clays or dispersible cellulose, which impart less viscosity for a given level of structure. Dispersible cellulose provides thixotropy with low viscosity. The viscosity profile can be increased by formulator if desired. A soluble polymer may only give thixotropy at high viscosity, which cannot be decreased by the formulator.

Clays and dispersible cellulose impart additional suspended material to the formulation, the clay crystals themselves, or residual partially attrited microcrystalline cellulose. Suspensions might also contain titanium dioxide, used as an opacifier/whitener. However modern image and microspectral analytical techniques allow drug particle size and polymorphic purity to be tracked even against a background of suspended insoluble excipient. Automated analysis can rapidly capture individual images of tens of thousands of particles, which can then be analysed to generate statistically relevant descriptors of size, shape and transparency. Kippax *et al.*[26] used morphological screening to eliminate some excipient particles, followed by Raman spectroscopy to isolate the drug particles for size analysis ("morphologically directed Raman spectroscopy"). The composition of an example of a commercial structured vehicle suspension using dispersible cellulose is given in Table 5.1.

Although dimensionally inconsistent, % w/v is preferred for liquid formulae because all the individual % w/w figures will vary with each change to the formulation. The summation of % w/v figures will total one hundred times

Table 5.1 An example of a commercial structured vehicle suspension using dispersible cellulose[27].

Ingredients	% w/v
Cimetidine base (B polymorph)	2.00
Avicel RC591	1.50
Water	25.00
Propylene glycol	5.00
Glycerol	5.00
Butylparaben	0.10
Propylparaben	0.05
Sodium saccharin	0.04
Vanilla (Firmenich 54.286C)	0.05
Cream (FDO FC 900772)	0.10
Titanium dioxide 50% in glycerol	0.40
Sorbitol 70% in water*	82.34
Total	121.58
*Sorbitol	57.64
*Water	24.70

the relative density. This can be determined from the volume and density of prototype batches.

Avicel® RC591 can be used at levels as low as 1.2% w/v[21] but a higher level was used to ensure formulation away from the network percolation threshold. There is no penalty for added structure, and pourability is not affected.

The formulation illustrates a common problem with developing suspensions in high-solids syrup vehicles. If high-solids syrup raw materials are used then there will not be a lot of free water in which to disperse the SVF. As will be discussed in the section on manufacture it is desirable to disperse or hydrate the SVF in water first, as other added materials may interfere. In this case, dispersing the Avicel® RC591 in the 25% w/v free water resulted in an in-process concentration of 6% w/v. Although semi-solid at this concentration the extremely shear thinning nature of an Avicel® RC591 gel allowed it to stir into a larger volume of sorbitol syrup using low-shear propeller mixing. In selecting a SVF the ability to process high in-process concentrations at scale should be taken into consideration.

The drug:excipient ratio will be much lower in high-solids suspension vehicles, in this case the sorbitol loading being 5.8 g per dose. High levels of non-absorbed sorbitol may impose an osmotic drag *in vivo*, potentially reducing bioavailability.[28] Because high-solids vehicles are already viscous, addition of soluble polymers as suspending agents is less feasible.

5.8 Manufacture

Continuous manufacturing processes are preferred to eliminate unexpected problems with scale-up, which are often very problematic for suspensions. It is very easy to develop a suspension with high-shear mixers in beakers but such a process will not scale-up. As a minimum, the high-shear incorporation of the SVF should be done in-line. Regardless of the suspending agent used, dispersion and/or hydration of the SVF is process-sensitive and may be affected by the presence of other ingredients. Ideally the SVF should be dispersed and/or hydrated in water first, before other ingredients are added, unless it has been validated during development that their presence does not interfere with finished product rheology.

Heat may be advantageous to accelerate hydration of clays or reduce polymer viscosity but, in general, the heating of structured vehicle dispersions should be avoided as thixotropes resist both mixing and convection, which could be problematical at scale.

Polymer solutions are difficult to prepare because of slow hydration and dissolution and because of the tendency of soluble polymers to clump on addition to water. Such clumps, on wetting, encapsulate agglomerates of dry polymer in a gelatinous outer layer, which retards water penetration and wetting of the core. Such "fish eyes" commonly result from adding the dry polymer too quickly. Once formed, fish eyes do not disperse without subsequent high shear, although they will slowly hydrate. They can block screens and, if they persist in varying degrees into subsequent processing stages, they represent a source of variability if their number is scale- or process-dependent. It is also important to avoid entraining air as is difficult to deaerate structured vehicles.

High shear requires a gap, either between a stator and a rotor, or the orifice of a high-pressure homogeniser. A propeller mixer is inadequate. High pressure homogenisation is an in-line method, but stator–rotor mixers can be either immersed (point) or in-line. The advantage of in-line processing is that the whole dispersion goes through a defined high-shear history, which can compensate for variation in the preceding low-shear dispersion. There is also less risk of air entrainment. The hydrodynamics of stator–rotors in viscous fluids have been reviewed by Doucet *et al.*[29]

A stator–rotor immersed in a beaker is not representative of industrial processing. The power is unrealistically high, giving simultaneous homogeneity and homogenisation. Unfortunately, the power does not scale-up and homogeneity and homogenisation are no longer coincident. A thixotrope resists mixing so extending the homogenisation time does not guarantee homogeneity throughout the tank. In the absence of additional bulk stirring paddles it is possible to obtain three regions of high, medium and low shear within the same tank. The high-shear unstructured region ("cavern")[29] around the stator–rotor is isolated from the low-shear, possibly undispersed, region by a shell of structured material which forms just outside the shearing range of the immersed stator–rotor. Such caverns are almost an unavoidable phenomenon[29] when mixing viscous and/or non-Newtonian fluids with rotor–stator impellers. Emptying the tank averages the three regions. The resulting finished product rheology may be satisfactory but any change in equipment or scale may alter the proportions of the three regions and change the finished product rheology.

In-line processing, by pumping the dispersion through the stator–rotor, is scale-independent, subject to throughput constraints. A 100 l batch can be scaled up to 1000 l without changing the shear history but it will take ten times longer to transfer. However, the effect of increasing the throughput pump rate is easily investigated. A transfer to a second tank is preferred to recirculation in and out of a single tank. Bulk circulation needs to be ensured and additional time is required to ensure all the contents have passed through the stator–rotor during recirculation. In-line processing is less convenient in the laboratory but will pay dividends on scale-up. The 6% w/v Avicel® RC591 dispersion in the example formulation[27] was subjected to in-line high shear, as was the final product homogenisation.

In-line processing also avoids entrainment of air. De-aeration is hindered by the structure in the same way suspension is facilitated. Aeration (flotation) is used industrially for ore extraction[11] but it will do the same for drugs, especially for micronized hydrophobic drugs, which can stabilise the foam by particles preferentially accumulating at the air–water interface. Any surface foam will therefore be enriched in drug content. The amount of drug trapped in the foam may be only a small fraction of the total, insufficient to affect bulk assay. However, if the enriched foam gets into an individual bottle it could significantly increase the drug concentration above the label claim on that bottle. For this reason, the final homogenisation of the suspension should also be in-line. Final in-line high-shear homogenisation minimises the influence of variability in preceding low-shear unit operations, similarly to the way in which the high-shear dispersion of the SVF minimises the influence of variability in the initial low-shear crude dispersion.

The presence of other ingredients in the dispersion medium may interfere with the dispersion or hydration of the SVF. Smectite clays require water to penetrate between the layers for delamination, so anything dissolved in the dispersion water can be osmotically counterproductive, including minor components such as preservatives and chelating agents.[13] The liberation of the colloidal fraction from dispersible cellulose is hindered by ions. The higher the ionic strength the higher the degree of shear required for liberation.[21] Dissolved salts and solids in the dispersion medium can affect the hydration and dissolution of polymers.[15] The ability of an immersed stator–rotor to disperse the SVF will be limited with increasing viscosity of dispersion media such as syrup. Circulation is hindered and heat is generated.[29]

In summary:

- In-line high-shear dispersion of SVF
- Disperse SVF in water, first.
- In-line high-shear homogenisation of finished product

5.9 Specification

Absence of sedimentation and drug content uniformity should be included in the CQAs of a structured vehicle suspension. Control of rheological attributes, such as gel strength or thixotropy, is desirable but often these are not suitable for routine third-party quality control, because they are method- and equipment-dependent, and time consuming. Rheometers are much less common than simple viscometers such as Brookfield.

With one exception, viscosity should never be a CQA for suspendability. The lower Newtonian viscosity (zero shear viscosity) is directly relevant to suspension efficacy but is experimentally inaccessible to simple viscometers. Values in the 10^6 mPa s range are associated with suspension.[15]

Given that the range of apparent viscosities in a thixotropic suspension may span several decades it is highly inappropriate to specify a single apparent viscosity (unknown shear) with the typical ±10% limits. This imposes an unnecessary compliance burden. Sworn[15] cites a range from 10^6 mPa s at 10^{-6} s^{-1} for suspension, to 1 mPa s at 10^6 s^{-1} for spraying. This range is representative of dispersible cellulose in nasal spray suspensions.

It might be possible to correlate an apparent viscosity with a product rheological profile, in which case a minimum apparent viscosity would be more appropriate. A maximum apparent viscosity is only required when pourability or sprayability needs to be ensured. It is unlikely that both upper and lower viscosities can be measured on the same speed/spindle combination.

As one of the defining characteristics of a thixotrope is viscosity growth after a sudden decrease in shear rate a two-point apparent viscosity test could be applied. A low apparent viscosity measured on a freshly destructured sample ("initial") can be followed some hours later by a second apparent viscosity measurement on the same sample left undisturbed to restructure ("set-up"). This could perhaps be specified as a minimum ratio.

5.10   Conclusions

The structured vehicle approach to suspensions is relatively simple and avoids many of the problems of designing a flocculated system tailored around a specific drug. A learn-once use-many-times vehicle for multiple drugs is the ideal QbD approach in terms of cumulative learning.

In-line high-shear dispersion of the SVF in water first will avoid most formulation and manufacturing problems. Even if not a fully continuous process, in-line processing of the SVF dispersion and finished product will give a high degree of scale-independence, with a known, consistent shear history.

References

1. *Remington: The Science & Practice of Pharmacy*, ed. L. V. Allen, Pharmaceutical Press, London, 22nd edn, 2013.
2. J. T. G. Overbeek, in *Colloid Science*, ed. H. R. Kruyt, Elsevier, New York, 1952, vol. 1, p. 80.
3. W. Ostwald, *Z. Phys. Chem.*, 1901, **37**, 385.
4. ICH Q8 R2, http://www.ich.org/fileadmin/Public_Web_Site/ICH_Products/Guidelines/Quality/Q8_R1/Step4/Q8_R2_Guideline.pdf.
5. FDA, *Request for Quality Metrics; Notice of Draft Guidance Availability and Public Meeting; Request for Comments, Federal Register/Vol. 80, No. 144, 434973–434977*, 28th July 2015.
6. F. Morrison, *Rheol. Bull.*, 2004, **73**(1), .
7. T. Tadros, *Dispersion of Powders in Liquids and Stabilization of Suspensions*, Wiley VCH, Weinheim Germany, 2012.
8. B. V. Derjaguin and L. Landau, Theory of the stability of strongly charged lyophobic sols and of the adhesion of strongly charged particles in solutions of electrolytes, *Acta Physicochim. URSS*, 1941, **14**, 633–662.
9. E. J. W. Verwey and J. T. G. Overbeek, *Theory of the Stability of Lyophobic Colloids*, Elsevier, New York, 1948.
10. J. Shi, 29 August 2002, http://muri.lci.kent.edu/References/NIM_Papers/stabilization_of_NP_suspensions/2002_Shi_steric_stabilization.pdf.
11. Q. Min, Y. Duan, X. F. Peng, A. S. Mujumdar, C. Hsu and D. J. Lee, *Drying Technol.*, 2008, **26**, 985–995.
12. *Handbook of Pharmaceutical Excipients*, ed. R. Rowe, *et al.*, Pharmaceutical Press, London, 7th edn, 2012.

13. Vanderbilt Minerals, *Veegum® Magnesium Aluminum Silicate Vanatural® Bentonite Clay for Personal Care and Pharmaceuticals*, http://www.vanderbiltminerals.com/ee_content/Documents/Technical/VEEGUM_VANATURAL_P_C_Pharma_Web.pdf.

14. B. Salopek, D. Krasic and S. Filipovic, *Rudarsko-geoloiko-naftni zbornik*, 1992, vol. 4, pp. 147–151.

15. G. Sworn, in *Gums and Stabilisers for the Food Industry*, ed. P. A. Williams and G. O. Phillips, Royal Society of Chemistry, 2004, vol. 12.

16. S. Fu, A. Thacker, D. M. Sperger, R. L. Boni, S. Velankar, E. J. Munson and L. H. Block, *AAPS PharmSciTech*, 2010, **11**(4), 1662–1674.

17. *KELTROL®/KELZAN® Xanthan Gum Book*, 8th edn, http://www.cpkelco.com.

18. Hercules, *Sodium Carboxymethyl Cellulose, Chemistry, Functionality, and Applications*, http://www.uta.edu/faculty/sawasthi/Enzymology-4351-5324/Class%20Syllabus%20Enzymology/carboxymethylcellulose.pdf.

19. United States Pharmacopeia 40/National Formulary 35.

20. G. H. Zhao, N. Kapur, B. Carlin, E. Selinger and J. T. Guthrie, *Int. J. Pharm.*, 2011, **415**, 95–101.

21. FMC Corporation, Philadelphia USA.

22. *British Pharmacopoeia 2017*, Stationery Office, London, 2016.

23. H. Green and R. N. Weltmann, *Ind. Eng. Chem., Anal. Ed.*, 1943, **15**, 201–206.

24. J. Mewis and N. J. Wagner, *Adv. Colloid Interface Sci.*, 2009, **147–148**, 214–227.

25. T. Dapčević, P. Dokić, M. Hadnađev and V. Krstonošić, *Food Processing, Quality and Safety*, 2008, vol. 35, (1), pp. 33–39.

26. P. Kippax, D. Huck, A. Virden, C. Levoguer and J. Suman, *Pharm. Technol. Eur.*, 2011, **23**, 2.

27. B. Carlin, J. Healey, G. S. Leonard and G. D. Tovey, US Patent 4,996,222, 1991.

28. M.-L. Chen, A. B. Straughn, N. Sadrieh, M. Meyer, P. J. Faustino, A. B. Ciavarella, B. Meibohm, C. R. Yates and A. S. Hussain, *Pharm. Res.*, 2007, **24**(1), 73–80.

29. L. Doucet, G. Ascanio and P. A. Tanguy, *Chem. Eng. Res. Des.*, 2005, **83**(A10), 1186–1195.

Excipients: Kano Analysis and Quality by Design

BRIAN A. C. CARLIN[*a] AND C. G. WILSON[b]

[a]Lawrenceville, NJ, USA; [b]University of Strathclyde, Glasgow, Scotland, UK
*E-mail: brianac.carlin@gmail.com

6.1 Excipients

Almost without exception, drugs are administered to the patient as a formulation, which provides advantages compared with administration of the active pharmaceutical ingredient alone. The creation of a pharmaceutical product is an engineering and chemical process with a long history, stretching back to the beginning of rational medicine, the creation of herbals describing formulae and the generation of the characteristic tools of pharmacy. The objective is to achieve a stable, definable dosage form, which is identifiable, easy to use and which releases the active pharmaceutical ingredient (API) appropriately. To construct a dosage form, improve the handling, and then release the API in or on the body, we need excipients.

The definition of excipient belies the range of possible uses. "Excipients are substances, other than the active drug or prodrug, which are included in the manufacturing process, or are contained in a finished pharmaceutical dosage form".[1] They are often referred to as the so-called "inactive" or "inert" ingredients in a product but they can exert indirect and direct physiological effects which are additional to the required action of the API. For

Drug Discovery Series No. 64
Pharmaceutical Formulation: The Science and Technology of Dosage Forms
Edited by Geoffrey D. Tovey

Published by the Royal Society of Chemistry, www.rsc.org

example, nasal formulations need preservatives to prevent contamination. However, preservatives cause suppression of ciliary beat frequency of the nasal mucosa,[2] reducing clearance. In the gut, the inhibition of P-glycoprotein efflux from intestinal cells by some surfactants and block copolymers[3] used in formulations is another example of a direct physiological effect. Similarly, water fluxes in the gut lumen are dynamic and high levels of polyols in a formulation can indirectly exert a physiological effect by imposing a high osmotic load in the gut.[4] By design, rate-controlling polymers in controlled-release formulations affect the dissolution and pharmacokinetics of the active drug.

The rationale for including an excipient in a formulation is its functionality, such as a tablet disintegrant or a suspending agent. However, most excipients are multifunctional and their performance in a specific application may depend on the process and formulation. Magnesium stearate is often more significant in terms of side effects on dissolution and compactability than its nominal functionality of lubrication. As discussed in the United States Pharmacopoeia (USP) chapter (1078) on good manufacturing practices for bulk pharmaceutical excipients,[5] excipients may be included in the formulation to:

- aid in the processing of the drug delivery system during its manufacture,
- protect, support or enhance stability, bioavailability or patient acceptability,
- assist in product identification,
- enhance any other attribute of the overall safety, effectiveness or delivery of the drug during storage or use.

Excipients, which often constitute the major proportion of the formulation, essentially enable the finished pharmaceutical dosage form. It would be difficult to dispense microgram doses of potent drug in isolation. Very low doses are sometimes encountered, and content uniformity during production requires staged addition of ingredients under appropriate engineering control. Drug solubility may be too high or too low, or excretion may be too fast, requiring solubilisers and/or rate-controlling compounds to be added to the API. The pharmacodynamics of modified-release dose forms are therefore controlled by the excipients.

A simple overview of the multiple functionalities built into typical oral solid dosage forms is summarised in Figure 6.1. The four main operands are manufacturing, storage, release and modification of absorption. For other routes, other excipients, including air-displacement agents, propellants, oily bases, adhesives *etc.*, become important. The USP[5] provides a useful list of excipients grouped by functional category, which may be applicable to multiple dosage forms.

Functional categories of excipients are formally described in the USP Chapter (1059) on Excipient Performance.[5] Each category is described, the functional mechanism discussed, together with relevant physical and chemical

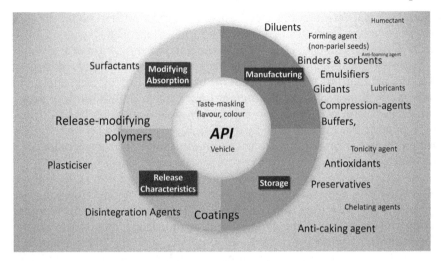

Figure 6.1   The spectrum of excipient functionality in an oral dosage form.

properties. Sheehan and Amidon[6] discuss the background and rationale for this non-mandatory guidance chapter, highlighting that it can be more difficult to set meaningful specifications for excipients than for drug substances. Excipients are not made for specific applications, may be multi-sourced, and are often manufactured for multiple industries with specifications wider than the higher grades needed for pharmaceutical applications. The quantity sold for pharmaceutical applications may be too small a proportion of total sales to justify the cost of setting and maintaining appropriate compendial standards.

> "Excipients are a chemically diverse group of materials, with examples encompassing all states of matter (solid, liquid, gas and semi-solid). Some may be manufactured using batch processing, but many are manufactured using continuous processing. The scale of excipient manufacture is often much larger than that used for active pharmaceutical ingredients (API) or finished drug products. In addition, very few materials sold as excipients are manufactured exclusively for pharmaceutical use. Very often, the pharmaceutical usage of an excipient may represent a minor fraction of the total output of the material".[7]

Excipients should always be sourced from the original manufacturer or their authorised distributors. If a vendor is unable or unwilling to identify the original manufacturer it is unlikely they will be able to provide the data on the excipient required for quality by design (QbD), in addition to the increased supply chain risk.

Quality control of an excipient in a medicinal product has wider importance beyond purity and identity. Making a medicine is an engineering process and other parameters need to be controlled.

"In some applications, excipient suppliers and users will need to identify and control material attributes in addition to monograph specifications. Manufacturers should anticipate lot-to-lot and supplier-to-supplier
variability in excipient properties and should have in place appropriate
control measures to ensure that CMAs [Critical Material Attributes] are
maintained within the required limits."[5]

Serial experimentation—step-wise change of a single parameter or quality
attribute—is an inefficient method of finding a position of robustness. Better
matrix statistical treatments, such as design of experiments (DOE), can be
used to map plateau regions for operational and ingredient specifications. In
addition, process analytical technologies can be used to measure the manufacturing process and deviations. This understanding underpins QbD, as
discussed later.

The USP Chapter (1059) on excipient performance[5] warns that reliance on
pharmacopoeial compliance alone is not consistent with QbD:

"Excipients are used in virtually all drug products and are essential for
product manufacturing and performance. Thus, the successful manufacture of a robust product requires the use of well-defined excipients and
manufacturing processes that consistently yield a quality product. Excipients used in drug products typically are manufactured and supplied in
compliance with compendial standards. However, the effects of excipient
properties on the critical quality attributes (CQAs) of a drug product are
unique for each formulation and process and may depend on *properties
of excipients that are not evaluated in USP or National Formulary (NF) monographs*. The effects of variations in excipient material attributes depend on
the role of an excipient in a formulation and the CQAs of the drug product."

"Good product development practices, which at times are termed QBD principles, require understanding excipient CMAs that contribute to consistent
performance and are the foundation of a control strategy that accommo-
dates excipient variability, consistently achieving final product CQAs."

The Handbook of Pharmaceutical Excipients[8] provides more detailed
monographs on a wide range of the commonly used excipients. Aulton's
Pharmaceutics[9] includes the science of formulation and drug deliv-
ery, designed and written for newcomers to the design of dosage forms.
Formulation of a wide range of pharmaceutical products is covered by
Remington.[10]

It is extremely unlikely that a new or novel chemical entity excipient will be
used because a new excipient incurs the same regulatory safety burden as a
new drug substance. The safety of an excipient relates to the level of human
exposure *via* a specified route of administration and a regulatory safety
assessment, as part of a finished-product marketing application, is necessary
for pharmacopoeial listing. The absence of a separate regulatory approval

mechanism for excipients has long been a barrier to the introduction of new-chemical-entity excipients. Moreton[11] highlighted this double jeopardy: a new chemical entity excipient incurs the cost of safety studies, but there is no regulatory mechanism for review and approval as a pharmaceutical excipient. The developer of a new excipient must persuade a pharmaceutical company to incorporate the new excipient into their new product so that the excipient is reviewed as part of the drug product marketing application. Because pharmaceutical companies seek to minimize regulatory risk, few will be willing to incur the added risk of incorporating a new excipient. Even if not optimal, pharmaceutical companies will favour existing excipients rather than complicate their regulatory filings. Consequently, the regulatory environment does not encourage new excipient development. Only three new chemical entity excipients have been launched in the last two decades:

- Sulfobutylether β-cyclodextrin (Captisol®), which enabled a pharmaceutical product, so development costs and inclusion in the marketing application were supported by the pharmaceutical company.
- Hydroxystearic acid PEG ester (Solutol® HS15). The regulatory risk was reduced by use of the International Pharmaceutical Excipient Council (IPEC) Novel Excipient Safety Evaluation Procedure.[12,13] This allows review by an independent panel of toxicologists which can be shared with FDA. It is not a regulatory approval but highlights at an early stage any safety issues, reducing the risk that the novel excipient will delay the finished product approval.
- Polyvinyl caprolactam–polyvinyl acetate–polyethylene glycol graft copolymer (Soluplus®), which was developed for hot-melt extrusion and launched in 2009.

There are benefits of using well-established excipients that have already been administered to humans by the intended route in similar dosage forms, have been manufactured to an acceptable standard, and obtained from reputable quality suppliers *via* a secure supply chain. If not subject to Good Manufacturing[14] and Distribution[15] Practices and change control[16] there is the potential for contamination, adulteration, substitution, undeclared additives and/or degradation. However, these factors by themselves do not eliminate the effects of excipient variability. Quality by Design seeks to minimize the risk that raw material variability will adversely affect the finished product quality. This is now also a requirement for current Good Manufacturing Practice (GMP) under section 711 (Enhancing the Safety and Quality of the Drug Supply) of the Food and Drug Administration Safety and Innovation Act 2012.[17]

"the term 'current good manufacturing practice' includes the implementation of oversight and controls over the manufacture of drugs to ensure quality, including *managing the risk of and establishing the safety of raw materials*, materials used in the manufacturing of drugs, and finished drug products".

Most pharmacopoeial monographs are primarily concerned with chemical identification, purity and quality with few requirements for physical or chemical properties that relate specifically to excipient function or performance. Excipient manufacturers and users must therefore independently identify and control excipient CMAs that go beyond monograph specifications.[6] Pharmacopoeial specifications are insufficient for some applications, such as biologics development, which require ultra-pure grades of excipients.

6.2   Quality by Design (QbD)

QbD brings an integrated science- and risk-based approach to the design and quality of pharmaceutical products,[18] providing improved understanding of the interplay between material attributes and process parameters, and how they influence the quality attributes of the finished product. The International Conference on Harmonization of Technical Requirements for Registration of Pharmaceuticals for Human Use (ICH) Q8 (R2)[19] provides an overview of the application of QbD to pharmaceutical development.

The quality of a design depends on how well the product performs against predictions based on the design inputs. Design inputs are knowns. Unknowns which subsequently adversely affect product performance are not criteria against which to assess (with hindsight) the quality of the design (unknown unknowns).

> "There are known knowns; there are things we know that we know.
> There are known unknowns; that is to say there are things that we now know we don't know.
> But there are also unknown unknowns—there are things we do not know we don't know".[20]

Excipient unknowns fall into several categories and may be unknown to the user, unknown to the excipient manufacturer or unknown to both.[21] Risk assessment requires that unknowns (not unknowable) be addressed with all stakeholders, including the excipient suppliers. Carlin[22] reviewed the categories of excipient unknowns, including composition, functionality/performance, limited utility of pharmacopoeial attributes, non-pharmacopoeial attributes, variability and criticalities.

The designer can be criticised if the design subsequently falls victim to an excipient effect unknown to the designer but known to the excipient manufacturer, who was not consulted during the design. Effects from an unknown can be likened to a 'black swan', a highly unexpected event for a given observer (the designer), which carries large consequences (product failure), and is subject to *ex-post* rationalization (why did no-one see it coming?).[23] The risk from 'black swans' is accentuated by the tendency to place too much reliance on what we know and ignore or underestimate what we don't know. The less we have to input, the easier it is to design.

Pharmaceutically, there tends to be over-reliance on the Certificate of Analysis (CoA), focused on pharmacopoeial parameters, which may be of limited relevance to determining excipient fitness for purpose in a specific application. Other unspecified excipient attributes may vary uncontrolled in the background, but will be unknown to the user unless discussed with the excipient manufacturer.[22]

In QbD, product critical quality attributes (CQAs) are selected to meet a quality target product profile (QTPP), the "prospective summary of the quality characteristics of a drug product that ideally will be achieved to ensure the desired quality, taking into account safety and efficacy of the drug product".[19] Critical material attributes (CMAs) and critical process parameters (CPPs) can then be identified which need to be controlled to ensure the finished product CQAs. The desired quality may also include manufacturing robustness (right first time) as stated in the proposed FDA quality metrics initiative.[24] Quality risk management in pharmaceutical development has been reviewed by Charoo and Ali.[25]

A design space can be proposed, which is "the multidimensional combination and interaction of input variables (*e.g.*, material attributes) and process parameters that have been demonstrated to provide assurance of quality".[19]

QbD also looks beyond the development and asks how quality will be maintained throughout the product lifecycle, *via* the control strategy, "a planned set of controls, derived from current product- and process-understanding, that assures process performance and product quality".[26]

In theory, the concept of incorporating excipients into QbD is deceptively simple:

$$QTPP = \Sigma CQA = \int (\Sigma CPP + \Sigma CMA)$$

In practice, this approach is overly simplistic, for several reasons. It assumes a simple linear relationship between an excipient attribute or set of attributes and performance of that excipient in a finished product. It ignores interactions and assumes that the relationship identified during development will hold during scale-up and throughout the product life-cycle.

6.2.1 Criticality

The term "Critical" is used in QbD for anything which affects the safety or efficacy of the finished product. MIL-STD-1629A uses the word "critical" as a severity classification, one stop short of catastrophic, and defines a criticality as a relative measure of the consequences of a failure mode and its frequency of occurrences.[27] The ICH defines criticality in terms of severity, probability and detectability.[19] This is reflected in the common QbD practice of assigning numeric scores of severity, probability and detectability to each excipient, so that the excipients can be ranked in terms of importance, and classified as either critical or non-critical. Such scoring is often arbitrary.

The use of the term "non-critical" when applied to excipients is problematical, as discussed later in the section on "Critical" *vs.* "Non-critical" Excipients. O'Keeffe *et al.*[28] have criticized the binary approach of classifying process parameters and quality attributes as being critical or non-critical as being overly simplistic, and not consistent with science and risk-based thinking. They suggest that a better approach would be to rank attribute and parameter criticalities with the relative risks of not achieving the desired quality attributes, a "spectrum of importance" with respect to process parameters and material attributes.

In theory a non-critical excipient cannot have a CMA, but do all critical excipients have to have CMAs? A common pitfall is to assign one or more pharmacopoeial attributes as CMAs without full understanding of whether other unspecified attributes may be more important, as discussed later in the section on factoring excipients into QbD during development.

A major problem with the ICH definition of criticality is that it does not include the definition, common in physics or mathematics, as a transition between two states. A criticality or critical transition is defined as being in a state, or at a point, where some quality, property or phenomenon undergoes a definite change.[29] A criticality, a point of transition from one state to another, can be critical if encountered during production. An example of a criticality is the critical micelle concentration, where the properties of the dilute solution below the critical concentration are not predictive of the micellar system above.

Percolation thresholds are examples of criticalities in pharmaceutical systems. Percolation theory deals with clusters in random systems and long-range connectivity. An example would be increasing the water content in a water-in-oil emulsion. The oil is the continuous phase in which the water is dispersed but at a critical water concentration the system may invert to give an oil-in-water emulsion, where the water is now the continuous phase. Powder mixes may exhibit similar behaviour with particles of one component (A) dispersed in another (B). As A increases beyond a critical concentration, or percolation threshold, the mix becomes a dispersion of B in A. If the properties of A and B are different there may be a marked discontinuity in the properties of the mix. In his review of the application of percolation theory to powder technology Leuenberger[30] warns that:

"formulations which contain a component with a critical concentration, *i.e.* close to the percolation threshold p_c, may lead to non-robust conditions during scale-up and during subsequent large-scale production activities."

An example would be a disintegrant in a hydrophobic tablet matrix at a level just sufficient to provide a contiguous network for water wicking.

Even a slight variation in content uniformity (or the disintegrant properties, inter-batch or inter-supplier) could render parts of the tablet batch non-disintegrating.

Leuenberger[30] also gave an example of a percolation threshold in tablet hardness *vs.* relative density. Below the criticality the granules disintegrate, above it they do not. It should also be remembered that force transmission and the resultant densities within a compact are not homogeneous, and may be dependent on factors such as the tablet geometry.[21,31]

Conflicting technological objectives are another source of criticalities. The closer the formulation is to a performance margin or point of failure, the greater the effects of excipient variability. Ranging studies during development are useful: if you can vary the level of an excipient by ±50% and maintain product performance, then the effect of variability of that excipient is generally going to be less than that associated with a ±5% titration that has greater effects. However, if you are trying to balance too many multiple competing objectives, then you will have a very narrow operating margin with much greater susceptibility to excipient variabilities and unknowns. Good examples can be found with design-critical rate-controlling polymers in modified release. For example, the higher the level of gelling-matrix-former in a hydrophilic matrix tablet formulation, the lower is generally the influence of variability in the excipient attributes. If faster drug release is required, it is generally advisable to maintain a high level of a "weaker" polymer rather than reduce the original polymer to a level where the influence of excipient variability is greater. Similarly, maintaining a high loading of a rate-controlling controlled release film-coating is preferable to reducing to a level where it is subject to the influence of both the coating precision and variability of the excipient attributes.[21]

Excipients may disproportionately affect CQAs, if their variability interacts with a criticality in the application. In that case, a hitherto unremarkable excipient variability now starts to govern the transition from one state to another. The term "explosive percolation" refers to the characteristic binary step function where the system goes from one state to another with little or no warning. Criticalities are also known as latent conditions since they are unknown to the designer.[32]

6.2.2 Complexity

Complexity theory attempts to explain phenomena not explainable by traditional (mechanistic) theories, to treat systems *as they are, and not by simplifying them* (breaking them down into their constituent parts). It recognizes that complex behaviour emerges from a few simple rules, and that all complex systems are networks of many interdependent parts which interact according to those rules.[33] This is the opposite of the bottom-up approach in traditional pharmaceutics, which extends from the excipients, through the formulation, to product performance.

A definition very relevant to excipients in QbD is that of Christensen and Moloney,[34] who define complexity as:

> "the repeated application of simple rules in systems with many degrees of freedom that gives rise to emergent behaviour not encoded in the rules themselves".

QbD requires greater understanding of material effects and excipients should hence not be regarded as simple "inert" ingredients to be combined in a fixed recipe. Their behaviour in the formulation requires careful evaluation during product development, but, given the complexity, there may still be excipient-related surprises during the product lifecycle.

Based on both authors' personal experience in this field, pharmaceutical product development, control and regulation has become too dependent on repeated application of simple rules, such as fixed formulae and fixed processes, and there is an (over)reliance on pharmacopoeial compliance, potentially rendering finished product quality more vulnerable than is necessary.

The complexity of both raw materials and finished products is often underestimated, leading to systems with many degrees of freedom, such as variability and unspecified attributes in the excipients, and criticalities or latent conditions in the finished product. The more unknowns, the more degrees of freedom.

Emergent behaviour not encoded in the rules is inevitable, usually manifesting as special cause variation in the finished product, often correlating with variability in an excipient attribute, previously thought "non-critical" and within historical norms.

6.2.3 Special Cause Variation

The term "special cause" was coined by Deming[35] and is characterized by:

- New, unanticipated, emergent behaviours
- Inherently unpredictable
- Outside historical experience
- Inherent change in the system?

Special cause variation is attributable to a specific component of the system. Removing all the special causes leaves the intrinsic noise of the system, common cause variation. Deming's reference to inherent change in the system is consistent with criticalities in the finished product.

Due to the complexity of the excipients, and the products into which they are formulated, excipients represent a reservoir of special cause variation in the finished product which must be addressed by the designer. As special cause variation is, by definition, unpredictable it is not experimentally accessible during development and must be factored into the control strategy. Paradoxically the more rigid, or fixed, the system the more susceptible it is to the effects of excipient variability. The flexibility built into the system to cope with special cause variation then becomes a criterion of design quality. There is little benefit having a product which works perfectly so long as nothing changes. Products are subject to cumulative changes throughout their lifecycle, often subject to univariate change control, until multivariate failure ensues. The failure need not correlate with a critical excipient. It could also correlate with an attribute of a so-called "non-critical" excipient, within its norms of variability.

6.2.4   "Critical" *vs.* "Non-critical" Excipients

In QbD the (sometimes arbitrary) binary classification of excipients between critical and non-critical poses the question as to why the so-called "non-critical" excipients continue to be associated with finished product special cause variation, adversely affecting CQAs. It is better to regard all excipients as critical. If truly non-critical then such excipients would be optional, in which case why add them?

A better approach is to focus on the design-critical excipients during late stage design of experiments (DOE). Design-critical means that there is some reason to expect an effect on finished product performance, such as the type and level of a modified release polymer. Do not assume that inclusion of so called "non-critical" excipients during development without incident confirms that they are not critical. If there is no reason to expect effects, then absence of effects does not prove absence of potential effects: just that it had not occurred during the experiments. Even if not design-critical, all excipients in the product are potentially performance critical: appropriate contingencies should be built into the control strategy. Adoption of Kano analysis can be useful in the assessment of excipients during the design phase of product development.

6.3   Kano Analysis

Kano models derive from an analysis of customer satisfaction with goods and services and emerged during the late 1980s.[36] The terminology seems a little odd at first but it was adopted as management speak and has been influential in constructing models that should lead to product improvement and customer satisfaction. The basic model has three elements relating the degree of sufficiency of a quality attribute to the degree of customer satisfaction with that attribute, as shown in Figure 6.2:

- basic (must have or in Kano's terms must-be),
- one-dimensional or performance—the more of a good feature the better,
- 'attractive' (exciter) features which the client sees as 'really great' and unlike the one-dimensional feature, exerts a disproportionate effect on satisfaction. Sadly, the opposite would lead to customer dissatisfaction and disproportionate dislike.

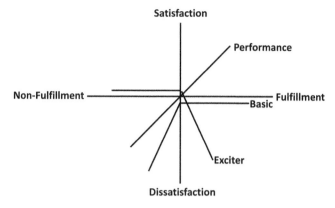

Figure 6.2 Kano analysis: basic, exciter and performance.

As an example, consider the smart phone. Provision of a back camera was once an exciter, enabling "selfies." Now that the ability to take "selfies" is taken for granted the same back camera is now a basic. One would now not market a smart phone without this feature.

The *x*-axis represents the physical sufficiency of a given attribute and the resultant satisfaction is represented on the *y*-axis. A basic is a minimum requirement or "entry ticket" without which a product is not feasible. Such product attributes are taken for granted by customers and only result in dissatisfaction if not present or insufficient. Another commercial example would be the cleanliness of a hotel room; not a purchase criterion but grounds for complaint if not clean on check-in.

A performance attribute has proportional satisfaction, "the more the better". Speed of check-in, to continue the hotel example. Attractive or exciter attributes normally provide satisfaction when present, but no dissatisfaction if absent, the customer being unaware. A complimentary gift in the hotel room would be an example, a pleasant surprise. In Figure 6.2 the excitement is shown as negative (dissatisfaction) because in pharmaceutical product quality surprises are generally unwelcome.

Applying the Kano model to excipients, the *x*-axis represents the expression of a particular excipient attribute in a formulation and the *y*-axis reflects the quality of the finished product. Variability in the excipient attribute will be a range on the *x*-axis but the product response on the *y*-axis will vary dependent on the Kano type used to map to the *y*-axis, namely basic, exciter or performance.

6.3.1 Kano Basic

Excipient attributes included in the basic category fall into two types. The first are compliance and compatibility. Non-compliance with specification should be detected by the quality system, but lack of GMP in the manufacture of an excipient can result in the finished product being deemed adulterated.

Chemical compatibility with the API is another basic requirement identified during preformulation studies. These minimum-compliance and compatibility standards are independent of specific product design requirements.

The second type of excipient attributes in the basic category relate to the so-called "non-critical" excipients. The lack of effects of variability is illustrated by the flat-line response, where variability is well away from the level below which there is dissatisfaction. A good test of "non-criticality" is to range the concentration downwards. If there is no effect on CQAs, it suggests a margin or reserve of performance. Variability in an excipient close to a minimum level or quality will render the finished product more susceptible to that variability, in which case the excipient becomes critical and the attribute becomes a CMA. Basic attributes are known and specified. Statistically relevant information on their variability may be available from the excipient manufacturer.

6.3.2   Kano Exciters

Exciter attributes are unknown to the designer, so when they are subsequently discovered during the product lifecycle they come as a surprise. The absence of expression of an exciter attribute in the formulation contributes to satisfactory product quality. This is in contrast to a basic attribute, which must be expressed in excess of a minimum. Surprises in a pharmaceutical product are generally negative, which is why the exciter response curve is shown as negative in this excipient context. (The traditional concept of the exciter is to delight or at least attract the customer). It is possible to have a positive outcome, such as a previously unspecified attribute leading to process or yield improvements. However, in most cases a hitherto unimportant variability starts to correlate with a finished product out-of-trend or out-of-specification result. Absence of excitement is a characteristic of the so-called "non-critical" excipients. An exciter attribute cannot be a CMA, since it is unknown at the time of design. It is important to recognize that unknown (to the designer) does not always mean unknowable. What do the excipient manufacturers know?

6.3.3   Kano Performance

Ignoring interactions, the effects of a performance excipient in a finished product will be a function of its concentration (c), and expression of the relevant excipient attribute (x), *e.g.* strength, potency or efficacy. Variability in x can directly affect performance, therefore such excipients can be regarded as critical, or more specifically design-critical.

$$\text{Performance} = \int(c,x) \text{ or Performance} = \int x \text{ for fixed formulae}$$

Interactions will result in dependency on other formulation or process variables. Design-critical excipients deliver a specific functionality to the product and require titration in the formula. A rate-controlling polymer in a sustained-release matrix or coating is a good example, due to release being faster the lower the rate-controlling polymer concentration, and *vice versa*. For design-critical excipients, their attribute *x* would be considered a CMA. Most pharmaceutical formulae are fixed. It would make more sense to allow a range rather than a single concentration, in order to offset variability in *x*, in which case *x* becomes less critical.

In the Kano model (Figure 6.2) the performance attribute is traditionally shown as a line of proportionality ("more is better"), but in practice there will be constraints. For example, you cannot keep adding an excipient, no matter how beneficial, without making a tablet too big to swallow, or running into side-effects. Magnesium stearate is an excellent lubricant, but unconstrained addition renders most formulations unfeasible. More realistic profiles are shown in Figure 6.3 where satisfaction levels off (plateaus) or decreases, being constrained by some other limiting property. Such a constraint is analogous to the therapeutic index of a drug, which is the range between the maximum tolerable dose and the minimum effective dose. A narrow effectiveness index for an excipient in a formulation represents a finished product criticality. Any variability in the excipient will result in sub-optimal performance. The plateau in Figure 6.3 is hence a more attractive operating region. The wider the formulation concentration range in which an excipient is effective, the less susceptible the formulation will be to variability in the excipient performance.

The logic of Figure 6.3 can be illustrated by the choice of suspending agent for pourable or sprayable aqueous suspensions. If a soluble thixotropic polymer, such as xanthan, is used, there is conflict between suspension and viscosity (constraint), whereas the use of dispersible cellulose BP[37] gives suspension without viscosity (plateau). In this example we have uncoupled viscosity and suspension which frees the designer from the constraint of having to balance viscosity *versus* suspension. If such a performance- or design-critical excipient is in a plateau operating region does it cease to be a critical excipient with no need for a CMA? Variability in the attributes of a suspending agent in the plateau region will have less effect than those of a

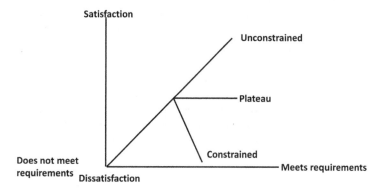

Figure 6.3 Kano analysis: performance, constrained *vs.* plateau.

constrained suspending agent close to the critical concentration needed to balance competing objectives. It should also be remembered that a CMA is not always intrinsic to an excipient, but may depend on the application. For example, morphology and particle size distribution are not relevant to an excipient in solution.

6.4 Factoring Excipients into QbD During Development

Potential CMAs can be identified *a priori* for the design-critical excipients and confirmed experimentally, which may also identify significant interactions. A major caveat is that a CMA might not be a specified attribute. For example, many excipients are polymeric and specified by a dilute solution apparent viscosity. This reflects an average molecular weight and it is hence important to ask the excipient manufacturers how they meet the viscosity specification. If one manufacturer offers a particular molecular weight while another one blends higher and lower molecular weight fractions to meet the same viscosity specification, then their grades may not be interchangeable in a particular application. Also, because the apparent viscosity is measured at a low concentration, it may not be predictive of rheologies at higher concentrations. For example, above a certain critical concentration C^*, corresponding to the onset of polymer coil overlap, a gel network may form.[38,39] This can be problematic in controlled-release matrix applications, where the effective concentration of the rate-controlling polymer in the hydrating–swelling–gelling barrier layer may be an order of magnitude higher than the typical dilute solution concentrations (<5%) used for viscosity specification and grade differentiation. Fu *et al.*[40] found that dilute solution viscosities of several grades of sodium alginate were not predictive of viscoelastic properties at higher concentrations.

Apparent particle size is another common attribute which can be misleading in a particular application. For example, laser scattering methods are common but their assumption of sphericity is not applicable to most excipients. Laser scattering is dominated by larger particles and may miss multimodal particle size distributions, as demonstrated for microcrystalline cellulose by Gamble *et al.*[41]

It is the responsibility of the designer to fully characterise the excipients selected for a particular application. If additional attributes need to be controlled, then the excipient should be specified as such, after confirming with the excipient manufacturer(s) that it is economically and technically feasible to do so. Discussion with excipient manufacturers early in the development process is recommended as they may be able to identify additional failure modes in a particular application and suggest potential CMAs not on existing specifications. They can also supply data on their excipients, which can be subjected to multivariate analysis (MVA), to guide selection of appropriate batches for inclusion in experiments. Additional characterisation results, and data from the excipient manufacturer, can also be used to avoid the increasing regulatory requests to justify reliance on supplier or pharmaco-poeial specifications.

In the absence of discussion, the only data on the excipient will be from the CoAs accompanying each purchased batch of excipient. This is likely to be a statistically unrepresentative sampling of the historical variability.

> "When only a small number of excipient lots are examined without know-ing how they compare to the overall variability of the excipient, there is a high probability of obtaining erroneous results, either due to the omission of a critical physico-chemical property of the excipient in the study that has a high impact on the drug product performance or due to the examination of only a small domain of the overall variability of the excipient".[42]

Requesting historical CoA data from the excipient manufacturer covering multiple years will give a much better assessment of the excipient variability, and the manufacturer's process capability. MVA can identify what type of batches to factor into the DOE. The use of MVA in the pharmaceutical indus-try to enable process understanding and improvement has been reviewed by Ferreira and Tobyn.[43]

The excipient manufacturer may also be able to share in-process data, which usually involves higher frequency testing of certain attributes. This is particularly valuable for the many common excipients manufactured con-tinuously on kiloton scales. A CoA may represent a week's production or several hundred tons, so some CoA results will typically be a composite or an average. This will tend to smooth the data. Higher frequency in-pro-cess data hence provides a more realistic picture of the true variability. The excipient manufacturer may also have data on attributes not on the specification.

6.5   Excipient Samples for QbD

Provision of excipient samples for QbD has been sufficiently problematic to merit the issue of a QbD Sampling Guide by IPEC.[7] This was partly in response to the misconception that QbD requires batches of product incor-porating excipient lots manufactured at the extremes of specifications.

> "Simply evaluating excipient lots which have been manufactured at the extremes of specifications, even if it were possible, is not a valid method of applying the principles of QbD. QbD requires enhanced understand-ing of the interplay of CMAs and Critical Process Parameters (CPPs) and how they influence product CQAs, not merely experimental experience. Similarly, evaluating samples of equivalent grades from different excipient suppliers is also not QbD. It is extremely unlikely there will be sufficient enhanced understanding of the interactions between CMAs and CPPs with respect to product CQAs from such an approach".[7]

Excipient manufacturing plants, especially continuous configurations, typically produce excipient at mid-specification (or well below/above target for one-sided specifications). Generally, these manufacturing plants are not designed to produce materials at the extremes of specification. Operating at a specification limit would potentially put 50% of the material outside specification. Given the scale of operation, the costs of disposal or reprocessing for such an approach are prohibitive. Excipient samples at the extremes of specification are therefore essentially unavailable. Even if it is possible to manufacture an excipient lot at an extreme of specification for one particular attribute, trying to simultaneously target multiple attributes at their extremes of specification is virtually impossible, as there may be interdependencies between attributes.

Alternative methods of simulating material at the edge of specification for a specific attribute include:

- Fractionation: *e.g.* sieve cuts to evaluate the effect of particle size. Milling and granulation are less preferred as other material attributes may be altered.
- Level of incorporation: *e.g.* in the case of viscosity, more or less excipient can be added to simulate lots at or outside the viscosity limits.
- Conditioning: *e.g.* equilibrating excipient at higher humidity to increase the moisture content, or drying it to decrease the moisture content.
- Spiking with known concomitants, process aids and additives to simulate extremes of excipient composition
- Grade bracketing, where multiple grades are available reflecting ranges of the attribute(s) of interest beyond the original specification limit(s).
  - If there are no unanticipated effects of bracketing, then the original specification limits for the attribute(s) of interest may be relied upon.
  - If bracketing has an unacceptable effect on finished product performance, then bracketing with blends of the grades may be used to set limits for that attribute in that application.

A better approach than targeting unrealistic specification extremes is to perform MVA on the excipient manufacturer's data and select batches on the 95% confidence limits of the Hotelling T2 plot. If the data shows clustering, then batches can also be chosen from the different clusters. Kushner[42] and Thoorens *et al.*[44] showed clustering corresponding to manufacturing site differences in the physicochemical properties of Avicel® PH102. Unless the excipient manufacturer's data is current, it is unlikely that historical batches of interest will still be in inventory, unless there are retained samples. This is another reason for starting discussions with the excipient manufacturer as early as possible, because you may have to wait for future batches falling into the multivariate regions of interest.

"MVA methods allow the extraction of information contained in large, complex data sets, thus contributing to increased product and process understanding".[43]

6.6 All Excipients Are Critical

From the earlier Kano discussion it can be seen that all excipients are potentially critical in terms of affecting finished product CQAs. Some excipients are design-critical, with direct cause–effect relationships with finished product performance, with the potential for their variability to affect product performance. The "non-critical" excipients and their variabilities have no apparent effect on CQAs, but the absence of evidence of a problem is not evidence of the absence of a problem. The question then arises as to how can so-called "non-critical" excipients be associated with special cause variation at some stage in the product lifecycle? Their effects are indirect, due to interaction with finished product criticalities. If there was no such interaction during development, then something must have changed, either process or product drift. Such drift will not be detectable by univariate change control, hence the value of continuous multivariate monitoring.

Change control of pharmaceutical products is nearly always univariate. The product performance is checked before and after the change to confirm that the product remains within specification and that the change has had no effect on CQAs. This may be formalized as a comparability protocol, to cover foreseeable events such as switching suppliers of an excipient. The weakness of this approach is that other attributes of the product may change, which are not reflected in the specified parameters or the CQAs. After several changes of supplier and a few process tweaks, sequentially qualified one step at a time, the system may have drifted and the next change triggers an unexpected problem. Past performance is not always predictive of future performance. A hundred white swans do not guarantee that the next swan will not be black. Ideally, monitoring of specified parameters should be complemented with multivariate monitoring.

$$\text{Variability} + \text{Drift} + \text{Criticality} = \text{Effect}$$

It is important to note that excipient variability may not be causative. The variability may not have changed, but is now within range of a criticality and is starting to correlate with finished product quality excursions. The effect of drift can be illustrated by a simple process capability model as shown in Figure 6.4.

Process capability is essentially the ratio of the specification range relative to the range of variability, usually quantified as ±3 standard deviations ($\pm 3\sigma$), assuming only common cause variability. It is a measure of the ability of the process to yield product within limits. For the purposes of illustrating the combined effect of drift, variability and criticality, only a single limit is needed. This corresponds to an unknown limit within the product beyond which an excipient variability will affect product CQAs.

$$\text{Capability Index} = \min\left[\frac{(\text{upper limit} - \text{mean})}{3\sigma}, \frac{(\text{lower limit} - \text{mean})}{3\sigma}\right]$$

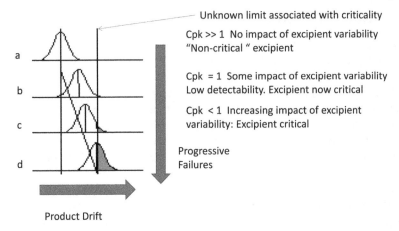

Figure 6.4   Effects of product or process drift.

In Figure 6.4(a) the criticality limit is on the right-hand side. The effect of the excipient variability is not seen as it is well to the left of the criticality limit. The process capability is >1 and the product can tolerate some drift, with the 3σ limit being well away from the criticality. The excipient would hence be regarded as non-critical, as there is no discernible effect on product CQAs.

Figure 6.4(b) shows the 3σ limit drifting to the criticality limit. The process capability now has a value of one, and the previously non-critical excipient is now critical. The excipient variability is starting to correlate with product quality excursions. Detectability is low as the incidence of excursions will still be parts per million, assuming a normal distribution. Further drift (Figure 6.4(c) and (d)) will increase the incidence of quality excursions, and the process is no longer capable of producing product within limits. Note that the severity of an excursion may be disproportionate, if the drift straddles a percolation threshold between two different regimes. An analogy is the significance of a misstep when you are at the cliff-edge, compared with being well away from the edge.

6.7   Control Strategy and Changes in Excipient Criticality During the Lifecycle

CMAs are identified during development, but if variability in a hitherto non-critical excipient attribute starts to correlate with effects on the finished product CQAs, then there are three options:

1. Alter the process
2. Alter the formulation
3. New CMA

A variation may need to be filed if the necessary process/formula changes fall outside the scope of the product approval or applicable waivers (*e.g.* scale-up and post approval changes (SUPAC)[45]). This could cause a significant delay and expense. It is often difficult for the excipient manufacturer to address, as the problem may not be replicable in the laboratory (user or manufacturer). This can be the sign of a criticality specific to a product, at scale, on a particular site. Assuming that the excipient manufacturer has not changed something (or that the user has not precipitated the crisis by changing excipient supplier) it is worth asking the excipient manufacturer what process or raw material variants are covered by their specification. This could include different sites or equipment trains/scales. A representative selection of excipient batches can then be introduced into the users production to identify "good" or "bad" batches in terms of the presenting quality issue. Specifying preferred excipient batch types should be accompanied by appropriate characterization to identify a CMA to distinguish "good" from "bad". The advantage of adding a CMA is that no prior regulatory approval is required. In the longer term (the next window of regulatory opportunity), a variation can be filed containing corrective formulation and/or process options, together with justification for retiring the now redundant added CMA.

To pre-empt the special cause variation associated with too many raw material and product degrees of freedom, the control strategy must include continuous multivariate monitoring during the product lifecycle. Multivariate monitoring is preferred because quality depends on multiple variables, which individually may remain within univariate limits but interact to cause a quality problem.

With the quality metrics initiative[24] and current good manufacturing practice (cGMP)[16] it is essential to monitor for out-of-trend (OOT), instead of waiting for out-of-specification (OOS), results. MVA can contribute to an early warning system that enables effective management review (quality assurance instead of reliance on quality control). Kushner[42] found the domain of prior experience for Avicel® PH102 during a development project to be only a small part of the overall variability observed in lots of Avicel® PH102 manufactured over an eight-year period. This type of potential risk from excipient variability can be addressed by further development batches to expand the excipient experience domain before production, if the risk is deemed unacceptable. Alternatively, if the risk is deemed acceptable, comparability protocols (including MVA) in production can allow expansion of the excipient experience domain during the product lifecycle.

Another advantage of applying MVA to excipient data is that it provides a convenient means of demonstrating user oversight of excipient quality, as required by cGMP.[17] Excipient users should anticipate lot-to-lot and supplier-to-supplier variability in excipient properties and therefore should have appropriate controls in place to ensure consistent excipient performance.[6]

Access to excipient manufacturers' data can accelerate product development and increase product robustness with respect to the effects of excipient variability. Such access will require collaboration, under a confidentiality agreement, which should be a two-way agreement, as pharmaceutically-aligned excipient manufacturers cannot help if they are unaware of the application. The excipient manufacturer can help to identify CMAs for design-critical excipients during development and provide CMAs to counter special cause variation during the finished product life cycle. Continuous MV monitoring of raw material data for drift may pre-empt drift and quality problems in the finished product.

6.8   Conclusions

The pharmaceutical industry relies on a limited range of pharmacopoeial excipients but official specifications may not be adequate to control excipient performance in a specific application. Residual degrees of freedom will contribute to quality excursions if they interact with latent finished product criticalities.

Experimental results during development are snapshots and the criticality ranking of individual excipients may change throughout the product lifecycle. Continuous multivariate monitoring throughout the product lifecycle is therefore essential to the control strategy.

The Kano approach illustrates why the effects of a particular excipient can change during the product lifecycle. Only the critical or Kano-performance excipients provide proportional responses amenable to experimentation during development. Kano basic or exciter excipients may be deemed "non-critical" due to absence of experimental response in their plateau region, only to later prove critical due to product or excipient drift.

Given the complexity of excipients greater collaboration with excipient manufacturers reduces the effects from unknowns and is consistent with cGMP.

"If a man will begin with certainties, he shall end in doubts; but if he will be content to begin with doubts he shall end in certainties."[46]

References

1. FAQs, *What Are Pharmaceutical Excipients?* http://www.ipecamericas.org/what-ipec-americas/faqs#question1.
2. T. Hofmann, M. Gugatschga, B. Koidl and G. Wolf, *Arch. Otolaryngol., Head Neck Surg.*, 2004, **130**, 440–445.
3. K. M. R. Srivalli and P. K. Lakshmi, *Braz. J. Pharm. Sci.*, 2012, **48**(3), 353–367.
4. M.-L. Chen, A. B. Straughn, N. Sadrieh, M. Meyer, P. J. Faustino, A. B. Ciavarella, B. Meibohm, C. R. Yates and A. S. Hussain, *Pharm. Res.*, 2007, **24**(1), 73–80.
5. *U.S. Pharmacopoeia-National Formulary [USP 40 NF 35]*, United States Pharmacopeial Convention, Inc., Rockville, MD, USA, 2017.

6. C. Sheehan and G. Amidon, *Am. Pharm. Rev.*, 2011, **14**(6), 10–18.

7. International Pharmaceutical Excipient Council, *The Excipient QbD Sampling Guide*, 2016.

8. *Handbook of Pharmaceutical Excipients*, ed. R. C. Rowe, P. J. Sheskey, W. G. Cook and M. E. Fenton, Pharmaceutical Press, UK, 7th edn, 2012.

9. *The Design and Manufacture of Medicines*, ed. M. Aulton and K. Taylor, Elsevier, 5th edn, 2017.

10. *Remington, the Science & Practice of Pharmacy*, ed. L. V. Allen, Pharmaceutical Press, London, 22nd edn, 2013.

11. R. C. Moreton, Tablet excipients to the year 2001: a look into the crystal ball, *Drug Dev. Ind. Pharm.*, 1996, **22**(1), 11–23.

12. http://www.ipecamericas.org/content/ipec-novel-excipient-evaluation-procedure.

13. S. Ku and R. Velagaleti, Solutol HS15 as a novel excipient, *Pharm. Technol.*, 2010, 108–110.

14. NSF International, *NSF/IPEC/ANSI 363-2014 Good Manufacturing Practices (GMP) for Pharmaceutical Excipients*, http://standards.nsf.org/apps/group_public/download.php/26765/NSF%20363-14%20-%20water-marked.pdf.

15. *The IPEC Good Distribution Practices Guide*, 2006, http://ipecamericas.org/system/files/IPEC_GDP_Guide_final.pdf.

16. *The IPEC Significant Change Guide for Pharmaceutical Excipients*, 2014, http://ipecamericas.org/system/files/IPEC_Significant_Change%20_Final_2014.pdf.

17. *S.3187 — 112th Congress: Food and Drug Administration Safety and Innovation Act (FDASIA)*, 9th July 2012, https://www.govtrack.us/congress/bills/112/s3187, Accessed 1 Mar 2018.

18. *Pharmaceutical Quality for the 21st Century a Risk-based Approach*, U.S. Food and Drug Administration, May 2007, https://www.fda.gov/aboutfda/centersoffices/officeofmedicalproductsandtobacco/cder/ucm128080.htm.

19. ICH Q8 R2, http://www.ich.org/fileadmin/Public_Web_Site/ICH_Products/Guidelines/Quality/Q8_R1/Step4/Q8_R2_Guideline.pdf.

20. D. Rumsfeld, 2002, http://www.defense.gov/Transcripts/Transcript.aspx?TranscriptID=2636.

21. B. Carlin, in *"The Role of Excipients in QbD" in "Pharmaceutical Quality by Design"*, ed. W. Schlindwein and M. Gibson, John Wiley And Sons Ltd, Chichester, UK, 2018, in press.

22. B. Carlin, *J. Excipients Food Chem.*, 2012, **3**(4), 143–153.

23. N. N. Taleb, *The Black Swan: The Impact of the Highly Improbable*, Random House Publishing Group, NY, 2nd edn, 2010.

24. FDA, *Request for Quality Metrics; Notice of Draft Guidance Availability and Public Meeting; Request for Comments, Federal Register/Vol. 80, No. 144, 434973–434977*, 28th July 2015.

25. N. A. Charoo and A. A. Ali, *Drug Dev. Ind. Pharm.*, 2013, **39**(7), 947–960.

26. ICH Q10, https://www.ich.org/fileadmin/Public_Web_Site/ICH_Products/Guidelines/Quality/Q10/Step4/Q10_Guideline.pdf.

27. US Dept Defense, *MIL-STD-1629RevA*, 1980, http://src.alionscience.com/pdf/MIL-STD-1629RevA.pdf.

28. D. O'Keeffe, C. Campbell and K. O'Donnell, *J. Valid. Technol.*, 2016, **22**(1), http://www.ivtnetwork.com/article/spectrum-importance%E2%80%94 challenging-concept-criticalnon-critical-qualification-and-validation-ac.

29. http://www.merriam-webster.com/dictionary/critical.

30. H. Leuenberger, *Adv. Powder Technol.*, 1999, **10**(4), 323–352.

31. B. Eiliazadeh, B. J. Briscoe, Y. Sheng and K. Pitt, *Part. Sci. Technol.*, 2003, **21**, 303–316.

32. J. Reason, *Br. Med. J.*, 2000, **320**(7237), 768–770.

33. Business dictionary, http://www.bu6sinessdictionary.com/definition/complexity-theory.html.

34. K. Christensen and N. R. Moloney, *Complexity & Criticality*, Imperial College Press, 2005.

35. W. E. Deming, *Out of the Crisis*, MIT Center for Advanced Engineering Study, Cambridge, Mass, 1986.

36. M. Löfgren and L. Witell, *Qual. Manag. J.*, 2008, **15**(1), 59–75.

37. *British Pharmacopoeia*, Stationery Office, London, 2017.

38. G. Sworn, in *Gums and Stabilisers for the Food Industry*, ed. P. A. Williams and G. O. Phillips, Royal Society of Chemistry, 2004, vol. 12.

39. T. Tadros, *Dispersion of Powders in Liquids and Stabilization of Suspensions*, Wiley VCH, Weinheim Germany, 2012.

40. S. Fu, A. Thacker, D. M. Sperger, R. L. Boni, S. Velankar, E. J. Munson and L. H. Block, *AAPS PharmSciTech*, 2010, **11**(4), 1662–1674.

41. J. F. Gamble, W.-S. Chiu and M. Tobyn, *Pharm. Dev. Technol.*, 2011, **16**(5), 542–548.

42. J. Kushner, *Pharm. Dev. Technol.*, 2013, **18**(2), 333–342.

43. A. Ferreira and M. Tobyn, *Pharm. Dev. Technol.*, 2015, **20**(5), 513–527.

44. G. Thoorens, F. Krier, E. Rozet, B. Carlin and B. Evrard, *Int. J. Pharm.*, 2015, **490**, 47–54.

45. https://www.fda.gov/downloads/drugs/guidances/ucm070636.pdf.

46. F. Bacon, *The Advancement of Learning*, 1605, Book I, v, 8.

CHAPTER 7

Film Coating of Tablets

MARSHALL WHITEMAN

GlaxoSmithKline, Priory Street, Ware, Hertfordshire, SG12 0DJ, UK
*E-mail: marshall.x.whiteman@gsk.com

7.1 Introduction

The majority of tablet coatings are purely cosmetic, to improve the appearance and aid identification for branding and patient safety. These coatings should have minimal effect on the therapeutic efficacy of the product, although certain colours have been associated with psychological response to medicines.[1]

Film coatings can also be used to confer a range of functional attributes to tablets, including protection of the product from light or moisture, protection of the patient from unpleasant-tasting products and modifying the site or rate at which drugs are released in the body. These aspects will be covered in this chapter.

The coating of oral solid dose forms has been performed for centuries,[2] but modern tablet film coating is usually traced back to Abbott Laboratories in 1953, when a fluidised bed apparatus was used to apply coatings from organic solvents.[3] The technology used for tablet film coating has evolved over time from fluidised bed to solid drum to perforated pans, the current standard technology.

Drug Discovery Series No. 64
Pharmaceutical Formulation: The Science and Technology of Dosage Forms
Edited by Geoffrey D. Tovey
© The Royal Society of Chemistry 2018
Published by the Royal Society of Chemistry, www.rsc.org

Almost all tablets are now coated and while the dominant technology is film coating other techniques are available:

- Enrobing: uses a process similar to soft gelatine capsule manufacture, where tablets and polymer (*e.g.* gelatine) sheets are fed between die rollers that press the polymer around the tablet. Requires specific equipment. Can yield a capsule-like appearance with high gloss.
- Gel dipping: has been used for OTC (over-the-counter) products in particular as it permits a capsule-like two-tone appearance and high gloss. Relatively complex to manufacture as specialist equipment is required to hold, dip and dry the tablets.
- Hot-melt coating: used for pellet coating to provide taste masking or modified release. Low-melting waxes can be sprayed using conventional equipment with heated lines to prevent the melted wax from congealing in the equipment.
- Compression coating: uses standard tableting materials to compress a coating around a core, on a special tablet press. It is mostly used to provide a modified release profile of some sort *e.g.* a fast-release outer coat with a prolonged-release inner core.
- Sugar coating: uses inexpensive materials and a simple, solid coating pan to apply a subcoat, a bulking layer, pigments and polish, followed by printing with identifying text.
- Film coating: pigments are suspended in a solution or dispersion containing polymer and plasticiser in a solvent or continuous phase. This suspension is then sprayed onto tablets, typically in a perforated, side-vented coating pan.

The popularity of film coating stems from the flexibility, reproducibility and ease of control of the process together with the ready availability of equipment and materials. Sugar coating, though in decline for many years, is still common and is discussed briefly here.

7.2   Sugar Coating

Initially, a sealing sub coat (*e.g.* polyvinyl acetate phthalate (PVAP)) may be applied to prevent erosion or dissolution of the core. "Grossing up" is then performed, where an inert material (*e.g.* calcium carbonate) is applied with syrup to build up the characteristic smooth discoid shape. When the tablet is the right size and has a smooth, rounded surface, a thin layer of colour is applied (colourants are usually the most expensive ingredients in a sugar coat, so the thinner the better), then a final polish layer is applied (wax or shellac). The coating is usually carried out in a simple solid pan and the coating liquid may be sprayed or ladled onto the tablet bed. The steps of the process are illustrated in Figure 7.1. Product name or identification codes are printed onto the polished tablets as an additional operation, since it is not feasible to deboss sugar-coated tablets.

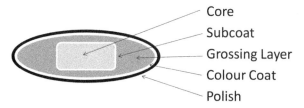

Figure 7.1 Layers in a typical sugar coating process.

Historically, sugar coating was the dominant coating technology, combining elegant, glossy appearance with the ability to mask the unpleasant taste of some drugs. Since the mid twentieth century, however, film coating became more and more common and is currently the technology of choice for pharmaceutical tablets. The remainder of this chapter is dedicated to film coating.

7.3 Film Coating Formulation and Materials

Film coat formulations generally have a polymer and plasticiser to form the film and a solvent or, in the case of dispersions, a continuous phase. For most pharmaceutical products, one or more pigments are also included to render the coating opaque and to impart a colour. In the majority of cosmetic coatings, these components represent the entire formulation, but occasionally additional materials may be added, for example, a surfactant to improve wettability and spreading of the coat on the tablet surface, flavourings to enhance taste masking of the product, or anti-tack agents to prevent the tablets sticking together during coating.

7.3.1 Tablet Cores

The first step in developing a film coating formulation and process is to develop a suitable tablet core. Other chapters cover the fine detail of formulation and manufacture, but here we can at least say that an ideal tablet for coating should have good mechanical strength (tensile strength >1.7 MPa[4]), and not be prone to capping, lamination (solid fraction *ca.* 0.85[4]) or abrasion (friability <1%),[5] should be dimensionally stable to temperature variation (avoid mineral excipients by preference) and have the surface properties to ensure good adhesion with coating polymers. Pandey *et al.*[6] showed that inclusion of the surfactant SLS (sodium lauryl sulfate) and the lubricant magnesium stearate can reduce the adhesion between a tablet and coating to the point that significant logo bridging occurs and that this was worse for tablets made using roller-compacted granules than for either wet-granulated tablets or those made by direct compression (logo bridging is an appearance defect where adhesion of the coat to the tablet is weak enough that the tensile strength of the film causes it to pull away from the logo and form a 'bridge' over the intagliations, obscuring the embossing). The shape of the tablet should avoid flat surfaces and sharp corners, to facilitate movement during the coating process and minimise potential for damage or adherence between tablets. Aulton[7] made indentation hardness profiles across a variety of tablet shapes and concluded that the best shape overall was normal concave, which showed an even hardness profile across the tablet, with no weakening at the edges.

Failure to take these requirements into account when formulating the tablet core can lock in issues that it may not be possible to overcome by changing the coating formulation and process.

7.3.2 Polymers

The film-forming polymer is the most important constituent of a coating—literally, the glue that holds the coat together. Polymers usually constitute roughly 50–75% of the coat formulation and so the properties of the coating depend very heavily on the properties of the polymer. In his seminal work on film coating theory and practice, Banker[8] described how the mechanical properties and solubility of film coatings were determined by polymer chemistry and structure, plasticisation, presence of dispersed solids and solvent effects. We will now briefly consider some of these factors.

7.3.2.1 Polymer Properties Important in Film Coating

7.3.2.1.1 Solubility. The solubility of a polymer in gastrointestinal fluids is of prime importance for coating, determining whether the polymer is appropriate for immediate-release, prolonged-release or delayed-release applications. The polymer solubility in the coating vehicle determines whether the coating can be sprayed as a polymer solution with pigments suspended in it, or needs to be delivered as a latex or pseudolatex dispersion.

7.3.2.1.2 Permeability. The permeability of polymer films is a key parameter for some prolonged-release products and also low permeability is vital for moisture protection and other barrier applications, although the intrinsic permeability may be reduced by additional formulation components to enhance barrier properties.

7.3.2.1.3 Viscosity. Viscosity of polymers in solution is important for coating, as it dictates whether the coating suspension can be sprayed reliably and also influences the appearance of the coated tablet. A lower viscosity coating suspension will spray more evenly and tend to give a smoother surface finish. A rule of thumb for viscosity is that suspensions become difficult to spray reliably above 400–500 mPa s. Coating suspensions may well have viscosities around 100 mPa s or less, so it is important to ensure that the suspension is stirred continuously during coating, to prevent settling out of dispersed pigments.

7.3.2.1.4 Tensile Strength. Films with a high tensile strength are more resistant to cracking or splitting and generally provide more protection to the tablet than weaker films. Tensile strength and Young's modulus increase with increasing molecular weight of polymer.

7.3.2.1.5 Adhesion. Aside from the obvious benefit of the film sticking to the tablet better, strong film adhesion means that the coat will adhere well into the debossing on a tablet and is less likely to show the phenomenon of logo bridging, where poor adhesion within the debossing causes the film to pop out and form a bridge across the logo, making it difficult or impossible to read. Poor adhesion in combination with low tensile strength can lead to split coats peeling back from the tablet surface.

7.3.2.2 Polymer Choice for Specific Applications

Table 7.1 lists the most common film-forming polymers, indicates their solubility or permeability characteristics and states whether they are commonly used for immediate, delayed or prolonged release coatings.

Table 7.1 Coating polymers and their uses.

Polymer	Solubility[a]	Uses[b]
Cellulose acetate	Semipermeable	PR
Cellulose acetate phthalate	>pH6	DR
Ethylcellulose	Insoluble	PR
Hydroxypropyl cellulose	Soluble	IR
Hypromellose	Soluble	IR
Hypromellose acetate succinate	>pH5.5–7	DR
Hypromellose phthalate	>pH5.0–5.5	DR
Polyvinyl acetate dispersion	Insoluble	PR
Polyvinyl acetate phthalate	>pH5	DR
Polyvinyl alcohol	Soluble	IR
Shellac	>pH7	DR, PR
Zein	>pH11.5	DR, PR
Ammonio methacrylate copolymer (type a)	Insoluble, high permeability	PR
Ammonio methacrylate copolymer (type b)	Insoluble, low permeability	PR
Basic butylated methacrylate copolymer	<pH5	IR
Methacrylic acid–ethyl acrylate copolymer (1:1)	>pH5.5	DR
Methacrylic acid–ethyl acrylate copolymer (1:1) dispersion 30%	>pH5.5	DR
Methacrylic acid–methyl methacrylate copolymer (1:1)	>pH6.0	DR
Methacrylic acid–methyl methacrylate copolymer (1:2)	>pH7.0	DR
Polyacrylate dispersion	Insoluble, low permeability	PR

[a]All solubility information taken from *The Handbook of Pharmaceutical Excipients*.[9]
[b]IR = immediate release; PR = prolonged release; DR = delayed release.

7.3.2.2.1 Immediate Release. For immediate-release coatings, good aqueous solubility is required to facilitate release in the gastrointestinal tract. Consequently, most immediate-release coatings are formulated as aqueous polymer solutions with insoluble components, such as pigments, suspended in them.

The most commonly used immediate-release coating polymer is hypromellose (hydroxypropylmethylcellulose). It is available in several different grades, usually distinguished by their solution viscosity. Rowe[10] showed that as the viscosity grade increases, molecular weight increases and the breadth of the molecular weight distribution increases. All grades have a certain amount of high-molecular-weight material. It has also been shown[11] that as the viscosity grade increases, hypromellose films become harder, less elastic and more resistant to abrasion. Hypromellose shows quite a steep increase in tensile strength with molecular weight, making molecular weight a useful formulation parameter to fine tune film properties—either by selection of a particular grade with specific properties or by combining grades to modulate the properties of the film. Hypromellose does suffer from relatively poor adhesion, so in products where adhesion is an issue, additional formulation components may be needed. Hydroxypropylcellulose has greater adhesion than hypromellose, but poorer tensile strength, so the two can be combined to give a coat with good strength and adhesion. Addition of a solid component, such as lactose or microcrystalline cellulose, to the coat formulation can also improve adhesion.[12]

Originally used as glue on postage stamps and more recently as a moisture barrier coating, polyvinyl alcohol (PVA) is increasingly being used as an immediate release coating polymer due to its excellent adhesion and the smooth finish it imparts to tablets. Its low viscosity means it can accommodate a higher proportion of pigments in suspension than hypromellose and the same weight of coating can be applied more quickly.[13] PVA–polyethylene glycol (PEG) copolymers are also becoming common in immediate release coating formulations. The incorporation of PEG onto the PVA backbone makes a very flexible, self-plasticising film and the low solution viscosity permits a high solids loading and hence faster coating.

7.3.2.2.2 Delayed Release. Delayed-release products are those which do not release drug immediately on ingestion, but are subject to some sort of trigger event. For most products, the delaying mechanism is the pH-dependent solubility of the film-forming polymer and the trigger is the tablet entering a part of the gastrointestinal tract where the local pH is at or above the point where the polymer is soluble. Polymers are more or less completely unionised (and hence insoluble) when the pH of the gastrointestinal tract is 2 or more units below the pK_a of the polymer and more or less completely ionised (soluble) at 2 or more pH units above the pK_a.

A broad approximation to pH in the gastrointestinal tract would be to say that it is acidic in the stomach and gets progressively more alkaline along the length, although there is considerable variation in pH, particularly in the colon.[14,15]

Acrylic polymers are the most commonly used for delayed release, followed by cellulosic or vinyl esters (*e.g.* hypromellose acetate succinate, hypromellose phthalate, cellulose acetate phthalate, polvinylacetate phthalate).

The most common type of delayed-release product is an enteric coated tablet or capsule. Enteric coating is usually done either to protect acid-labile drugs from stomach acid or to deliver drugs that are preferentially absorbed in the small intestine to their favoured absorption site. Since gastric pH is rarely above 3 and intestinal pH is typically over 6, we can use a number of different polymers that dissolve in the pH range 5–6 to provide a satisfactory enteric barrier which will release drug once out of the stomach (see Table 7.1). The most commonly used enteric polymer is methacrylic acid–ethyl acrylate copolymer (1:1), which can be applied as an organic solvent solution or as an aqueous pseodolatex dispersion. Hypromellose acetate succinate and polyvinyl acetate phthalate are also common, as both are available in forms that dissolve at pH 5.5 and are generally more stable than other non-acrylic polymers.[16]

Delayed-release coatings can also be used for colonic delivery and there are several ways to approach formulation of coatings for release in the colon:

1. Using standard enteric coatings or coatings that dissolve at higher pH and modulating delivery using the thickness of the coating (*e.g.* methacrylic acid–methyl methacrylate copolymers are available in grades soluble at pH 6 and 7, while hypromellose acetate succinate is available in grades soluble at pH 6 or 6.5).[17]
2. Using synthetic compounds, such as azo polymers,[18] or polysaccharides, such as galactomannans,[19] that can be degraded by colonic bacteria or that are cleaved by enzymes only present in the colon.
3. Using combinations of insoluble polymers and colon-degradable materials (*e.g.* ethylcellulose and pectin[20]).

7.3.2.2.3 Prolonged Release. Prolonged-release coating polymers are typically insoluble in water, with a range of permeability characteristics. Historically, these tended to be applied from organic-solvent systems, but the ready availability of latex or pseudolatex dispersions has allowed aqueous coating to become the dominant process.

For release to occur the drug must be taken (however slowly) into solution and so films must be permeable to some extent. This permeability may arise from the properties of the film-forming polymer, from the inclusion of soluble (pore-forming) materials along with an insoluble film former, or from application of the coat in such a way as to leave tortuous channels through which gastrointestinal fluids may pass to dissolve the drug. Release rate may be modulated either by varying the thickness of coat applied (increasing the tortuosity and reducing the number of channels), or by varying the proportion of soluble material in the coat.[21]

The most common prolonged release polymers are ammonio methacrylate copolymer types A (high permeability) and B (low permeability), and ethylcellulose. These polymers are most often applied as aqueous latex or pseudolatex dispersions.[22] In practice, it makes little or no difference whether they are true latexes or pseudolatexes.

7.3.2.2.4 Moisture Barrier Coatings. Many different polymers or combinations have been used as moisture barrier coatings. The issue with moisture barrier coatings is that they should protect the core from moisture, yet at the same time be capable of releasing the drug when swallowed. For this reason moisture barrier formulations typically require a compromise. Polymers most often used for moisture barrier coating include acrylics and polyvinyl alcohol. Studies on a variety of moisture barrier coatings have shown inconsistency in the rank order of barrier performance and that stability can be worse for coated tablets than for uncoated, although there does appear to be some advantage in using a moisture barrier coating when the tablet cores are hygroscopic.[23]

7.3.3 Plasticisers

The majority of polymers used in film coating form brittle films without addition of plasticisers—an undesirable quality in a tablet coat. Consequently we typically include a plasticiser in a film coating formulation.

Plasticisers act by intermingling with the polymer chains, weakening their intermolecular attractions, allowing the chains to move past each other more easily—transforming the polymer into a more pliable material[22] and lowering the glass transition temperature (T_g). T_g is the point at which the polymer changes from a hard, glassy material into a softer, more rubbery one. The extent to which a plasticiser lowers the T_g is often used as a measure of its effectiveness.[27]

Some suggested selection criteria for plasticisers in pharmaceutical products include:[24,25]

- Biocompatibility—clearly protection of the patient from potential toxicity must be the prime consideration;
- Compatibility with the polymer—there is no single plasticiser that can be used with all polymers, although it has been established which plasticisers work best with the common coating polymers;
- Effect on drug release—in particular, the type and level of plasticiser can affect the performance of modified-release coatings;
- Effect on mechanical properties—just as drugs can have undesirable side effects, plasticisers will influence other properties than the one we are concentrating on, T_g;
- Processability—for example, if the plasticiser leads to increased tackiness it could be more difficult to apply the coat.

Table 7.2 Plasticisers used in film coatings.

Hydrophilic	Hydrophobic
Glycerin (P)	Acetyl tributyl citrate (OE)
Mannitol (P)	Acetyl triethyl citrate (OE)
Polyethylene glycol (P)	Benzoyl benzoate (OE)
Polyethylene glycol 3350 (P)	Castor oil (OG)
Polyethylene glycol monomethyl ether (P)	Chlorobutanol (alcohol)
Propylene glycol (P)	Diacetylated monoglycerides (OG)
Pullulan (polysaccharide)	Dibutyl sebacate (OE)
Sorbitol (P)	Diethyl phthalate (OE)
Sorbitol sorbitan solution (P)	Tributyl citrate (OE)
Triacetin (OE)	Triethyl citrate (OE)
	Vitamin E (OE)

Plasticisers are sometimes categorised according to their chemical nature, *e.g.* whether they are polyols (P), organic esters (OE) or oils/glycerides (OG),[22] but it is more common to classify them according to whether they are hydrophobic or hydrophilic. In Table 7.2 the plasticisers identified in USP 40-NF35 [26] have been identified by both methods.

The most commonly used plasticisers in film coating are polyethylene glycol of various molecular weights and triethyl citrate, though propylene glycol, triacetin, dibutyl sebacate and glycerine are used. Phthalates and mineral oils, once common, are no longer used on grounds of toxicity. Generally vendors will recommend plasticisers for their polymers or else include them in preformulated systems at an appropriate level. The optimum amount of plasticiser for a coating formulation will vary according to the plasticiser and polymer combination, but a starting point for new formulations is around 10–20% of plasticiser by weight of polymer. Occasionally, levels as low as 5% or as high as 40% may be needed.

As well as reducing T_g, plasticisers lead to a reduction in film tensile strength, so there is always a trade-off between flexibility and strength. Increasing plasticiser level also tends to cause films to become more tacky, which can lead to appearance defects as tablets stick together and are pulled apart during the process.[28]

7.3.4 Colours

Colours are primarily used for branding and identification of products, but to the extent that they are opaque (do not transmit light, but reflect it), they can contribute to the stability of products that are photosensitive[29] and may reduce the permeability of films to a degree. Pigments are essentially solid inclusions within the film structure and, hence, increasing pigment concentration will reduce film tensile strength.[30] The critical pigment volume concentration (CPVC) is a parameter sometimes used to describe pigments. It is the pigment level in a formulation beyond which there is insufficient polymer to completely envelope the pigment.[31]

Insoluble pigments are preferred to water-soluble dyes as there are colour migration and stability issues with the latter.[32] Pigments also tend to be more opaque than dyes and so, when complete coverage of the tablets has been achieved, the colour will remain the same however much more coating is applied. With transparent dyes, the perceived colour is affected by the amount of coat applied.

Opacity depends on the reflection, absorption, scattering and refraction of light. In particular, the difference in refractive index between the polymer and the pigment can significantly affect the amount of light reflected and hence the opacity. The amount of light reflected at the polymer–pigment interface, R, assuming normal incidence and no absorption, is approximated by Cooper's equation:[33]

$$R = [(n_1 - n_2)/(n_1 + n_2)]^2$$

What this means in practice is that coatings will be transparent if the refractive indices of the pigment (n_1) and polymer (n_2) are the same, as $R = 0$. If the refractive indices are different, then the greater the difference between them, the more opaque the film will be. Most coating polymers have refractive indices close to 1.5 and aluminium lake pigments are also at around this level. Iron oxides and titanium dioxide have refractive indices in the range of 2–3 and so their presence in a film confers greater opacity. Some materials are anisotropic (have different refractive indices depending on their orientation) and this has been used to create a bicolour effect using the logo to create a different orientation of the pigment to that on the surface of the tablet.[34]

In addition to their greater opacity compared with aluminium lake pigments, the oxides also have a broader regulatory acceptability and so are the pigments of choice for products likely to be commercialised internationally. Colour regulations[35,36] change frequently, and while a formulator could, in time, find out all of the relevant regulatory details for pigments, in practice consulting suppliers of the colourants is the quickest and easiest approach.

While using a single pigment is feasible, the range of colours would be very limited, so the common practice is to use blends of two to three pigments to give the desired hue. This can lead to problems when trying to match an existing colour due to colour metamerism. This is a phenomenon whereby colours observed under one light source appear identical, but when observed under a different light source, they are different. This arises because the same apparent colour can be created from different combinations of pigments, which appear identical under certain lights, but different according to the illumination. When matching colours, it is established practice to do so under standard lighting conditions *e.g.* using a D65 artificial daylight lamp. Of course, if a colour is matched using the same pigments in the same relative quantities, the colour will be identical under all light conditions.

7.3.5   Solvent or Continuous Phase

Water is the most commonly used solvent or continuous phase for coating as it has none of the cost, safety or environmental issues associated with organic solvents. There are disadvantages to water, however: the viscosity of polymer solutions in water is usually higher than those in organic solvents, so they are harder to pump effectively and there is a lower polymer concentration at which the viscosity will become limiting for pumping. Water also has a lower latent heat of vaporisation than organic solvents, so more energy is needed to evaporate the solvent in a given time. A comparison between water and ethanol as a solvent for HPC[37] found that the ethanol solution was better wetting and the viscosity is lower, so pumping and atomisation were easier and drying was faster. A separate study[38] showed that the improved wetting with organic solvent gave rise to better film adhesion. In most cases, the superior performance of organic solvents is not sufficient to offset the cost, safety and environmental downside.

7.3.6   Other Components

A variety of other ingredients may be used,[31] depending upon the requirements of the individual formulation. Surfactants may be added to promote wetting of the tablet by the coating and also ensure rapid dissolution of the coating in the stomach. Many coat formulations, especially functional coats, can be tacky and materials may be added to reduce this stickiness during the coating process, talc and high molecular weight polyethylene glycols are examples of anti-tack agents. Flavours can be incorporated into the coating if there is a problem with taste masking, but this is unusual. For some formulation types, materials may be added to increase coat adhesion *e.g.* lactose.

7.4   The Coating Process

7.4.1   Film Coating Equipment

The most commonly used equipment for commercial film coating is the side-vented pan, or perforated pan coater. Similar designs are available from multiple manufacturers but the operating principle is generally as depicted in Figure 7.2.

Heated and, ideally, dehumidified air is fed to the coating drum. It passes through the tablet bed, heating the tablets and evaporating the solvent or continuous phase and is removed *via* an exhaust plenum. The coating liquid is delivered by an array of guns mounted on an arm positioned within the coater. The perforated drum rotates to form a moving bed of tablets, allowing all tablets to pass through the spraying zone.

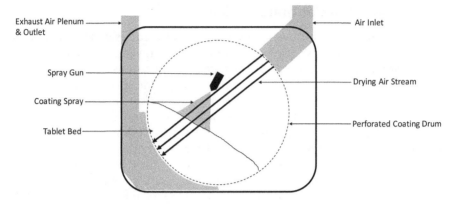

Figure 7.2   Schematic of perforated pan coater.

While most coaters are of a fixed size, several manufacturers offer coaters with interchangeable drums so that a wide range of batch sizes can be accommodated in a single piece of equipment.

There are a number of coaters available that have been designed for continuous manufacturing. The most common type of design is to simply elongate the drum and spraying arm and raise the front relative to the back so that tablets can feed in a continuous stream at the front and then tumble down the slope to exit at the lower rear end. A disadvantage of this approach is that there can be a wide range of tablet residence times in the coater. A design that claims to dramatically reduce residence time variation is the Bohle coater, which features multiple interconnected drums such that product is held in each drum until sufficient product has accumulated to pass over a weir into the next drum.

Different approaches to continuous coating have been taken be GEA, who have adopted the concept of coating multiple small batches very quickly, to give a quasi continuous process. The Consigma™ coater comprises multiples of drums, which rotate very rapidly but nevertheless handle tablets gently and coat 3 kg at a time. The number of coating drums can be adapted to match the output of a continuous tableting line.

7.4.2   Preparation of the Coating Liquid

This will vary slightly depending on whether the coating liquid is to be based on a solution of a polymer or a dispersion and whether it is to be made from the individual components or, as is the norm in the pharmaceutical industry, from a ready-formulated product. The fundamentals remain the same however—ensure even distribution of the components, ensure the polymer is dissolved completely where appropriate and avoid or eliminate drawing air into the liquid during mixing, as the presence of bubbles or foam will lead to erratic pumping and spraying of the coating liquid. A low-shear propeller mixer is preferred, to minimise aeration of the liquid and give better fine control over the mixing. For polymer solutions, the solid ingredients should be added to the solvent while stirring sufficiently to create a vortex, which will draw the solid into the body of the liquid and facilitate dissolution and dispersion. If the mixer head is too small relative to the size of the vessel, it will be very difficult to achieve a sufficient vortex without drawing excessive air into the liquid. A good rule of thumb is for the mixer head to be approximately 1 to 1/3 of the diameter of the vessel. The vessel itself should have either a flat bottom or a very shallow angle, so that the mixer had can get close to the bottom (to keep the pigments in suspension during processing, it may be necessary to move the mixer lower into the vessel). Ideally the shaft of the mixer should be parallel to the sides of the vessel—if it is at an angle, it will tend to draw in more air and it will be difficult to get the mixing head close to the bottom of the vessel. Usually a coating suspension will be allowed to stand (with continuous stirring to keep the solids in suspension) for around 1 hour prior to use, but it can be much longer depending on manufacturing schedules *etc.*

For dispersions, the excipients are typically homogenised with the water and this is added to the dispersion using a low-shear mixer.

7.4.3   Application of the Coating Liquid

Coatings are applied by pumping the coating suspension through a spray gun fitted with a nozzle to atomise the coating liquid. Any appropriately sized pump may be used, though it is most common to use a peristaltic pump as it is easier to clean (the interior of the tubing is the only part that comes into contact with the coating). The pump speed must be calibrated for the suspension and tubing being used. Each gun should have its own supply of coating liquid and they should be adjusted so that the flow rate through each gun is similar and the liquid is sprayed evenly across the tablet bed. If adjustment is needed this is done using the needle adjustment screw at the back of the gun (Figure 7.3).

Suspension is pumped through the nozzle of the spray gun. The nozzle can be closed or open depending on the position of an air-actuated needle (this airstream is termed the operating air). The fluid is atomised at the nozzle tip by the action of an annular air stream (atomising air). The shape of the atomised spray pattern is adjusted by a third air stream delivered from holes either side of the nozzle (pattern air). The exact configuration of these elements may differ somewhat between types of spray gun, but they are all required in every case. Traditional spray gun designs suffer from build-up of dry coating around the nozzle, which can drop into the product and spoil it. This phenomenon is called bearding. Manufacturers have been moving to so-called anti-bearding designs, wherein the pattern air jets are positioned closer to the nozzle tip, so that air is blown away from the nozzle and bearding is much reduced or eradicated.

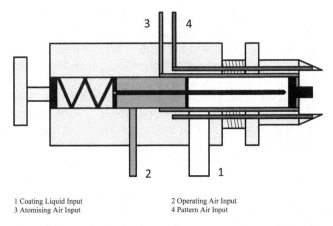

1 Coating Liquid Input   2 Operating Air Input
3 Atomising Air Input   4 Pattern Air Input

Figure 7.3   Schematic showing the operation of a two-fluid spray gun.

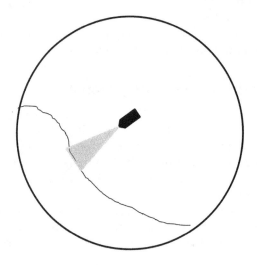

Figure 7.4   Typical positioning of the spray gun.

As the coater drum turns, the tablet bed will ride up the drum in the direction of rotation until at the leading edge of the bed begins to cascade back down into the drum. Spray guns are usually placed so they spray at the top third of the bed, near the foot of the cascade, as illustrated in Figure 7.4. For small-scale coaters, a single spray gun may suffice, but for most commercial-scale coaters, multiple guns will be needed and it is important to set these up properly for optimum performance. The distance between the guns and the distance of the guns from the tablet bed should be such that the spray does not dry before it can deposit on the tablet surface and the pattern from each gun should only minimally impinge on that from neighbouring guns. The individual spray patterns should merge into a single

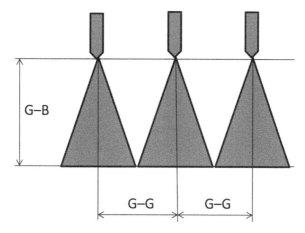

Figure 7.5 Illustration of gun to gun (G–G) and gun to bed (G–B) distances.

spray pattern covering the width of the tablet bed, but not going beyond the bed width, which would cause coating suspension to be sprayed onto the walls of the coater. The gun to gun and gun to bed spacings are shown schematically in Figure 7.5. Failure to get a single, uniform, broad pattern over the width of the bed can lead to poor coverage of the bed and the coating process will take much longer to produce an even colour on all tablets within the batch.

The atomising air pressure controls the size of the spray droplets. Insufficient atomisation leads to over-wetting of the tablet bed (since larger droplets need more heat for evaporation, larger droplets lead to a 'wetter' process), causing defects to the coating, while too high a pressure can lead to spray drying of the coat before it reaches the tablet surface (dryer process).

The pattern air must be chosen to produce the desired spray pattern. Ideally, this should have an even droplet size and droplet density across the whole pattern. Increasing the pattern air pressure gives a flatter cone of spray. If the atomising pressure is low, the cone will be rounder and will have larger drops in the centre than at the edge (wetter in the centre). There will be an optimum level of atomising air pressure that gives an even, elliptical spray pattern with uniform droplet size. Increasing the pressure beyond this point will force the spray cone into a dumbbell shape with a fine droplet region in the middle and the two lobes having large droplets. Either too low or too high an atomising pressure can thus lead to areas of over-wetting in the tablet bed and uneven coating.

An extensive study of spraying parameters found that coating efficiency (amount of coating applied to the tablets as a percentage of the actual amount sprayed) was greatest when atomising air and pattern air velocities were about the same and when spray rate was relatively high.[39] This implies that a wetter process is more efficient than a dryer one, which is logical, since excessive drying would increase coating lost by spray drying, though spraying too quickly would lead to overwetting and a different set of problems.

7.4.4 Distribution and Mixing

It is important to have good mixing of tablets within the bed in a coater, to ensure that each tablet gets about the same number of passes under the spraying zone. If some tablets pass through the coating zone less often, there may be variability in colour.

One of the key parameters for controlling coating is the rotational speed of the drum. If the speed is too slow, then mixing may be inadequate, resulting in variable colour, and individual tablets will be in the spraying zone for longer, which can lead to over-wetting and coating defects. Conversely, if the drum speed is too high, tablets may suffer mechanical damage, such as abrasion and logo erosion. It has been shown that a combination of a high pan speed and slow spray rate, using an adequate number of guns, leads to low weight gain variability on coated tablets.[40]

Coating pans usually have certain design features to assist with distribution and mixing. Anti-slide bars, as the name suggests, prevent tablets from sliding, *en masse*, down the wall of the pan as it rotates. As well as hindering mixing, sliding of the tablet bed can lead to abrasion against the pan wall and scuffing, where grey marks are created by abrasion between the steel drum and the titanium dioxide in the coat. Baffles are also a standard feature, though the design of the baffles may vary between equipment manufacturers. The baffles act to turn the powder bed from the edges back to the middle of the drum and so promote a uniform distribution of coating liquid across and through the bed. The influence of baffle design on mixing effectiveness has been studied[41] and different designs can have different effectiveness. In general, tubular baffles are better than ploughshares, which are better than rabbit ear baffles.

When transferring between different coaters (*e.g.* during scale up) the drum speed must be adjusted according to the pan diameter – tablets in a 60″ pan will be travelling much faster than tablets in a 16″ pan if both are rotating at 5 rpm. The appropriate drum speed should be determined experimentally by observing the tablets in the drum, but a useful rule of thumb to choose a starting point when changing equipment is to divide the diameter of the smaller drum (D_S) by that of the larger (D_L) and multiply the speed of the smaller drum (ω_{cycS}) by the result. Hence the speed of the larger drum (ω_{cycL}) is given by:

$$\omega_{cycL} = \omega_{cycS} \times (D_S/D_L)$$

This is based on the linear speed of the drum at the periphery, but does not take the specific motion of the tablets in the bed into account, which is why direct observation is required.

Other factors that influence the mixing are the size and shape of the tablet cores, the shape of the pan and the volume of tablets in the pan.

7.4.5 Drying

The key factors that influence drying in a coater are the flow rate, temperature and humidity of the drying air and the spray rate of the coating liquid. Increasing air flow or air temperature make for a hotter, dryer process, whereas increasing the drying air humidity or the coating spray rate will make the process cooler and wetter. Too dry a process leads to spray drying of the coat and infilling of the debossed logo; too wet a process leads to sticking and picking. It is widely accepted that the temperature of the tablet bed is a key parameter and this is closely related to the outlet air temperature. The inlet air temperature is usually automatically adjusted so as to maintain a steady outlet temperature. Air flow rate is usually held at a fixed set point and not varied during processing and inlet air humidity is typically controlled to a low level. Consequently the major variable parameters influencing drying are bed temperature (with outlet temperature as a surrogate).

7.4.6 Control

The essence of controlling the coating process is to ensure the process is wet enough to avoid spray drying while simultaneously ensuring it is not so wet that over-wetting occurs. Various thermodynamic models have been developed to aid in controlling the process.[42] These models combine the key variables to give a value for bed or outlet temperature and can be readily applied to commercial coating operations. In the absence of such models, coating processes can be well controlled using bed or outlet temperature, spray rate and drum speed.

7.4.7 Effects of Process Parameters on Product Quality

There are ample illustrations of defects and their causes readily available from supplier websites. Table 7.3 presents the potential effects on the product of changes to the major process variables.

7.5 Evaluating Film Coats

This section describes some of the more common tests used to assess film coatings. Many of the measurements done to assess film coating formulations are done on free film samples, which can be made by a casting technique or by spraying onto a surface and removing the sample for subsequent testing. While this approach is almost universal, it does raise the question of whether a free film behaves in the same way as a film sprayed onto a tablet

Table 7.3 The effects of process parameter changes on the product.

Process parameter	Increase		Decrease	
	Effect on process	Effect on product	Effect on process	Effect on product
Inlet air temperature	Increases bed temperature	Faster drying; may lead to spray drying of the coat before it reaches the tablet; may lead to 'orange peel'	Decreases bed temperature	Slower drying; may lead to over-wetting of the tablets and hence picking, sticking or twinning
Inlet air humidity	Reduces drying capacity of air	If drying capacity becomes limiting, may lead to over-wetting of the tables and hence picking, sticking or twinning	Increases drying capacity of air	Usually not an issue
Air flow rate	Increases bed temperature	Faster drying; may lead to spray drying of the coat before it reaches the tablet; may lead to 'orange peel'	Decreases bed temperature	Slower drying; may lead to over-wetting of the tablets and hence picking, sticking or twinning
Pan speed	No effect on bed temperature or humidity	If speed is too high erosion of the tablet surface, edge chipping or tablet breakage may occur	No effect on temperature or humidity	If pan speed is too low, may lead to over-wetting of the tablets and hence picking, sticking or twinning
Spray rate	Decreases bed temperature, increases humidity	May lead to over-wetting of the tablets and hence picking, sticking or twinning	Increases bed temperature, decreases humidity	May lead to spray drying of the coat before it reaches the tablet, may lead to 'orange peel'
Atomising pressure	May reduce bed temperature slightly	Reduces the size of spray droplets, may lead to spray drying of the coat before it reaches the tablet, may lead to 'orange peel'	May increase bed temperature slightly	Increases size of spray droplets, may lead to colour variation, may lead to over-wetting of the tablets and hence picking, sticking or twinning, may lead to 'orange peel'
Pattern air pressure	If the pattern air pressure is not adjusted correctly, there may be localised areas where the droplets are too large or too small, so effects related to over-wetting or over-drying may occur. Usually local over-wetting is more problematic due to sticking, picking and twinning			

surface. Only the colour and surface appearance of coatings are likely to be used as routine tests for commercial products—the other tests are mostly limited to research and product development.

7.5.1 Appearance

Tablet specifications typically include 'description' rather than appearance and this is usually limited to a broad statement of colour, shape and markings. In most cases, though, appearance will be designated an in-process CQA and assessed by inspection after film coating, but prior to packing. Appearance is assessed in terms of acceptable quality levels (AQL—the maximum level of defective tablets in a specified sample size) or limiting quality levels (LQL—the level below which the customer will not accept the product). Details of appropriate sampling plans for various desired quality levels are available from published sources, for example ISO 2859.[43] Different AQLs are commonly defined for different classes of defect (*e.g.* critical, major, minor) so that there is zero to very low tolerance for more serious defects, but a greater tolerance for minor defects, such as minor picking or pitting of the tablet that has been coated over.

7.5.2 Colour

Colour may be assessed by either visual or spectrophotometric comparison with a reference sample. The latter typically uses a reflectance spectrophotometer to measure colour difference (ΔE) in the Hunter Lab Colour Space:[44]

$$\Delta E = [\Delta L^2 + \Delta a^2 + \Delta b^2]^{1/2}$$

where L ranges from 0 (Black) to 100 (white), a is a green (negative values) to red (positive values) continuum, and b is a blue (negative values) to yellow (positive values) continuum. Generally, colours should be within ± 5 ΔE units of the reference, but ideally a visual comparison is done in addition to the spectrophotometric test.

7.5.3 Gloss

Gloss may be assessed visually or by reflectance measurements, or may be inferred from surface roughness measurements—rougher surfaces are obviously less glossy than smoother ones, all else being equal. Glossier tablets are more visually appealing to the patient and give an impression of quality, but they also tend to cause fewer problems in terms of packaging, as they move well in automated equipment. Gloss is measured in gloss units (GU) and is based on the specular reflection (mirror-like reflection) at the same angle to the reflective surface as the incident light (Figure 7.6). Typically for coatings it is measured at 60°, but if the sample is highly glossy, a 20° angle may be used, or if it has very low gloss, 85° can be used.

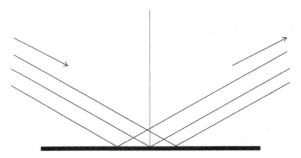

Figure 7.6   Gloss meter measures specular reflection.

Glossmeters are readily available and the methods for their use are given in ISO 2813.[45]

7.5.4   Roughness

Surface roughness of tablet cores or coated tablets can be quantified using laser profilometry.[46] Rougher tablet core surfaces indicate the likelihood that adhesion of the coat to the core will be stronger, as there are more surface features for bonding. The average roughness, R_a, is the most widely used measure of roughness. It is the average absolute deviation from the mean of amplitude measurements ($|y_i|$) across a surface and is given by:[47]

$$R_a = \frac{1}{n}\sum_{i=1}^{n}|y_i|$$

7.5.5   Tensile Strength

Film strength for free film samples can be conveniently measured using standard techniques. ASTM D638 or ISO 527[48] methods may be used interchangeably since both tend to give very similar results. A template is used to prepare samples of identical shape. The ends of the sample are clamped in a crosshead press set up to pull the sample at a constant rate, while measuring the force and the sample length. Stress (force/area) and strain (elongation/original length) may be calculated and plotted to give Young's modulus, E (E = Stress/Strain). Typically, the tensile strength at breaking is quoted, along with Young's modulus.

7.5.6   Adhesion

Adhesion may be measured using a crosshead press. Ideally, flat-faced tablets are used and the sidewall coating is removed from the tablet, which is fixed to the lower and upper platens of the press with double sided adhesive

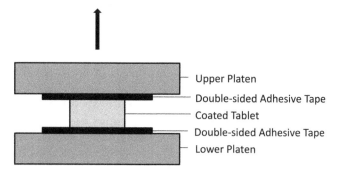

Figure 7.7 Adhesion testing rig.

tape (Figure 7.7). The upper plate is moved upward at a constant rate and the force is measured. The force at the point of film separation from the tablet is taken as a measure of the adhesion.

7.5.7 Moisture Vapour Transmission Rate (MVTR)

The simplest way to estimate film permeability to water is to put a saturated salt solution in a vial, to deliver a known vapour pressure, fasten a free film sample over the mouth of the vial and place the vial in a chamber at a different known vapour pressure. The different vapour pressures create a driving force for moisture transmission through the film. The amount of water transmission is determined from the weight loss (or gain) over time. If the weight change is plotted *versus* time, the slope of the line gives the MVTR. If desired, a permeability constant, P, can be calculated from:

$$P = \text{MVTR} \times L/A \times \Delta P$$

where L is the thickness of the film, A is the area of film available for transmission and ΔP is the vapour pressure gradient.[49]

7.5.8 Glass Transition Temperature

T_g is conveniently determined using differential scanning calorimetry (DSC). Film material is heated at a constant rate and the phase transition points are determined from the temperature at which endothermic or exothermic events occur.

7.5.9 Minimum Film Forming Temperature (MFFT)

MFFT is measured by heating a polymeric dispersion to evaporate solvent or continuous phase. Below the MFFT a powder residue will remain, while at or above the MFFT, a clear film will be formed.

7.5.10   Enteric Tests

USP40-NF35, General Chapter 701,[50] Disintegration, stipulates testing six tablets in simulated gastric fluid for 1 hour, then transferring them to simulated intestinal fluid for a period stated in the product monograph. The tablets should be intact after the first stage and disintegrate fully within the specified time for the second stage. If any of the six fail to disintegrate, a further 12 tablets are tested and 16 of the 18 must disintegrate within the allotted time. In practice, companies making enteric products often choose to carry out much more stringent testing, using tens or even hundreds of tablets.

7.6   Conclusions

It is impossible, within a few pages, to give a complete treatment of film coating, but this chapter has touched on the key formulation components of polymer, plasticiser, pigment and solvent, and how they contribute to the properties of the film coating and the overall product. It has also given an overview of the coating process and the effects of the various process parameters thereon. The chapter is concluded by a discussion of the various measurement techniques that are most commonly used to assess coatings in development and commercial manufacture. Those wishing to delve deeper will find the references cited to be a useful start.

References

1. A. J. M. de Craen, P. J. Roos, A. L. de Vries and J. Kleijnen, *BMJ [Br. Med. J.]*, 1996, **313**, 1624.
2. K. H. Bauer, K. Lehmann, H. P. Osterwald and G. Rothgang, *Coated Pharmaceutical Dosage Forms*, Medpharm Scientific Publishers, Stuttgart, 1998, ch. 1, p. 13.
3. G. C. Cole, in *Pharmaceutical Coating Technology*, ed. G. C. Cole, J. E. Hogan and M. Aulton, Taylor & Francis, London, 1995, ch. 1, p. 4.
4. M. Leane, K. Pitt and G. Reynolds, *Pharm. Dev. Technol.*, 2015, **20**, 12.
5. USP39-NF34 S1, *United States Pharmacopeial Convention, General Chapter <1216>*, 2016, p. 1609.
6. P. Pandey, D. S. Bindra, S. Gour, J. Trinh, D. Buckley and S. Badawy, *J. Pharm. Sci.*, 2014, **103**, 3666.
7. M. E. Aulton, *Pharm. Acta Helv.*, 1981, **56**(4–5), 133.
8. G. S. Banker, *J. Pharm. Sci.*, 1966, **55**(1), 81.
9. *The Handbook of Pharmaceutical Excipients*, ed. P. Sheskey, W. Cook and C. Gable, Pharmaceutical Press, London, 2016.
10. R. C. Rowe, *J. Pharm. Pharmacol.*, 1980, **32**, 116.
11. R. C. Rowe, *Pharm. Acta Helv.*, 1976, **51**, 330.
12. H. Khan, J. T. Fell and G. S. Macleod, *Int. J. Pharm.*, 2001, **227**, 113.

13. L. A. Felton, *Int. J. Pharm.*, 2013, **457**, 423.
14. C. G. Wilson, in *Advances in Delivery Sciences and Technology, Controlled Release in Oral Drug Delivery*, ed. C. G. Wilson and P. J. Crowley, Controlled Release Society, St. Paul, MN, 2011, ch. 2, p. 27.
15. A. W. Basit, *Drugs*, 2005, **65**(14), 1991.
16. K. Thoma and K. Bechtold, *Eur. J. Pharm. Biopharm.*, 1999, **47**(1), 39.
17. M. Ashford, J. T. Fell, D. Attwood and P. J. Woodhead, *Int. J. Pharm.*, 1993, **91**, 241.
18. C. Tuleu, A. W. Basit, W. A. Waddington, P. J. Ell and J. M. Newton, *Aliment. Pharmacol. Ther.*, 2002, **16**(10), 1771.
19. S. Hirsch, V. Binder, V. Schehlmann, K. Kolter and K. H. Bauer, *Eur. J. Pharm. Biopharm.*, 1999, **47**(1), 61.
20. Z. Wakerley, J. T. Fell, D. Attwood and D. Parkins, *Pharm. Res.*, 1996, **13**, 1210.
21. D. Palmer, H. Vong, M. Levina and A. R. Rajabi-Siahboomi, *Poster Presentation at AAPS Annual Meeting and Exposition*, San Diego, 12–16 Nov 2007, Poster T2056.
22. J. E. Hogan, in *Pharmaceutical Coating Technology*, ed. G. C. Cole, J. E. Hogan and M. Aulton, Taylor & Francis, London, 1995, ch. 2.
23. E. Mwsegiwa, G. Buckton and A. W. Basit, *Drug Dev. Ind. Pharm.*, 2005, **31**(10), 959.
24. M. Rahman and C. S. Brazel, *Prog. Polym. Sci.*, 2004, **29**, 1223.
25. E. Snejdrova and M. Dittrich, in *Recent Advances in Plasticizers*, ed. M. Luqman, Intech, Rijeka, 2012, ch. 3.
26. US40-NF35, *Front Matter: Plasticizers*, Rockville, MD, 2017.
27. V. Y. Senichev and V. V. Tereshatov, in *Handbook of Plasticizers*, ed. G. Wypych, ChemTech Publishing, Toronto, 2004.
28. M. Wesseling, F. Kuppler and R. Bodmeier, *Eur. J. Pharm. Biopharm.*, 1999, **47**(1), 73.
29. R. Teraoke, Y. Matsude and I. Sugimoto, *J. Pharm. Pharmacol.*, 1988, **41**, 293.
30. A. O. Okhamafe and P. York, *Int. J. Pharm.*, 1984, **22**, 273.
31. L. A. Felton and S. C. Porter, *Expert Opin. Drug Delivery*, 2013, **10**(4), 421.
32. L. A. Felton and J. W. McGinity, *Drug Dev. Ind. Pharm.*, 2002, **28**(3), 225.
33. A. C. Cooper, *J. Oil Colour Chem. Assoc.*, 1948, **31**, 343.
34. R. C. Rowe and S. F. Forse, *J. Pharm. Pharmacol.*, 1982, **35**, 205.
35. 21 CFR, *Parts 73 & 74, Color Additives Approved for Use in Drugs*, 2012.
36. Regulation (EC) No.1333/2008 on Food Additives.
37. J. Bajdik, G. Regdon, T. Marek, I. Eros, K. Suvegh and K. Pintye-Hodi, *Int. J. Pharm.*, 2005, **301**, 192.
38. P. K. Nadkarni, D. O. Kildsig, P. A. Kramer and G. S. Banker, *J. Pharm. Sci.*, 1975, **64**(9), 1554.
39. J. Wang, J. Hemenway, W. Chen, D. Desai, W. Early, S. Paruchuri, S.-Y. Chang, H. Stamato and S. Varia, *Int. J. Pharm.*, 2012, **427**, 163.
40. S. Just, G. Toschkoff, A. Funke, D. Djuric, G. Scharrer, J. Khinast, K. Knop and P. Kleinebudde, *Int. J. Pharm.*, 2013, **457**, 1.

41. G. W. Smith, G. S. Macleod and J. T. Fell, *AAPS PharmSciTech*, 2003, **4**(3), 37.
42. M. T. am Ende and A. Berchielli, *Pharm. Dev. Technol.*, 2005, **10**(1), 47.
43. ISO 2859-10:2006, *Sampling Procedures for Inspection by Attributes – Part 10: Introduction to the ISO 2859 Series of Standards for Inspection by Attributes*, International Organisation for Standardisation, Geneva, Switzerland, 2006.
44. R. S. Hunter, *J. Opt. Soc. Am.*, 1958, **48**(12), 985.
45. ISO 2813: 2014, *Paints and Varnishes – Determination of Gloss Value at 20 Degrees, 60 Degrees and 85 Degrees*, International Organisation for Standardisation, Geneva, Switzerland, 2014.
46. P. Seitavuopio, J. Heinamaki, J. Rantanen and J. Yliruusi, *AAPS PharmSciTech*, 2006, 7(2), 31.
47. E. P. Degarmo, J. Black and R. A. Kohser, *Materials and Processes in Manufacturing*, Wiley, 9th edn, 2002, p. 223.
48. ISO 527-1:2012, *Plastics – Determination of Tensile Properties – Part 1: General Principles*, International Organisation for Standardisation, Geneva, Switzerland, 2012.
49. L. A. Felton and S. C. Porter, *Drug Dev. Ind. Pharm.*, 2010, **36**(2), 128.
50. USP40-NF35, *General Chapter 701, Disintegration*, Rockville, MD, 2017.

CHAPTER 8

Oral Controlled Release Technology and Development Strategy

CHRISTIAN SEILER

Pharma XP Consulting Ltd., Hertford, Hertfordshire, UK
*E-mail: christian.seiler@pharmaxpconsulting.co.uk

8.1 Introduction

Most oral drug products are designed for immediate release (IR), with the dosage form primarily a means to deliver the correct quantity of active pharmaceutical ingredient (API) and expose it to gastrointestinal (GI) fluid. To perform *in vivo* as intended, the following three steps are generally required:

  i. API is physically released from the dosage form
  ii. API released from the dosage form dissolves in GI fluids
  iii. API dissolved in GI fluids diffuses through the GI membrane into the systemic circulation

Any API in the systemic circulation is gradually eliminated, resulting in lower blood levels unless further API enters the system.[1] The relationship between API levels in the blood and time post ingestion is termed the

Drug Discovery Series No. 64
Pharmaceutical Formulation: The Science and Technology of Dosage Forms
Edited by Geoffrey D. Tovey
© The Royal Society of Chemistry 2018
Published by the Royal Society of Chemistry, www.rsc.org

Table 8.1   Key clinical issues and observations with immediate-release dosage forms that require mitigation strategies such as controlled-release formulation development.

Clinical issue	Mitigation options	Risks of mitigation options
Toxicity associated with high or rapidly increasing plasma concentration	IR dose reduction	- Insufficient trough → May require frequent dosing
	Extended release	- More complex development - Once daily dosing (QD) requires adequate colonic absorption
Lack of efficacy due to low plasma concentration	IR dose increase	- Toxicity associated with peak plasma levels
	Frequent dosing	- Lower market uptake
	Extended release	- More complex development - QD requires adequate colonic absorption

pharmacokinetic (PK) profile. For IR products, the PK profile is largely governed by API properties (see Section 8.2), with clinical issues, as listed in Table 8.1, often linked to the PK profile.

As highlighted in Table 8.1, using higher doses to tackle low API concentration is often not possible, due to the risk of unacceptably high API concentrations. In such cases, the only option for IR products to achieve safe and efficacious concentrations is frequent dosing. Reduced patient adherence is the main reason for avoiding frequent dosing, and a key driver for controlled release (CR) development. The main advantage of CR dosage forms is that they represent levers for API release, and hence also for the API input rate into the systemic circulation. When IR products require frequent dosing, *e.g.* two or three times daily, this is typically due to the following scenarios:[1]

i. Only the small intestine allows adequate API absorption, with drug release in the lower GI tract (*i.e.* colon) not contributing to the therapy
ii. The API diffuses rapidly through the GI membrane and is rapidly metabolised and eliminated

The time window for API input into the systemic circulation is significantly shortened in both scenarios, resulting in declining blood levels shortly after product ingestion (Figure 8.1).

CR formulations can be designed with different API release profiles (dissolution examples are shown in Figure 8.2), whose choice is generally governed by the API properties and the clinical need.[1]

Different CR technologies offer different ranges of release profiles, and the aim of this chapter is to discuss these technologies in sufficient detail to make them accessible to the reader. What all oral CR products have in common is that the release-controlling unit remains largely intact while travelling through the GI tract. In many cases, the API can thus be delivered to

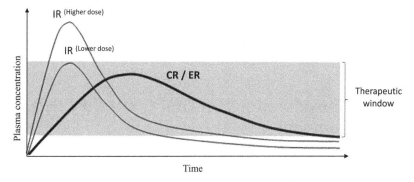

Figure 8.1 PK profile illustrations for (i) immediate-release (IR) formulations failing to maintain plasma concentrations within the therapeutic window and (ii) a controlled-release/extended-release (CR/ER) formulation whose plasma concentrations remain within the therapeutic window.

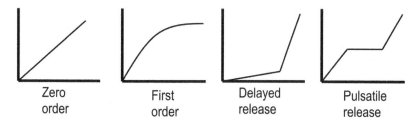

Figure 8.2 Examples of dissolution profiles achievable by oral controlled release formulations.

the body throughout most of the day, exploiting the fact that it typically takes a day for the product to travel through the GI tract. It is hence important to consider the interplay between CR dosage forms and the GI environment (see Section 8.2), together with key considerations such as API attributes, manufacturability and quality-by-design (QbD).[1]

Oral CR dosage forms are generally either monolithic (single-unit) or multiparticulate (multi-unit), with the options listed below covered in Sections 8.3 and 8.4.

   i. Hydrophilic matrix tablets
   ii. Inert matrices
  iii. Multiparticulate or pellet formulation technologies
   iv. Osmotic drug delivery systems
   v. Proprietary and other technologies

Since human PK is generally the most important attribute, the release profile is a critical formulation deliverable. A sound strategy for product performance evaluation is hence essential for successful CR development. *In vivo* and *in vitro* studies generally differ significantly in terms of costs, lead times,

API requirements and strategic relevance, which can make choices difficult. A staged approach, from *in vitro* to *in vivo* testing, is hence generally a good compromise, especially if it is governed by the clinical stage of the program (see Section 8.5).

8.2 Key Considerations for Oral Controlled Release Development

CR formulations are generally developed to simplify therapy. As discussed in Section 8.1, CR formulations shift drug release control from the API to the dosage form. This has important implications for development, with key considerations discussed in this section.

8.2.1 Gastrointestinal Environment

All oral CR technologies have in common that the release-controlling unit remains largely intact while travelling through the GI tract. The interplay between dosage form and GI environment is hence very important, especially the effect of physiological conditions such as motility, residence time, ions, pH and enzymes.[2] Figure 8.3 shows how parameters like environmental

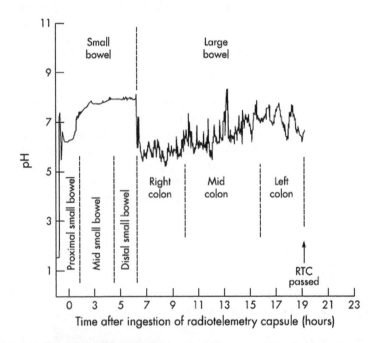

Figure 8.3   Gastrointestinal pH profile, and residence times in different regions of the gastrointestinal tract, as measured using a radio-telemetry capsule (RTC).[66] Reproduced from *British Journal of Sports Medicine*, K. A. Rao, E. Yazaki, D. F. Evans and R. Carbon, **38**, 482–487, 2004, with permission from BMJ Publishing Group Ltd.

pH and residence time differ for the stomach, small intestine (small bowel) and colon (large bowel).

Since CR dosage forms spend most of their time in the colon, a once-daily dosing regimen generally requires APIs with good colonic absorption (see Section 8.2.2). This particularly applies to CR formulations, which release API over an extended period of time, since released API serves little purpose if it can not be absorbed through the GI membrane. APIs with poor colonic absorption and a need for extended drug release have historically prompted the evaluation of so-called gastroretentive (GR) formulations. Their design principle is increased dosage form residence time in the stomach, which in turn extends the time window for drug release and absorption (see Section 8.4.5.1).

8.2.1.1 Food Effects

The presence of food can significantly alter the GI environment and is generally associated with (i) increased stomach hydrodynamics, (ii) retention of larger particles within the stomach (due to a contracted pylorus) and (iii) compositional changes of the gastric juices, including elevated pH (*e.g.* pH 4–5 *versus* 1–2 for fed *versus* fasted administration).[3] The influence of food on the GI environment affects its interplay with oral dosage forms, the extent of which is governed by food type and quantity and dosage form attributes such as physical integrity, size and release characteristics. When pharmacokinetic PK differences for fed *versus* fasted administration require a fed or fasted product label, one generally refers to a "food effect". In contrast, nonconsequential performance differences linked to food administration are generally termed "an effect of food". Food effects for IR dosage forms are generally governed primarily by API attributes, whereas food effects for CR dosage forms can also be attributable to the dosage form. Food susceptibility can differ significantly between technologies, especially when physical attributes differ significantly. Administration without regard to food is much preferred, since it simplifies the dosing regimen and increases patient adherence, with food effect risks, hence, being an important consideration when selecting CR technologies. Reliable prediction of food effects, however, has been amongst the most challenging tasks in CR development.

8.2.2 API Attributes

Many measures characterising API exist, with Table 8.2 showing key attributes for oral dosage forms.

The attributes most important for CR formulations are solubility, half-life and absorption, which are hence discussed further in this section.

8.2.2.1 API Solubility

API solubility is a key variable for CR development, since most technologies require API to dissolve within the dosage form (for diffusion-based release). This is in contrast to IR dosage forms, which primarily require

Table 8.2 Key API attributes for an oral solid dosage form.

	Compound attributes
Physicochemical	- Solubility
	- Wettability
	- Particle size
	- API phase/form
	- Molecular weight/volume
Biopharmaceutical	- Permeability
	- Half-life

API dissolution in the surrounding fluid, where the solid-to-liquid ratio is much more favourable. Some CR technologies are hence designed such that no dissolution within the dosage form is required for drug release (see Section 8.4).

8.2.2.2 API Half-life

The typical goal of CR formulations is to maintain therapeutic blood levels over a prolonged period of time, which requires API to enter the systemic circulation at similar rates to its elimination. The half-life thereby describes the rate of drug elimination, incorporating processes such as metabolism and urinary excretion. The half-life in the earlier parts of the PK profile is generally referred to as alpha half-life. This is most relevant for CR formulations and should not be confused with the terminal, or beta, half-life. Compounds with short alpha half-life (<8 hours) are prime candidates for oral CR development, while no established CR technology can generally accommodate very short alpha half-lives (<2 hours), at least not for once-a-day administration.[4]

8.2.2.3 API Absorption

Since release control *via* the dosage form is imperative for CR formulations, the rate of API release has to be lower than that of API absorption. However, this assumes that drug absorption occurs uniformly throughout the GI tract, which is rarely the case, since it consists of discrete regions of rather different characteristics. If API absorption is limited to a specific region, extended drug release can be detrimental, since only parts of the dose would be available for absorption. A key example is the colon, where CR formulations generally reside for the longest time and which is often associated with poor absorption, due to limited API solubility and/or permeability (see Section 8.2.1). Since this can determine the success or failure of CR development, it is important to assess the risk of poor colonic absorption early in the development cycle.[4] Commonly used methodologies are (i) *in vitro* evaluation of API transport across colonic cell membranes (*e.g.* Caco-2, LLC-PK1), (ii) rat perfusion studies or (iii) *in vivo* regional absorption studies in animals (*e.g.* dogs) or humans (*e.g.* Enterion™, IntelliCap® *etc.*) (Section 8.4.5).

8.2.3 Manufacturability and Quality-by-Design (QbD)

Manufacturability and quality-by-design (QbD) are important considerations for CR development, since CR products need to maintain drug release within well-defined limits, while remaining physically intact. This can require a substantial body of work, especially if design-space understanding is to be developed. The scope of this section is limited to QbD considerations for polymeric raw materials, since they represent the key difference between CR and IR products.

In contrast to IR products, CR formulations generally have upper and lower dissolution limits, across multiple time-points describing the profile. It is hence important to demonstrate that the manufacturing process consistently meets all product performance criteria. Since CR technologies commonly use higher viscosity polymers for release control, it is important to consider their batch-to-batch variability during product design. Since all functional polymer attributes (*e.g.* viscosity, substitution levels, particle size) have acceptance ranges, it is important to test CR product performance across them. This is where a QbD approach exploring possible raw material variations, within supplier or pharmacopoeial specifications, is advisable. It is thereby important to not rely on certificate-of-analysis (CoA) data, since these are often average values from very large production runs, which are typically filled continuously into standard-sized drums, ready for distribution. Due to this approach and raw material manufacturing process variability, the characteristics of a specific drum of material can hence be quite different to those quoted on the CoA. It is hence advisable to ask suppliers for samples of material with known characteristics and use these for QbD-type formulation work, or analyse representative samples of an excipient drum.

8.3 Process Technologies Utilised for Controlled Release Dosage Forms

8.3.1 Conventional Manufacturing Technology

CR technologies utilise established conventional unit operations, as listed below, which are discussed in other chapters and hence not explored further here.

- Granulation (roller compaction/high-shear wet granulation/fluid bed granulation)
- Drying (fluid bed drying)
- Tablet compression
- Capsule filling
- Tablet coating

8.3.2 Rotary Granulation Technology

Rotary granulation is a single-pot technique involving spheroid production, drying and coating, with the equipment commonly known as rotary processor, rotary fluidised bed, rotary fluid bed granulator, rotor fluidised bed

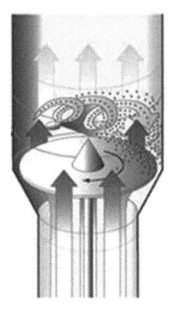

Figure 8.4   Schematic depiction of rotary granulation equipment.[73]

granulator or fluid bed roto-granulator.[5] It is a hybrid of fluidisation and spheronisation, with centrifugal, fluidisation and gravitational forces acting on the product and interactions between equipment, formulation and process variables determining pellet characteristics. The schematic representation in Figure 8.4 shows that the equipment involves a rotating plate at the base of a cylindrical vessel, with an air gap through which fluidising air enters.

A spray gun is located in the lower portion of the equipment wall, immersed in the powder mass. Due to the tangential spray mode, the spray droplets travel concurrently to the powder.[6]

8.3.3   Extrusion–Spheronisation (E–S) Technology

E–S pelletisation is a process involving four steps, namely wet mass preparation (granulation), wet mass shaping (extrusion), extrudate breakage and rounding (spheronisation) and fluid removal (drying).[7] Sieving of dried pellets is a common additional step, to avoid highly varied pellet size.[8]

8.3.3.1   Blending and Granulation

Typical equipment includes high-shear wet granulators (HSWG), planetary mixers or sigma blade mixers, all of which can distribute liquid homogeneously.[7] The liquid, usually water, imparts rheological properties to the formulation, facilitating shaping during extrusion and spheronisation. This generally requires higher fluid levels than conventional granulation, *e.g.* 30–40% w/w dry basis. For water, this makes the process difficult to use for drugs undergoing hydrolysis.[9]

8.3.3.2 Extrusion

According to Newton,[9] the extruder type and operating conditions are important, as they often affect pellet characteristics, with extruders generally classified as follows:[8]

- Screw-fed
- Sieve and basket
- Gravity-fed
- Piston-fed (ram extruder)

All these extruder types force material from large to small cross-sections (the die), compressing the wet mass and removing air, before forcing the material through the aperture. Most pharmaceutical formulations use screen extruders, despite their risks of screen wear and distortion, which are associated with extrudate and pellets of varying quality.[9] Ram extruders are often used in early development, as they require little material and can measure rheological properties.[8] General rules of thumb for extrusion are as follows:

- Pellet size is proportional to hole diameter and often formulation-dependent for a given aperture
- The risk of liquid migration, and hence formulation heterogeneity, increases with decreasing extrusion speed, which, in turn, affects pellet formation during spheronisation

8.3.3.3 Spheronisation

Spheronisers consist primarily of a friction plate with a grooved surface, which rotates within cylindrical walls.[7] As a result of the rotation, the extrudate is carried towards the wall, where it rises and falls in "torus"-form, with pellet rounding as a result of interactions with the wall, plate and other extrudate strands.[9] The wet mass must have enough plasticity to deform, but must not adhere to the equipment or other particles.[8] If the formulation is too wet, fluid migration to the pellet surface increases agglomeration propensity and results in increased pellet size and wide size distributions.[9] Figure 8.5 summarises pellet formation mechanisms proposed in the literature.

Depending on the formulation, it can take from 1 to 30 minutes for changes to occur, with spheronisation times of 2–10 minutes being most common.[7] If round pellets are not obtained within 30 minutes, the formulation is generally too "dry", with longer spheronisation being unlikely to help. Once rounded, a good formulation can be spheronised for prolonged periods of

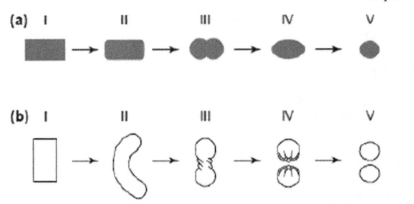

Figure 8.5 Pellet-forming mechanisms during spheronisation, according to (a) Rowe, and (b) Baert.[7] Reprinted from *International Journal of Pharmaceutics*, **116**, C. Vervaet, L. Baert and P. J. Remon, Extrusion-spheronisation a literature review, 113–146, Copyright 1995, with permission from Elsevier.[7]

time without further changes in size or shape.[9] According to Newton,[9] plate velocity (in terms of tip speed) and spheroniser load can both affect pellet attributes. Low velocities risk insufficient pellet rounding, while high velocities can significantly reduce pellet size, with low spheroniser loads producing too many interactions and *vice versa*.

8.3.3.4   Drying

The E–S liquid is primarily a processing aid and needs removing at the end of the cycle. Pellets can be dried at ambient or elevated temperatures, typically using fluidised bed or tray driers.[7,9] Less common alternatives are microwave and freeze-driers, with different studies having shown that the drying method can affect pellet microstructure. According to Dhandapani *et al.*,[8] microwave-dried pellets are generally softer, rougher and more porous, while freeze-dried pellets are generally weaker, larger and more porous than conventionally-dried pellets.

8.3.4   Hot-Melt Extrusion (HME)/Hot-Melt Pelletisation (HMP)

HME is a process involving thermal melting of material conveyed through a channel, using at least one thermoplastic polymer or low-melting wax. It is used in a variety of industries, most notably in the plastics industry. A key difference of pharmaceutical extruders is that they must comply with pharmaceutical regulations, including cleaning and validation. Screw extruders, most notably twin-screw units, are most commonly used in pharmaceutical applications and hence are the focus here.[10,11] Most commercial units are modular, to facilitate screw modifications, since the design allows high or low shear to be imparted. Screws can typically be divided into three sections, as follows:[10]

(i) Feeding section
The purpose is to transfer material from the hopper to the barrel. The channel depth is usually at its largest in this section, to facilitate mass flow.
(ii) Melting or compression section
The channel depth decreases in this section, hence increasing the pressure and removing entrapped air. The polymer also typically begins to soften and melt in this section.
(iii) Metering section
The primary function is to reduce the pulsating flow and ensure uniform delivery through the die. The extrudate flow rate is thereby highly dependent on channel depth and section length.

The die at the barrel exit dictates extrudate shape, whose size generally increases upon exit, a phenomenon known as "die swell" (the extent of which is mainly governed by the viscoelastic properties of the polymers). Auxiliary downstream equipment is usually for product cooling, cutting or collecting.[10]

The screws in twin-screw extruders can rotate in the same direction (co-rotating) or opposite directions (counter-rotating), imposing different conditions. The counter-rotating designs are most common and are utilised when high shear is needed, since the material is squeezed through the gap between the approaching screws. Counter-rotating extruders, however, generally suffer from air entrapment, high-pressure generation, low maximum screw speed and low output.[10]

Screw dimensions are described as L:D ratio, *i.e.* length divided by diameter, with ratios of 20:1 to 40:1 most typical. The size of an extruder is generally described by its screw diameter, *e.g.* 18 mm (pilot scale) or 60 mm (production scale).[10] Temperature-sensitive materials generally use shorter screws, and hence lower L:D ratios, while longer screws are generally for high throughput.[11] Extruder residence times are typically between 5 seconds and 10 minutes, governed by the L:D ratio, extruder type, screw design and operational settings.[10]

Materials for HME processing are selected based on their physicochemical properties and interaction potential. These can be quantified theoretically, using solubility parameters, or experimentally, using differential scanning calorimetry or hot-stage microscopy, with differentials in solubility parameters indicating the likelihood of material miscibility.[10,12] Thermal stability of compounds is a prerequisite for HME, but thermolabile compounds are not automatically precluded, due to the relatively short extrusion times. When preparing amorphous solid dispersions, the mixture is commonly heated above both the polymer glass transition temperature (T_g), to induce plasticity, and the API melting point, to facilitate dispersion and conversion to the amorphous state. API solubility in the carrier generally increases with temperature, resulting in crystalline APIs either melting or become solubilised in the carrier matrix.[10] It is important to note that transforming the drug into an amorphous form makes it thermodynamically unstable and thus susceptible to re-crystallisation on storage.[11] Extrudate stability can be assessed by thermo-analytical or non-thermal techniques.[12] Since the incorporation of amorphous APIs in polymeric carriers may stabilise them long enough to ensure stability throughout product shelf-life, adequate carriers and other excipients are needed for successful development. Adequate kinetic stability is often achieved for carriers with high T_g and functional groups that are hydrogen bond donors or acceptors.[11] As highlighted, the carrier properties often dictate the processing conditions, with their physical and chemical properties also potentially controlling drug release.[10] Amorphous dispersions normally consist of amorphous drug particles embedded within a hydrophilic carrier, with the API dissolution rate improved due to its molecular dispersion within the rapidly dissolving carrier, its inhibition of drug precipitation and API wettability enhancement.[11] Functional excipients, such as plasticisers, fillers, pH modifiers, release modifiers, stabilisers, surfactants, antioxidants and processing aids, can be included in the blend and extruded at relatively low temperatures.[12] Plasticisers are typically low-molecular-weight materials capable of softening polymers, due to reduction of the T_g and polymer melt viscosity, making them more flexible. Typical examples are triacetin, citrate esters and low-molecular-weight polyethylene glycols.[10]

For IR applications, HME is typically used to prepare milled extrudate for compression or capsule filling (after blending with excipients), while near-spherical pellets are generally required for CR applications. For this purpose, tailored shaping devices at the extruder exit have been developed, typically referred to as die face pelletisers. These cut the molten material emerging from the die plate into small particles, using a rotating knife. The spherical shape of these pellets is thereby a result of cutting the extrudate at a temperature above its softening point, where viscous forces allow particles to contract and become spherical.[13] One example of such a system is the Sphero-THA, developed by Maag Automatik GmbH, Grossostheim, Germany (formerly Automatik Plastics Machinery GmbH), in cooperation with the Research Center Pharmaceutical Engineering, Graz, Austria. This system is said to have been designed to satisfy GMP requirements and accommodate sticky materials.

8.3.5   Wurster-based Coating of Controlled Release Pellets or Multiparticulates

The Wurster process is the industry standard for coating of CR pellets, featuring a perforated base plate with a centrally-located spray nozzle and a cylindrical insert, all contained within standard fluidised bed equipment (Figure 8.6).

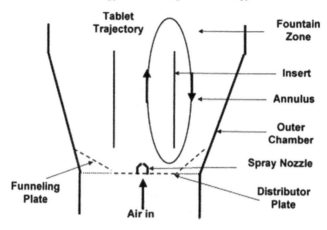

Figure 8.6 Schematic of the Wurster coating process.[53] Reprinted from *Powder Technology*, **110**, S. Shelukar, J. Ho, J. Zega, E. Roland, N. Yeh, D. Quiram, A. Nole, A. Katdare and S. Reynolds, Identification and characterisation of factors controlling tablet coating uniformity in a Wurster coating process, 29–36, Copyright 2000, with permission from Elsevier.

Fluidising air entering the equipment through the base plate causes a cyclic particle flow pattern, analogous to a water fountain, with the spray in the lower part of the chamber creating a distinct spray-zone. The regular flow pattern, with passage through the coating zone, facilitates the controlled build-up of coating on the particles. In the coating zone, the sprayed droplets collide with the fluidised particles, spread across their surfaces and dry rapidly. Wurster coating is quite a lengthy process, due to the relatively large total surface area available for coating and the low solids loading of coating suspensions or solutions (10–20%). This is also one of the reasons why the process is associated with high precision, manifesting itself in relatively small differences in coating thickness between pellets.

8.4 Key Oral Controlled Release Technologies

CR dosage forms are generally monolithic (single-unit) or multiparticulate (multi-unit) in nature and were historically developed for soluble APIs that diffuse rapidly through the GI membrane. These APIs were generally soluble enough to dissolve inside the dosage form and be released *via* diffusion. The most common CR technologies are hydrophilic matrices and CR pellets, but there are also other options, as described in this section.

8.4.1 Hydrophilic Matrix Tablets

Hydrophilic matrices have been used since the 1960s and have been the most popular oral CR technology to this day. They use the same established and cost-effective process technology as are used to produce conventional tablets,

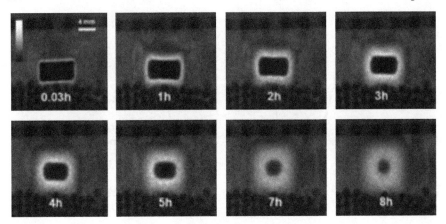

Figure 8.7 MRI images of an HPMC matrix tablet hydrating in water, with the hydrated portions of the tablet presented in white.[54] Reprinted from *Journal of Pharmaceutical Sciences*, **100**, H. D. Williams, K. P. Nott, D. A. Barrett, R. Ward, I. J. Hardy and C. D. Melia, Drug release from HPMC matrices in milk and fat-rich emulsions, 4823–4835, Copyright 2011, with permission from Elsevier.

such as blending, granulation (dry or wet), compression and film-coating, making them operationally largely indistinguishable. They can also deliver high unit doses of API and be used for APIs with different properties, which makes them so attractive to industry.[1,14] However, unlike IR tablets, CR matrices require at least one higher viscosity polymer that hydrates quickly and forms a pseudo-gel layer on the tablet surface. Rapid formation of this layer is critical to prevent disintegration of the core and to control water ingress and drug release rates.[15] According to Nokhodchi *et al.*,[2] different studies have shown that drug release from swellable hydrophilic matrices is governed by pseudo-gel layer thickness. Drug release can thereby occur *via* diffusion, erosion or a combination thereof (Section 8.4.1.2). Figure 8.7 shows magnetic resonance imaging (MRI) snapshots of a hydroxypropyl methylcellulose (HPMC) matrix hydrating in water, where the hydrated parts are indicated in white.

Figure 8.7 shows how the pseudo-gel layer forms, the tablet swells and the dry core size decreases with time. Since the release-controlling pseudo-gel layer only forms after tablet exposure to fluid, there are intrinsic robustness risks associated with this transition period. Specifically, any factors interfering with pseudo-gel layer formation can have undesired effects *in vivo* or *in vitro*, potentially altering the release profile significantly. The latter can be affected by factors such as tablet geometry, polymer type and concentration, API and excipient solubility and the fluid they are exposed to, most of which are discussed in Section 8.4.1.3. Matrix tablets are hence not as easy to design as they are to manufacture.[16] A study involving low polymer level (≤10% w/w HPMC) and a highly soluble compound nicely illustrates the transition from IR to CR in the early stages of dissolution testing (Figure 8.8).

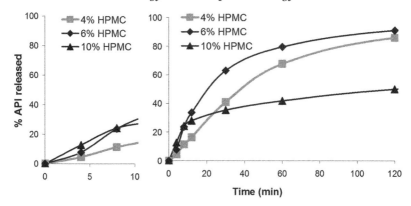

Figure 8.8 Dissolution data illustrating the transition from immediate to controlled release in the early stages of the dissolution test for a matrix tablet involving low polymer level (≤10% w/w HPMC) and a highly soluble compound (Note: dissolution method was USP 2 at 75 rpm with 0.25 M pH 6.8 phosphate buffer as the media; the Roller Compaction-based tablet formulations contained 80% API, 4–10% HPMC K100M, MCC, SiO$_2$ and SSF).[17]

Specifically, Figure 8.8 shows that the formulations with 10% HPMC exhibited enhanced drug release in the first minutes of dissolution, followed by a sharp decline thereafter, while the opposite trend was observed at 4 and 6% HPMC. Seiler *et al.*[17] attributed the more pronounced dissolution at the earlier time-points to HPMC swelling, which facilitates liquid penetration and API release, and the subsequent decline in release rate to pseudo-gel layer formation, which is more pronounced and rapid at higher HPMC levels. The various factors influencing drug release from matrix tablets are discussed in Section 8.4.1.3, with matrix tablets deemed robust if small changes in the manufacturing process or raw material attributes can be accommodated.[15]

8.4.1.1   Key Formulation Constituents

8.4.1.1.1   Release-controlling Polymer.   Different polymer types and grades are available for CR matrices, the respective advantages and disadvantages of which need to be considered in the context of a particular application. Examples are HPMC, hydroxypropyl cellulose (HPC), hydroxyethyl cellulose (HEC), xanthan gum, sodium alginate, polyethylene oxide (PEO) and cross-linked homopolymers and copolymers of acrylic acid (Table 8.3).[4]

The most popular polymer for matrix tablets is HPMC, which is associated with rapid hydration, good compression and gelling characteristics and very low toxicity. It is also globally available, exists in different grades and is non-ionic, minimising interaction risks in acidic, basic or electrolytic environments. Commercially available chemistries differ in their methoxyl and hydroxypropyl content (Table 8.4), which in turn affects polymer hydration and drug release.[2]

Table 8.3 Key controlled-release polymers used to prepare hydrophilic matrix tablets, including different viscosity grades available (as determined for aqueous solutions of specific polymer concentrations).[16,55–58,67,68]

Polymer type	Polymer grade	Molecular weight (Da)	Viscosity (mPa s) [% aq. sol.]
Hydroxyethyl cellulose (HEC)	G	300 000	250–450[a] [2%]
	M	720 000	4500–6500[a] [2%]
	H	1 000 000	1500–2500[a] [1%]
	HH	1 300 000	3400–5000[a] [1%]
Hydroxypropyl cellulose (HPC)	GF/GXF	370 000	150–400[a] [2%]
	MF/MXF	850 000	4000–6500[a] [2%]
	HF/HXF	1 150 000	1500–3000[a] [1%]
Hydroxypropyl methylcellulose (HPMC)	E50LV	91 300	40–60 [2%]
	E4M	400 000	2700–5040 [2%]
	E10M	746 000	7500–14 000 [2%]
	K100LV	164 000	80–120 [2%]
	K250	200 000	200–300 [2%]
	K750	250 000	562–1050 [2%]
	K1500	300 000	1125–2100 [2%]
	K4M	400 000	2700–5040 [2%]
	K15M	575 000	13 500–25 200 [2%]
	K35M	675 000	26 250–49 000 [2%]
	K100M	1 000 000	75 000–140 000 [2%]
	K200M	1 200 000	150 000–280 000 [2%]
Poly(acrylic acid) (Carbomer)	Carbopol 71 G[b,c]	–	4000–11 000[a] [0.5%]
	Carbopol 971P[c]	–	4000–11 000[a] [0.5%]
	Carbopol 974P[d]	–	25 000–40 000[a] [0.5%]
Polyethylene oxide (PEO)	WSR-1105	900 000	8800–17 600[a] [5%]
	WSR N-12K	1 000 000	200–400[a] [2%]
	WSR N-60K	2 000 000	2000–4000[a] [2%]
	WSR-301	4 000 000	1650–5500[a] [1%]
	WSR-303	7 000 000	7500–10 000[a] [1%]
Poly(meth)acrylates/ methacrylic ester copolymers	Eudragit RL100[e]	~150 000	–
	Eudragit RS100[f]	~150 000	–
Polyvinyl acetate/ polyvinyl-pyrrolidone	Kollidon SR	~500 000	–

[a]1 cP = 1 mPa s.
[b]1 Granular version of Carbopol 971 (for direct compression applications).
[c]Product trade name of the Lubrizol Corporation—USP/NF Compendial name is Carbomer Homopolymer Type A.
[d]Product trade name of the Lubrizol Corporation—USP/NF Compendial name is Carbomer Homopolymer Type B.
[e]Product trade name of Evonik—USP/NF Compendial name is Ammonio Methacrylate Copolymer Type A.
[f]Product trade name of Evonik—USP/NF Compendial name is Ammonio Methacrylate Copolymer Type B.

Table 8.4 Various grades of HPMC and their degrees of substitution.[15]

HPMC type	Methoxy (%)	Hydroxypropoxy (%)	Other names
K	19–24	7–12	Hypromellose 2208
E	28–30	7–12	Hypromellose 2910
F	27–30	4–7.5	Hypromellose 2906

The E- and K-series have been the preferred chemistries for matrix tablets for many years, especially the K-chemistry, which is associated with fast hydration and pseudo-gel formation, due to the highest hydroxypropoxyl-to-methoxyl ratio.[15]

8.4.1.1.2 Other Formulation Constituents. Since hydrophilic matrices use the same process technology as conventional tablets, they require similar functional excipients. These are hence not discussed in any detail here. However, their solubility characteristics can affect drug release. Specifically, soluble components will wet, dissolve and diffuse out of the matrix, while insoluble material generally remains in place until the surrounding polymer/excipient/drug-complex has eroded or dissolved away[15] (Section 8.4.1.3).

8.4.1.2 Drug Release Mechanisms

API release from matrix tablets can occur *via* diffusion, through the pseudo-gel, or *via* tablet surface erosion. Different drug release models have been devised, with the following model by Korsmeyer *et al.* being the most popular:[2]

$$M_t/M_\alpha = k \times t^n \qquad (8.1)$$

M_t/M_α = fraction of drug released at time t
k = diffusion rate constant
t = release time
n = release exponent (indicative of the drug release mechanism)

For $n = 1$, the release rate is independent of time, referred to as zero-order release (case II transport), while $n = 0.5$ indicates Fickian diffusion (case I transport). Values between 0.5 and 1 indicate that both diffusion and erosion contribute to the release kinetics (non-Fickian, anomalous or first-order release). As highlighted by Nokhodchi *et al.*,[2] the two extremes of $n = 0.5$ and $n = 1$ are only valid for slab geometries, with values of $n = 0.45$ and $n = 0.89$ describing cylindrical tablets.[2]

8.4.1.2.1 Diffusion-based Matrix Tablets. Diffusion has historically been the primary release mechanism of hydrophilic matrices, since they were developed predominantly for soluble APIs. A key limitation of diffusion-based matrices is that adequate release rates are only achieved if there is sufficient API dissolution within the tablet. The dissolution potential in the tablet is orders of magnitude lower than that in the surrounding environment (*e.g.* 900 ml dissolution media), due to the much higher solid-to-liquid-ratio within the tablet. In fact, API-to-liquid ratio is the primary reason why IR dosage forms require rapid tablet disintegration, since this exposes the API to the surrounding fluid and facilitates rapid dissolution and absorption. Dose, or drug loading, as well as API solubility, are hence generally more limiting

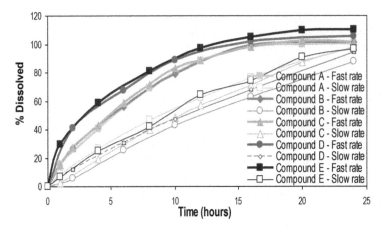

Figure 8.9 Polymer release profiles for a study involving five low-solubility compounds and two erosion-based hydrophilic matrix tablet platforms (targeting different release rates).[17]

for CR matrices than for their IR counterparts. The different formulation factors that affect drug release are discussed in more detail in Section 8.4.1.3.

8.4.1.2.2 Erosion-based Matrix Tablets. Classic hydrophilic matrices only offer limited opportunities for low-solubility APIs, due to the limitations of diffusion-based drug release. However, a shift of release mechanism from diffusion to erosion would, in principle, enable the use of matrix tablets for low-solubility APIs. A key difference between diffusion- and erosion-based matrices is that the latter require lower viscosity polymers, *e.g.* HPMC K100LV or K4M, since API release is governed by the rate of surface erosion (which exposes non-dissolved API to the surrounding fluid). Figure 8.9 shows polymer release profiles for a study involving five low-solubility compounds and two HPMC-based formulation platforms (each targeting a different release rate), which indicated that similar release rates could be achieved for the five compounds.[17]

A systematic screening study was performed for the matrix platforms, to evaluate the robustness risks associated with the erosion-based release. Unfortunately, the data from this study were never published or cleared for publication and hence are not reproduced here. However, what can be shared is that the release profiles of the erosion-based matrix platforms had shown considerable variability when using polymer lots with different degrees of substitution or viscosity (within compendial specifications). Based on this, the technical team of the company concluded that this formulation approach should not be considered for commercial product development.

8.4.1.3 Factors Affecting Drug Release

There are many factors that can potentially affect drug release from hydrophilic matrices, with this discussion focussing on matrices releasing API predominantly *via* diffusion.

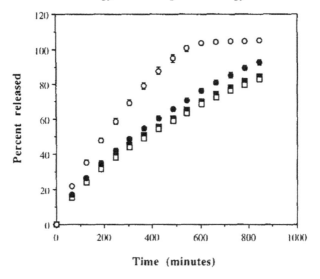

Figure 8.10 API release profiles of matrix tablets with different K-chemistry HPMC grade/viscosity (o KI00LV; ● K4M; □ K15M; ■ K100M). All tablets contained 2.5% API (adinazolam mesylate), 35% HPMC, 62% lactose and 0.5% magnesium stearate, with their dissolution and swelling profiles determined in 700 ml deionised water using a fully automated system.[69] Reprinted from *International Journal of Pharmaceutics*, **142**, K. C. Sung, P. R. Nixon, J. W. Skoug, T. R. Ju, P. Gao, E. M. Topp and M. V. Patel, Effect of formulation variables on drug and polymer release from HPMC-based matrix tablets, 53–60, Copyright 1996, with permission from Elsevier.

8.4.1.3.1 Tablet Size. The effect of surface-to-volume ratio on drug release is well-documented, having been demonstrated both theoretically and experimentally. Specifically, release from small tablets has been shown to be faster than release from their larger counterparts of the same composition, due to the differences in diffusional path-lengths.[4]

8.4.1.3.2 Polymer Viscosity. Viscosity is defined as a measure of resistance of a fluid to flow, with polymer solution viscosity dependent on molecular weight.[2] The texture and strength of a pseudo-gel varies with polymer type, viscosity grade and polymer concentration, with gel strength generally increasing with molecular weight. Figure 8.10 shows typical effects of polymer grade/viscosity on the drug release profile of HPMC matrices based on the K-chemistry.

Figure 8.10 clearly shows that the differences in release profiles for K4M, K15M and K100M are not proportional to their respective viscosities, with identical profiles obtained for K15M and K100M. This is a common observation, since gel strength tends to plateau at higher molecular weights.

8.4.1.3.3 Polymer Level. There must be sufficient CR polymer to form a uniform barrier and prevent immediate drug release. Increased polymer level

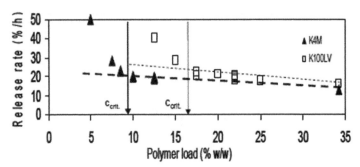

Figure 8.11   API release rate (first 2 hours) as a function of polymer load for formulations using different viscosity grades of HPMC.[17]

typically results in slower drug release, but there are often plateau regions where increased levels have little incremental effect. While the benefit of increased polymer level may not always be immediately apparent, it is worth noting that higher levels tend to decrease the sensitivity of formulations to variations in manufacturing process or raw material attributes. Percolation theory, as discussed by Caraballo,[71] can be used to (i) explain drug-release performance of hydrophilic matrices containing different types, levels or physical characteristics (especially particle size) of release-controlling polymer, and (ii) determine critical thresholds for CR polymer level. Above the threshold, a polymer is described as percolating through the system, similar to the outer phase of an emulsion, and as having hence a more pronounced effect on the whole system.[71] In the case of hydrophilic matrix tablets, the CR polymer level should be above its percolation threshold, to facilitate rapid transition from IR to CR behaviour and thus minimise release rate susceptibility to environmental factors.

For high API doses, it is important to balance tablet size with the need to include enough excipients to meet robustness criteria. This may necessitate deviations from recommended excipient levels in order to meet the target product profile, and there are commercial product examples where lower polymer levels had to be explored. Release rate robustness is the primary concern in this case, since pseudo-gel formation takes a finite time during which tablets display largely IR behaviour. Figure 8.11 shows the API release rate over the first 2 hours as a function of HPMC level, for USP 2 dissolution experiments performed at 50 rpm in 0.1 M phosphate buffer (pH 6.8). The formulations contained soluble API (>50% drug load), microcrystalline cellulose (MCC), lactose, magnesium stearate and HPMC K4M (4000 cps viscosity) or HPMC K100LV (100 cps viscosity), respectively.[17]

Figure 8.11 shows moderate and predictable increases in release rate below typical HPMC levels (20–30% w/w) until a critical concentration (c_{crit}) is reached. Below c_{crit}, release rate shows a sharp increase, attributed to the slower pseudo-gel formation, which allows immediate API release over a longer period of time. The other key point from Figure 8.11 is that c_{crit} is higher at lower polymer viscosity, attributed to a higher susceptibility to erosion at lower viscosity. While c_{crit} is likely to differ for different APIs or excipients, the presented trends and levels are deemed broadly representative of the polymers studied.[17]

8.4.1.3.4 Solubility.
Higher solubility drugs typically release at faster rates than less soluble ones, if the formulation is otherwise identical, due to the differences in diffusional driving forces.[15] However, API solubility needs to be viewed in the context of dose strength and drug loading, since their combination ultimately determines the API dissolution potential and kinetics. Although there are empirical models to determine release rates for different API solubilities, it is essential to verify their results experimentally. The reason is that drug–excipient interactions can have significant effects.

The solubility of diluents can also noticeably affect drug release from matrices, especially if their concentration is relatively high, since they affect the tablet microenvironment during drug release. For example, water soluble diluents like lactose can cause a marked increase in drug release rate, in contrast to insoluble components (*e.g.* MCC), which do not diffuse out of the matrix and hence help to maintain a solid tablet architecture that limits the mobility of penetrating liquid or dissolved API.[4] This is an important consideration when designing matrix formulations.

8.4.1.3.5 Particle Size.
The effect of API particle size on drug release from hydrophilic matrices is generally negligible, except for very large API particles in formulations with low levels of release-controlling polymer.[15] Furthermore, since most pharmaceutical companies control API particle size and avoid large API particles by default, this is generally not a significant parameter.

Polymer particle size, however, can greatly influence matrix tablet performance, since smaller particles have more total surface area, which provides for better polymer-to-water contact and increases polymer hydration and pseudo-gel formation kinetics, which are critical to performance.[15] Figure 8.12 shows an example of the effect of polymer particle size on drug release for a HPMC K4M matrix.

Figure 8.12 clearly shows that large HPMC size fractions failed to provide adequate release control for this model system. While IR products are primarily concerned with segregation of API, due to content uniformity concerns, manufacturers of CR products also need to be concerned about CR polymer segregation. HPMC CR grades hence come with well-defined particle size and are classified as "very fine" (with only 5–10% w/w exceeding 149 μm). However, while this is clearly beneficial to their function as CR polymer, there are implications for powder flow, since they should not be considered free-flowing.[15] This needs to be considered during formulation design.

8.4.1.3.6 pH and Ionic Strength.
For drugs with pH-dependent solubility, it is possible to increase solubility by modifying the tablet microenvironment. Ju *et al.*[18] studied the effect of matrix pH modification for acidic and basic drugs, using citric acid, *p*-toluenesulfonic acid, glycine and

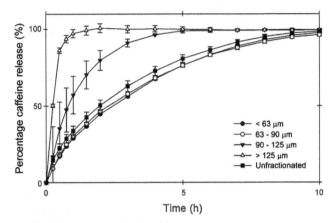

Figure 8.12 API release profiles of HPMC matrix tablets prepared using different HPMC particle size fractions. All tablets contained 10% API (caffeine anhydrous), 30% HPMC K4M, 39.3% lactose, 19.7% microcrystalline cellulose and 1% magnesium stearate and were studied for dissolution in 900 ml water using USP apparatus 1 at 100 rpm.[70] Reprinted from *International Journal of Pharmaceutics*, **401**, H. D. Williams, R. Ward, A. Culy, I. J. Hardy and C. D. Melia, Designing HPMC matrices with improved resistance to dissolved sugar, 51–59, Copyright 2010, with permission from Elsevier.

tris-hydroxymethyl aminomethane (THAM). They showed that pH-induced reductions in drug solubility could be overcome by addition of appropriate acidic or basic modifiers, resulting in faster release.[15] However, any such formulation levers need to be carefully assessed for robustness, especially with respect to product stability and manufacturability.

Due to its non-ionic nature, the viscosity of HPMC is generally stable over a wide pH range (*i.e.* pH 3–11). For drugs with pH-dependent solubility, the release from an HPMC matrix will hence also be pH-dependent.[2] This is an important consideration when selecting dissolution media for screening, especially in the case of formulation design for *in vivo* studies, considering the pH profiles in the GI tract.

Ionic strength can also affect the rate of drug release from HPMC matrices. Ionic strength of the GI tract generally ranges from 0 to 0.4 M, considering both fed and fasted states.[2] The effect of media with different ionic strengths on drug release can vary significantly between formulations and compounds and should hence be considered in the context of the other factors discussed. It is generally advisable to perform at least a basic robustness assessment for this risk, ideally by selecting conditions that are most reflective of the human GI tract.

8.4.2   Inert Matrix Tablets

According to Patel *et al.*,[4] CR matrix tablets involving hydrophobic or inert material were first reported in 1959. Such inert matrix tablets are non-eroding extended release (ER) formulations whose polymers do not interact with biological fluids and which do not exhibit the pseudo-gel formation of hydrophilic matrices. Like hydrophilic matrices, inert matrices utilise established and cost-effective process technology and can accommodate high drug loadings.[19] API release from inert matrices has been described as occurring *via* a leaching mechanism, with API particles in the matrix dissolving in the ingressing fluid and being subsequently released from the tablet, whereby liquid penetration into the matrix is generally the release rate-determining step. Drug release occurs *via* diffusion/migration through a network of pre-existing pores and pores created by dissolved API. At drug loadings in excess of 10–15% (v/v), there is typically a continuous structure connecting the API particles, often described as a percolating network. At lower loadings, API particles can be trapped completely in the insoluble matrix, resulting in incomplete API release.[4,20] According to Hughes,[19] the rate of drug release from inert matrices is generally dependent on API solubility, polymer type, polymer particle size, filler type and the mechanical strength of the tablets. Inert matrix tablets are generally pH-independent, with ethyl cellulose, acrylic polymers (*e.g.* polymethyl methacrylate) and polyvinyl acetate being the most popular excipients. Also reported in the literature are (semi-)synthetic polymers, like polyethylene, polyvinyl chloride, polystyrene and cellulose acetate, and lipophilic materials, like carnauba wax, hydrogenated castor oil and tristearin.[4,20] Due to the diffusion-based drug release, the application of inert matrices is generally limited to APIs with moderate to high solubility. Major drawbacks of most inert matrices are their inherent first order drug release and poor direct compression characteristics.[20]

8.4.3 Controlled Release Pellets/Multiparticulates

Multiparticulates/pellets are an important CR formulation strategy, with the advantage of supporting a wide range of dose strengths by delivering simply different pellet quantities, for adult and paediatric populations. This is generally not possible for monolithic dosage forms, like hydrophilic matrices, since their weight multiples result in differently sized dosage forms, which, in turn, affects drug release (due to mechanistic differences from pellets). There are a number of established processes for CR pellet preparation, most of which require barrier coatings to attain CR functionality. Technology selection can be governed by the particular application, available expertise/capability or economics. The following sections discuss the preparation of pellet core formulations (Section 8.4.3.1), membrane-coated pellets (Section 8.4.3.2) and finished dosage forms (Section 8.4.3.3).

Pellet formulations are generally associated with less *in vivo* variability than their monolithic counterparts, since they consist of a large number of release-controlling units. This is largely due to their different interplay with the GI tract, especially their spatial distribution. There are, however, unique challenges associated with CR pellets, mainly due to more complex processing, higher production costs and release barrier design criteria.

8.4.3.1   Pelletisation and Preparation of Multiparticulate Cores

Pelletisation and multiparticulate core preparation can generally be categorised as follows:

  i. Matrix architecture
  ii. Layered architecture

Cores with *matrix architecture* are characterised by API being distributed uniformly throughout the cores, which are generally formed by converting powders or granules into spherical or near-spherical particles with diameters of typically 0.5 to 1.5 mm and shapes that lend themselves to coating. Cores with *layered architecture* are prepared by depositing particulates onto seed particles, either *via* a liquid carrier or as dry powders, and converting them into API-containing solid layers. CR pellet cores with matrix or layered architecture generally require (i) application of a barrier membrane (Section 8.4.3.2), and (ii) API that can dissolve sufficiently within the core structure to facilitate diffusion-based release (which is the primary API release mechanism for these CR formulations).

Table 8.5 lists the main techniques used to prepare drug-containing pellets, which are discussed individually in the following sections.

8.4.3.1.1   Extrusion–Spheronisation.   Extrusion–spheronisation (E–S) originated in Japan, originally known as "Maurumerization". Introduced into Europe and the USA in the 1960s, it became one of the most widely used pelletisation processes in the pharmaceutical industry. One key advantage of E–S is its ability to produce pellets with high drug loading, up to 80% w/w.[9] As outlined in Section 8.3.3, the process involves granulation, extrusion, spheronisation and drying. E–S based pellets for CR applications generally also need to be suitable for film-coating, since the E–S cores do usually not control drug release (see Section 8.4.3.2 concerning pellet coating). Size and shape are important for coated E–S pellets, with their typical size range being 0.7–2.5 mm. The lower limit is determined by difficulties in providing extrusion screens with a uniform size of holes and sufficient rigidity to withstand the required forces, and formulations that extrude adequately through such dies.[9] Pellet shape is most commonly captured in terms of aspect ratio, as

Table 8.5   Processes for pelletisation and multiparticulate core preparation.

Technology	Matrix architecture	Layered architecture
Extrusion–spheronisation (E–S)	X	
Direct pelletisation (*via* rotary granulation)	X	
Dry powder layering		X
Spray layering		X
Hot-melt pelletisation	X	
Mini-tablets/Mini-matrices	X	

length over width. While Newton[9] accepts this as an adequate quality control measure, he challenges perceptions that aspect ratios of up to 1.2 are acceptable. His rationale is that such formulations are non-spherical and will show batch-to-batch variability in terms of shape.

In compositional terms, MCC is considered the gold standard for E–S, due to its unique combination of wet mass plasticity and cohesiveness. MCC is frequently described as a "molecular sponge", since it retains water in that fashion, with its water absorption and retention capacity attributed to its large surface area and high internal porosity. During extrusion, the "sponge" is compressed and water squeezed from the internal structure, acting as a lubricant. Following extrusion, the "sponge" expands and becomes more dry and brittle, which facilitates extrudate breakage during spheronisation.[8] Many formulations only require API and MCC, plus appropriate quantities of water. To reduce costs, especially when drug loading is relatively low, MCC can be substituted with fillers such as lactose, mannitol or calcium carbonate, which may affect the formulation performance.[9] Although MCC is a near-ideal spheronisation aid, there are some limitations, such as (i) prolonged drug release in the case of less soluble drugs (due to lack of pellet dissolution/disintegration), (ii) potential drug decomposition in the presence of MCC and (iii) drug adsorption onto MCC fibre surfaces.[8] It should also be noted that MCC exists as different brands and grades, differing in particle size or moisture content, which can affect their ability to produce pellets.[9] Although there are publications describing the evaluation of alternatives to MCC, such as powdered cellulose, starch and starch derivatives, k-carageenan, pectinic acid, chitosan, HPMC, HEC, crospovidone and PEO, none were found to be comparable to MCC.[8,9] Glycerol monostearate (GMS) has been highlighted as an extrusion aid, assisting the pellet formation of MCC-based formulations.

The most critical other component is the granulating fluid, which is necessary to impart plasticity to the powder. It has been shown that moisture content can be varied and still produce pellets of acceptable quality.[7] When levels are too low, excessive extrusion pressure is required to remove air voids, which, in turn, results in a brittle mass without the plasticity needed to form spheres. Such extrudates have the tendency to break during spheronisation and generate large amounts of fines. When fluid level is too high, pellets tend to agglomerate during spheronisation, due to excess liquid at the pellet surface, resulting in larger pellets than desirable.[7,8] It is also important to ensure that the process is not overly sensitive to fluid level, due to difficulties of controlling such formulations.[9] Water is the most widely used liquid, especially for MCC-based formulations. The high water content needed to form pastes of suitable consistency has prompted searches for alternatives, especially for APIs undergoing hydrolysis, but few studies have reported the use of non-aqueous solvents for MCC-based E–S pellets. Most of the non-aqueous liquids studied were alcohols, *e.g.* ethanol or isopropyl alcohol, where either (i) spheronisation was not achieved, (ii) pellets crumbled to powder upon drying or (iii) water addition (5% w/w or more) was required to meet pellet acceptance criteria. The literature indicates that solvents of high polarity, high surface tension and low viscosity are required to induce MCC plasticisation, which has been demonstrated for the solvent dimethyl sulfoxide (DMSO; which is not a suitable solvent for oral dosage forms, having been used for demonstration purposes only).[21]

MCC-water-based formulations have dominated the E–S landscape, attributable to MCC's unique "sponge"-like behaviour, and are likely to be required for robust E–S pellet product manufacture, especially for CR drug delivery.

8.4.3.1.2   Direct Pelletisation (*via* Rotary Granulation).   Direct pelletisation involves rotary granulation (Section 8.3.2), with spheroid formation occurring in the following steps:

- The powder mixture is moistened
- The moist mass is rounded into pellets
- The finished product is dried

In most of the literature, mixtures of MCC and lactose as well as water are used. The water content after addition of liquid has been found to be the key process parameter, which hence requires a high level of control.[5] Enough liquid is needed to yield suitably-sized spheroids, while too much liquid generally leads to oversized lumps, due to uncontrolled agglomeration.[6]

A study comparing rotary and conventional fluid bed granulation, using the same formulation and process variables, has revealed that the more intense and uniform powder movement in the rotary granulator allows for considerably higher spray rates than during conventional fluid bed granulation, despite lower air flow rates. Rotary granulation was found to be a good alternative to conventional fluid bed granulation, particularly when a fluid bed technique is needed to granulate formulations with poor flow or low drug content.[5] Table 8.6 shows details from a pilot-scale study by Kristensen and Hansen,[5] evaluating direct pelletisation using a Glatt GPCG-1.1 unit.

8.4.3.1.3   Dry Powder Layering.   Rotary granulation (Section 8.3.2) can also be used to manufacture pellets by dry powder layering. Seed particles,

Table 8.6   Parameters used for a pilot-scale rotary granulation/direct pelletisation study.[5]

Parameter	Parameter value
Quantity of powder (pre-blended and sieved)	825 g
Rotating plate speed	900 rpm
Binder spray rate	30–55 g min^{-1}
Inlet air temperature	25 °C
Atomising air pressure (Schlick 970/0-S3 nozzle)	1.0 bar
Air flow	40–60 m^3 h^{-1}

e.g. granules, pellets or mini-tablets, are thereby coated without the need for solvents or aqueous-based polymers, using just enough moisture to form liquid bridges and thus facilitate particle adhesion to the substrate.[15] Since the core preparation involves *layered architecture* and the same growth mechanism as dry-powder-based coatings, more details of this process can be found in Section 8.4.3.2.2.

8.4.3.1.4 Spray Layering. Pellet core preparation by spray layering involves Wurster processing (Section 8.3.5), with API-containing suspensions or solutions sprayed and dried onto fluidised particles. Although not essential, it is most common to use aqueous-based spray formulations. Curing, *i.e.* prolonged heat treatment, is a common process feature, to ensure continuous film formation after coating. Commercially available sugar or MCC spheres are the most common seed particles used in industrial practice. Since spray layering is a relatively slow process and there is generally a desire to minimise dosage form size, there are drivers to maximise drug loading. However, it is essential that API-containing layers adhere well to the substrate, to avoid API loss during downstream handling, including CR coating. Coating adhesion is governed by the type and number of bonds formed between two surfaces and is generally weakened by internal stresses. These can be the result of layer shrinkage (due to solvent evaporation) or thermal expansion differentials (*e.g.* on storage). A certain amount of binder, typically 10–20% of the API mass, is hence generally required.[22] Table 8.7 lists the most common water-soluble binders used to prepare pellet cores by spray layering, together with recommended solution concentrations.

It is standard practice to include antiadherents in the formulation [*e.g.* talc or GMS], to reduce tackiness (= stickiness) of the spray-layered beads during processing, handling or storage. However, since they can adversely affect mechanical and adhesive properties, it is important to strike an appropriate balance when selecting their type and level. Table 8.8 shows an example of a formulation prepared by spray layering.

Plasticisers are employed to increase coating flexibility and minimise film cracking risks, and hence avoid low API assays or uneven surfaces for

Table 8.7 Most common water-soluble binders used for core pellet preparation by spray layering, together with their recommended solution concentrations.[22]

Binder type	Recommended solution concentration
HPMC (*e.g.* 6 cps)	8–12%
HPC (*e.g.* Klucel EF)	8–10%
Povidone	10–20% (PVP K30)
	3–5% (PVP K90)
Starch	5–6% (corn starch)
	6–8% (Pregelatinised starch)
	8–10% (Partially pregelatinised starch)

Table 8.8 Example of a drug layer formulation prepared by spray layering, shown in terms of the solution/suspension as well as the resulting film composition.[22]

Formulation component	Solution concentration (% w/w)	Film concentration (% w/w dry basis)
API	25.0	81.2
Binder	4.0	13.0
Talc	1.0	3.2
PEG 4600	0.8	2.6
Water	To 100%	-

CR coating. For CR applications, core pellets prepared by spray layering generally need barrier-coating (Section 8.4.3.2).

8.4.3.1.5 Hot-Melt Pelletisation (HMP). As discussed in Section 8.3.4, HMP involves standard HME equipment, in combination with a pelletisation device that enables pellet preparation. According to Repka *et al.*,[12] the last two decades have seen increased interest in HME for drug delivery, as is also evident in the scientific literature. However, this is mainly for IR bioavailability enhancement, *via* amorphous solid dispersions, rather than CR applications.[23] HME can be performed such that it forms amorphous solid dispersions or maintains the API's crystalline state.[10] In either scenario, the API is embedded in a carrier system, which usually contains at least one thermoplastic polymer or low-melting wax. Upon cooling and solidification, these materials can act as thermal binders and/or drug release retardants, depending on whether IR or CR is intended. Materials are selected on the basis of their physicochemical properties and interaction potential (Section 8.3.4). The key benefits of HMP for CR applications are the continuous nature of the process, especially the associated cost benefits and batch size flexibility, as well as the potential for combining CR with solubilisation. For APIs with high solubility, CR formulations primarily require crystalline API to be uniformly distributed throughout the release-controlling polymer matrix, while the polymer matrix needs to form both an amorphous solid dispersion and control API release in the case of insoluble APIs. This is significantly more challenging, especially at higher drug loadings, but also significantly more attractive, due to limited CR options for insoluble APIs. One key challenge for ER applications, beyond operational considerations, is the relatively high surface area of the multiparticulates, considering diameters of typically 1–3 mm.

Many established pharmaceutical polymers can be used in HMP-based CR formulations, *e.g.* PEO, ethyl cellulose (EC), HPC, HPMC, acrylate copolymers (*e.g.* Eudragit RL/RS) or polyvinyl acetate–polyvinyl pyrrolidone blends (*e.g.* Kollidon SR). Since HPMC is most popular for conventional CR matrix tablets, it would also be potentially attractive for HME–HMP. However, the high glass transition temperature (T_g), low degradation temperature and high viscosity make it challenging for HME processing, with only few publications describing this use of HPMC.[23]

One HME CR study, by Ma *et al.*,[23] involved the following:

(i) Theophylline API, which is soluble (12–18 mg ml^{-1} aqueous solubility), has a high melting point (270 °C) and whose crystalline state was hence maintained throughout
(ii) Various grades of HPMC
(iii) Propylene glycol (PG), as a plasticiser, with a high boiling point (188 °C) that makes it amendable to HME processing.

Extrusion runs without plasticiser confirmed that they were needed, due to reaching the torque limit on the 16 mm twin-screw extruder (Prism Pharm-Lab 16, Thermo Fisher Scientific, New Jersey, USA). Extrusion runs with 10–40% PG resulted in the selection of 28% for the downstream evaluation of five HPMC grades. All extrudate samples contained 30% API and 42% HPMC and were cut into 1 cm strands for United States Pharmacopeia (USP) 2 dissolution testing in pH 4.5 buffer.[23] Extended release without evidence of burst release was achieved with most of the HMPC grades, with similar release profiles obtained for the higher viscosity grades (*i.e.* K4M, K15M and K100M). No API form change or significant dissolution changes were observed after 2 weeks open storage at 40 °C/75% relative humidity (RH).

Lian *et al.*[74] investigated HME as a means to simultaneously enhance the solubility of a poorly soluble API (nifedipine) and extend its release, up to 8 hours, as measured in fasted state simulated intestinal fluid (FaSSIF) using USP 2 apparatus. They employed a Leistritz ZSE 18HP twin-screw extruder combined with a Bay Plastic Machinery pelletiser BT25, utilising Plasdone S-630 copovidone and HPMC K15M, each at 40% w/w, as the key functional excipients. Pellets with drug loadings of 10–20% w/w were prepared successfully, some incorporating up to 5% polyethylene glycol (PEG) 3350 to reduce the torque and hence facilitate extrusion. According to the authors, API solubility was increased as much as sixfold to sevenfold, as a result of super-saturation, with the amorphous state maintained for pellet formulations stored for 3 months at 40 °C/75% RH, as determined by differential scanning calorimetry (DSC) and X-ray diffraction (XRD). Although only limited extended release (8 hours) was achieved in this study, it can nevertheless be considered a significant step forward, since it suggests that it is feasible to prepare HME pellets that deliver both solubilisation and controlled release.

8.4.3.1.6 Mini-tablets/Mini-matrices. Mini-tablets can be designed with diameters of 1–5 mm, with 2–3 mm being most common. Filling into capsules is the most common approach to prepare mini-tablet-based dosage forms. Due to the size and shape differences of mini-tablets, dosing wheels (which count the numbers into the capsule) are favoured over volume-based dosing (to ensure uniformity).[24] In the case of the commercial product Diclofenac Na 100 mg SR, each capsule contains 20 mini-tablets of 2 mm

Figure 8.13   Multi-tip punches and die, as used in the preparation of mini-tablets.[72]

diameter.[24] Mini-tablets are manufactured using the same unit operations as conventional tablets, with multi-tip compression tooling being the key difference (Figure 8.13).

Due to its tip-size, the tooling is one of the key vulnerabilities of mini-tablets, with maximum compression forces of 1.3 kN for a 2 mm tip quoted by Ghimire.[24] Another challenge when preparing mini-tablets is powder flow into dies, given their small diameter, which typically requires granulation of the powder blend. Careful flow assessments are hence necessary to ensure that robust tablets can be made at laboratory and commercial scale. Another parameter to consider carefully is granule size, since large granules can hinder efficient die filling.

Coating of mini-tablets can be performed using either a Wurster coater (Section 8.3.5), as is most common for CR coating, or a pan coater, *e.g.* with a perforated pan and mesh insert.[24] If mini-tablets are to be coated in Wurster equipment, special attention needs to be paid to tablet friability and tensile strength requirements, since they are subjected to very high impact forces, especially at commercial scale, where atomisation air can reach supersonic speed (approximately 340 m s^{-1}). Since atomisation air velocity is much lower in laboratory-scale coaters, successful coating at that scale is no guarantee for scale-up success. A sound scale-up strategy is hence required.

CR can either be achieved by adding a CR coat or by incorporating CR excipients into the core, as for hydrophilic matrix tablets (Section 8.4.1). Mini-tablets incorporating CR excipients in the core are often referred to as mini-matrices. A certain minimum polymer level is required for robust matrix tablet formulations, to avoid dissolution changes as a result of excipient lot-to-lot variability and/or stability storage. The high surface-to-volume ratio of mini-matrices generally reduces these risks, since they generally require higher polymer levels than conventional matrix tablets to achieve a certain release rate. However, it is important to choose the polymer grade carefully, to minimise the risk of burst release (assuming this is undesirable clinically). From experience, HPMC K100M is a particularly suitable polymer for mini-matrices.

8.4.3.2 Controlled Release Barrier Coating of Pellets/ Multiparticulates

While multiparticulate cores can be prepared *via* many different processes, there are only few options available for CR barrier coating. The following sections discuss the main coating approaches, namely spray layering (Section 8.4.3.2.1) and dry powder coating (Section 8.4.3.2.2).

8.4.3.2.1 Spray Layering. Wurster processing (Section 8.3.5) is at the heart of barrier-coated pellets prepared by spray layering, with pellets of layered or matrix architecture usually as the seed particles (Section 8.4.3.1). For all spray layering applications, film adhesion to the underlying particle surface is a key requirement, which is governed by the type and number of bonds formed between the two surfaces. Adhesion is generally weakened by internal stresses, which may arise from the shrinkage of the film, as a result of solvent evaporation, or from differences in thermal expansion (*e.g.* on storage).[25] Polymeric barrier films can be prepared *via* atomisation of organic solutions or aqueous dispersions (*e.g.* latexes or pseudolatexes). For either system, complete polymer particle coalescence and continuous film-formation are often not achieved at the end of coat application, with holes and/or channels still present. To minimize defects or imperfections, thermal post-coating treatment, referred to as "curing", is often required. The main advantages of aqueous systems over organic ones are lower environmental and health hazards (including explosion risks) and the potential for using higher solids loading in the coating formula (due to lower viscosity and reduced sticking tendency), resulting in reduced processing time. However, extensive curing requirements and the associated higher risk of drug release instability during shelf-life are a key downside of aqueous dispersions (since further polymer particle coalescence can lead to denser and less permeable films, and hence slower drug release).[25,26] According to Siepmann and Siepmann,[26] even optimal curing will not guarantee equivalent drug release for identical aqueous *versus* organic coatings, especially if polymer blends are used. One example of significant dissolution slow-down on storage, attributed to curing effects, is shown in Figure 8.14.

Drug Release Mechanisms. Two parallel transport processes have been proposed for barrier-coated pellets, namely diffusion through the intact coating and transport through pores or cracks. The lag-time prior to crack formation is thereby governed by (i) the coating permeability, (ii) the osmotic pressure differences across the coating, (iii) the pellet geometry, (iv) the coat thickness and (v) the mechanical properties of the coating.[27] The rate and extent of water uptake by the coating are other crucial parameters for drug release from the pellets.[28]

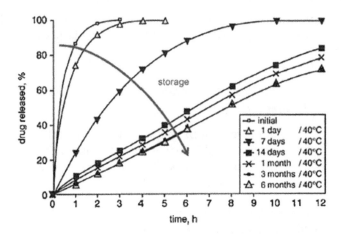

Figure 8.14   Dissolution changes for coated theophylline-containing pellets following open storage at 40 °C and 50% RH (Note: The pellets had been coated with an Eudragit RS 30D-based formulation containing 5% HPMC and 10% TEC).[26] Reprinted from *International Journal of Pharmaceutics*, **457**, F. Siepmann and J. Siepmann, Stability of aqueous polymeric controlled release film coatings, 437–445, Copyright 2013, with permission from Elsevier.

Key Formulation Constituents. To achieve a specific release profile, different parameters can be varied. Key examples are the types and levels of film-formers, plasticisers or water soluble polymers, *i.e.* so-called pore formers. The most common representatives of these constituents are listed in Table 8.9.

Plasticisers are employed to increase coating flexibility/integrity and hence minimise film cracking risks, which is particularly important for release-controlling films.[25] Pore formers are generally the key lever for drug release control and are discussed in more detail later in this section.

Water is the only solvent used for aqueous dispersions, while organic solutions typically utilise ethanol in combination with some water (*e.g.* 10–20%). While the solvent system is primarily the vehicle to deliver polymer to the pellet surface, solid–liquid interactions in the solution/suspension can potentially affect the characteristics of the resulting film. This needs to be considered carefully when designing CR pellets involving spray layering.

To reduce the tackiness (stickiness) of pellets during coating and subsequent storage, it is standard practice to include antiadherents such as talc or GMS. However, since these antiadherents can also adversely affect the mechanical and adhesive properties of the pellet coatings, it is important to strike an appropriate balance when selecting their type and level.[25]

Key Factors Affecting Drug Release. Pore formers are considered to be the key formulation levers for drug release from CR-coated multiparticulates.

Table 8.9 Most common functional excipients of barrier membranes prepared by spray layering.[28,52]

Ingredient class	Common materials
Film-former	- Acrylic acid derivates [*e.g.* poly(ethyl acrylate-*co*-methyl methacrylate) (EUDRAGIT-family, Evonik)]
	- Polyvinyl acetate (KOLLICOAT SR, BASF)
	- Cellulose derivatives [*e.g.* ethyl cellulose (ETHOCEL)]
Pore former	- Hydroxypropyl methylcellulose (HPMC) (*e.g.* METHOCEL)
	- Hydroxypropyl cellulose (HPC)
	- Polyvinylpyrrolidone (PVP)
	- Polyvinyl alcohol–polyethylene glycol (PVA–PEG) graft copolymer (KOLLICOAT IR, BASF)
Plasticiser	- Triethyl citrate (TEC)
	- Triacetin
	- Dibutyl sebacate (DBS)
	- Diethyl phthalate (DEP)
	- Fractionated coconut oil (FCO)
	- Oleic acid (OA)

HPC and HPMC are incompatible with ethyl cellulose (EC) and such films are hence generally inhomogeneous, presenting EC- and HPC/HPMC-rich domains [characterised by two glass transition temperatures (T_g)]. The control of their microstructure is hence important for robust product performance. Marucci *et al.*[29] used a release cell immersed in liquid to study EC/HPC films prepared by spraying ethanol:polymer solution (94:6 ratio) onto heated rolls (with the films peeled off and dried in a desiccator thereafter). Figure 8.15 shows the fraction of HPC leached as a function of HPC content of the film, while Figure 8.16 shows its water permeability for the same *x*-axis.

Figure 8.15 provides clear evidence that pore formers leach from inhomogeneous films, yet only above a certain level (here 22–24%), which represents a sharp transition-point (with <10% *versus ca.* 90% leached in this example). Figure 8.16 shows that leaching of the pore former significantly increases the permeability of the film, with permeability increasing with pore former level in the film, as long as the later is above the transition point discussed above. It is hence important to establish the latter for any system of interest and to avoid pore former levels near the transition point, since this region presents a significant risk to product robustness. In fact, it would be strongly recommended to change film components or use other levers, such as film thickness, rather than select a film formulation near its transition point. To minimise the risk of batch-to-batch differences or product changes during shelf-life, some opinions expressed in the literature hence favour systems without any such incompatibilities.

Another important parameter is coating thickness and it has been stipulated that the drug release rate from CR pellets is governed predominately by the thickness, uniformity and quality of the coating, which in turn can be affected significantly by the surface morphology of the substrate being

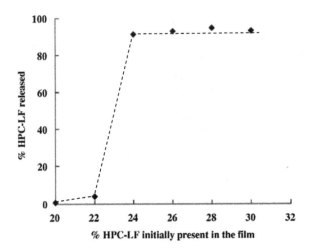

Figure 8.15 Fraction of HPC leached from free-standing EC/HPC-films after 24 hours immersion in pH 6.8 phosphate-buffer solution, as determined using a release cell under well-stirred conditions.[29] Reprinted from *Journal of Controlled Release*, **136**, M. Marucci, J. Hjärtstam, G. Ragnarsson, F. Iselau and A. Axelsson, Coating formulations: New insights into the release mechanism and changes in the film properties with a novel release cell, 206–212, Copyright 2009, with permission from Elsevier.

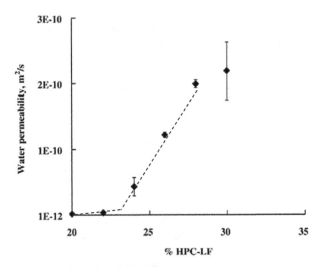

Figure 8.16 Water permeability of free-standing EC/HPC films containing different levels of HPC.[29] Reprinted from *Journal of Controlled Release*, **136**, M. Marucci, J. Hjärtstam, G. Ragnarsson, F. Iselau and A. Axelsson, Coating formulations: New insights into the release mechanism and changes in the film properties with a novel release cell, 206–212, Copyright 2009, with permission from Elsevier.

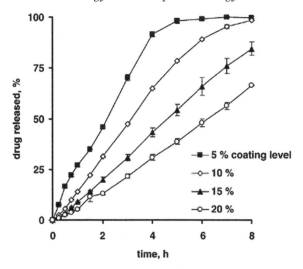

Figure 8.17   The effect of coating level on drug release for theophylline-containing pellets coated with 85:15 EC:PVA–PEG graft copolymer blends (Note: Dissolution media was 0.1 M HCl).[28] Reprinted from *Journal of Controlled Release*, **119**, F. Siepmann, A. Hoffmann, B. Leclercq, B. Carlin and J. Siepmann, How to adjust desired drug release patterns from ethylcellulose-coated dosage forms, 182–189, Copyright 2007, with permission from Elsevier.

coated. Certain minimum coating levels are required to compensate for uneven surfaces, with 20–50 μm being a typical target for CR pellets, according to Siepmann.[30] It is important to avoid coatings that are too thin, too thick, too brittle or too sticky, with different analytical techniques available to investigate the physical characteristics of the pellet coatings.[31] An example of the effect of coating thickness on drug release, here captured in terms of percentage of coating mass (relative to pellet core mass), is shown in Figure 8.17.

Figure 8.17 shows how release rate decreases proportionally with coating level (5–20%), which is the expected scenario for coatings of adequate robustness and homogeneity.

Another factor is the type of seed particle, especially for pellets with layered architecture, where sugar spheres and MCC spheres are most common. The reason is that MCC spheres, in contrast to sugar spheres, swell upon hydration and hence exert pressure on the coating. This is in addition to the osmotic pressure that is characteristic for both core types.

8.4.3.2.2   Dry Powder Coating.   Rotary granulation (Section 8.3.2) can be used for dry powder coating of granules, pellets or mini-tablets, involving the same steps as solvent-based coating (Section 8.4.3.2.1). The key difference is that dry powder coating only involves negligible amounts of liquid, hence lending itself to moisture-sensitive APIs and amorphous solid dispersions.[32] Since liquid saturation is generally low in the case of dry powder coating, API dissolution in the binding liquid is generally insignificant, irrespective of API solubility. Upon drying, the binder and other dissolved substances crystallise, with liquid bridges replaced by solid bridges.[32] According to Sauer *et al.*,[31] viscous flow, particle deformation and the resulting sintering of the polymer particles are the main driving forces for film formation. The time required for two powder particles to coalesce is thereby said to be directly related to coating viscosity, particle radius and surface tension of the coating. For elastic materials, the particle deformation is reversible and surface adhesion hence often poor.[32] Powder nozzles are generally used to charge the fine powder into the chamber, where it contacts the pre-wetted seed particles. Pre-wetting materials are aqueous solutions, most commonly plasticiser–water mixtures, which are partially mixed into the feedstock or injected into the chamber. According to Sauer *et al.*,[31] powder coating generally requires more plasticiser than liquid-based coating applications, with the plasticiser lowering the polymer T_g to levels that facilitate film-formation. The wetting agent and dry polymer are continually added to the seed particles, resulting in coating growth. This has been compared to the increase of a snow-ball rolling down a snow-covered mountain.[15] Although dry powder coating formulations often require higher coating levels than solvent-based processes, for similar release profiles, their coating times are said to be shorter.[32] Furthermore, the possibility of achieving hourly weight gains of up to 300% has been reported for dry powder layering. According to Sauer *et al.*,[31] the mechanism of film-formation can be summarised in the following steps:

- Particle coalescence and sintering, in a process involving partial polymer fusion
- Levelling/smoothing of the coating material, including layer densification
- Cooling and hardening of the coating

Since dry powder coatings are formed directly from powder particles, it is imperative that the latter have a controlled size distribution, especially with respect to the avoidance of large particles. This is particularly important since coatings typically have thicknesses in the order of only tens of micrometres to 100 μm, with powder particles of similar size risking compromising coating integrity.[15] According to Sauer *et al.*,[31] the diameter of the coating powder should be less than 1% of the substrate diameter, for acceptable uniformity, surface adhesion, appearance and processing times. Similar to spray layering, EC is commonly used for dry powder coating. Key advantages of dry powder coating are the elimination of the solvent removal stage and the corresponding processing time reduction. However, to obtain coatings with adequate functionality, a curing stage is typically required, which can be performed *in-situ* in the granulator (*dynamic curing*) or in tray ovens (*static curing*). In the case of *dynamic/in situ* curing, both heated air and frictional movement of the product bed contribute to the curing process.

Although either approach can be sufficient for film formation, dynamic curing is generally associated with smoother and more uniform films.[15] Sauer et al.[31] reported that lower curing temperatures generally require longer curing, whereas higher curing temperatures come with higher risks of pellet agglomeration. Curing is particularly important for release-controlling films, due to the importance of film integrity to material transport. A common issue with insufficient curing is that storage conditions facilitate polymer particle coalescence and thus change barrier functionality relative to the initial state. In the case of release-controlling films, this typically results in slower drug release, due to the formation of a more effective barrier.

Dry powder coating formulations also require anti-sticking agents, to prevent adhesion during processing. Even though their amount has not been studied extensively, blends with 10% talc have been successful in preventing the powder particle agglomeration during coating or storage. However, it should be noted that talc may also inhibit the adherence of the powder coating to the substrate, due to its anti-sticking properties. Colloidal silicon dioxide is an alternative, which (unlike talc) is not premixed with the polymer prior to coating, but applied thereafter (*e.g.* at 1–2% of polymer weight).[32]

Colorcon has reported a dry powder layering example involving a centrifugal fluid bed granulator (Glatt GPCG-1), sugar spheres of 840–1000 μm diameter, the API lansoprazole and a 3% HPC solution.[34] This had been chosen over spray layering or extrusion–spheronisation due to API incompatibilities with typical binders. Table 8.10 summarises the process parameters.

The final pellets were 1190–1410 μm in size, with Young et al.[33] describing them as spherical, dense, low in fines and suitable for downstream coating using solvent-based processes. Figure 8.18 contrasts the seed particles and final pellets, showing the snowball-like appearance of the latter.

In summary, it can be said that dry powder coating of pharmaceutical pellets has developed considerably over the past decade and has been able to differentiate itself from its solvent-based counterparts. However, commercial uptake has been slow, most probably due to the need for additional development and customised equipment.[31]

Table 8.10 Parameters used for the pilot-scale powder layering process.[33] Published with permission from Colorcon.

Parameter	Parameter value
Quantity of sugar spheres	2250 g
Rotor speed	200 rpm
Binder spray rate	20 g min^{-1}
Powder addition rate	15 g min^{-1}
Inlet air temperature	55 °C
Outlet air temperature	45 °C
Bed temperature	45 °C
Atomising air pressure	1.5 bar
Air flow	68–80 m^3 h^{-1}
Total processing time	113 min

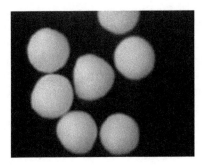

Figure 8.18   Images of uncoated seed particles (*left*) and dry powder-layered pellets (*right*).[33] Published with permission from Colorcon.

8.4.3.3   *Multiparticulate-based Finished Dosage Forms*

Multiparticulate-based products typically come as capsules, tablets or sachets, with the focus of this section on capsule filling and multiple-unit pellet system (MUPS) tablets.

8.4.3.3.1   Hard Capsules.   Hard capsules are the most common dosage form for CR pellets, due to their simplicity and flexibility. However, their production costs are generally higher than those for tablets, due to (i) the lower throughput of capsule filling machines, (ii) capsule procurement costs, and (iii) costs associated with capsule integrity control after filling.[25,34,35] The capsule filling equipment for CR pellets is the same as for powdery formulations, except for their dosing configurations. The reasons for the differences in the latter are that pellets are usually larger (*e.g.* 800–1400 µm mean diameter), cannot be compressed like powders, and are typically coated. Other considerations for pellet filling are (i) the surface roughness of CR coatings, due to the effects of higher friction and electrostatic charging on filling performance, and (ii) pellet shape, since fewer filling problems are associated with highly spherical pellets.[36]

8.4.3.3.2   Multiple-Unit Pellet System (MUPS) Tablets.   Tablets can generally be prepared at lower cost than capsules and come with lower tampering risks and fewer oesophageal transport difficulties.[35] There are hence commercial drivers for delivering CR pellets in tablet format, which are referred to as multiple-unit pellet system (MUPS) tablets.[37] MUPS tablets use conventional compaction technology and can be designed as scored tablets, without affecting drug release, since each pellet is a release-controlling unit whose properties are maintained during compaction.[25,34] A single MUPS tablet can hence potentially support multiple dose strengths, as long as pellets are distributed homogenously within the tablets. However, pellet compaction is generally challenging, due to the need to maintain pellet functionality in the final tablet. Key risks are (i) multiparticulates fusing during compression, yielding monolithic rather than multiparticulate performance, or

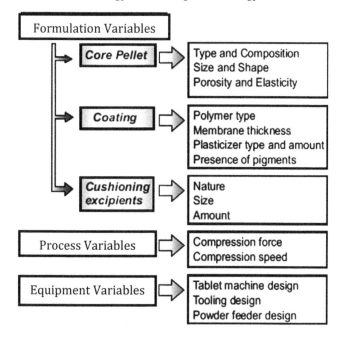

Figure 8.19 Key factors influencing the design of MUPS tablets.[25] Reprinted from *Journal of Controlled Release*, **147**, S. Abdul, A. V. Chandewar and S. B. Jaiswal, A flexible technology for modified-release drugs: Multiple-unit pellet system (MUPS), 2–16, Copyright 2010, with permission from Elsevier.

(ii) damaged functional coatings, compromising barrier properties in the case of CR coatings. As long as the film has sufficient strength and elasticity to prevent rupture, its deformation is generally of little consequence. Lehmann *et al.*[38] suggested that coatings need to achieve at least 75% elongation.[34,37] To prevent pellets from being altered, cushioning excipients are used, which rearrange themselves between the pellets and reduce the void space, thus preventing direct contact between pellets during compression. Other challenges for MUPS tablets are weight variation, low hardness and poor friability.[37] Various material- and process-related parameters thus need to be optimised to obtain MUPS tablets with physical integrity and similar drug release to the incorporated pellets. To minimise problems during development, it is important to know how material- and process-related parameters affect formulation performance, with Figure 8.19 summarising the key factors.[25]

According to Abdul *et al.*,[25] pellet compaction occurs in the following stages:

(1) Pellet bed volume reduction due to rearrangement of pellets (filling inter-particle voids)
(2) Pellet bed volume reduction due to local surface deformation (involving pellet surface flattening)

(3) Bulk deformation of pellets (*i.e.* change in pellet dimensions), parallel to densification of pellets

(4) Cessation of the volume reduction, due to low inter- and intra-granular porosity

Stages 1 and 2 are characterised by significant volume reduction and low bed strength, due to insufficient intergranular bonding. Stage 3 results in stronger inter-granular bonding, which is further increased in stage 4, despite little further volume reduction.[25]

As highlighted previously, MUPS tablet preparation is challenging, since pellet fusion into a non-disintegrating matrix needs to be avoided. For MUPS tablets containing coated pellets, it is imperative that the functional coating withstands compaction, since coating cracks are likely to affect drug release. The coating strength, ductility and thickness are hence important characteristics, influencing rupture and deformation capacity of the compressed pellets.[25] Coating polymers for pellets can be categorised as either (i) cellulosic (*e.g.* EC), or (ii) acrylic (*e.g.* methacrylic acid and methacrylic acid ester copolymers). These can be formulated as aqueous colloidal dispersions (*e.g.* latexes or pseudolatexes) or organic polymer solutions.[34] According to Gothoskar and Phale,[37] solvent-based coatings have been shown to be affected less by compression than their aqueous-based counterparts, due to their higher flexibility and mechanical stability. According to Bodmeier[34] and Abdul *et al.*,[25] most compaction studies for EC-coated pellets revealed coating damage and loss of extended-release properties. This can be attributed to the poor mechanical properties of EC. Films prepared from acrylic polymers are more flexible than EC and therefore generally more suitable for MUPS tablets. Eudragit NE 30D films were found to be very flexible, without the addition of plasticisers, showing elongation in excess of 365%. Bodmeier[34] attributed this to their lack of strong inter-chain interactions (*e.g.* hydrogen bonds). Flexible films with elongation in excess of 125% were obtained for plasticised Eudragit RS and RL 30D, whereas Eudragit L 30D produced weak and brittle films (elongation <1%). Pellets coated with Eudragit NE 30D or plasticised Eudragit RS/RL 30D, with elongation in excess of 75%, were compressed successfully into fast-disintegrating tablets, without significant coating damage or changes in drug release.[34] Since some traditionally used CR coatings are not particularly suitable for MUPS tablets, it is important to consider this upfront during the pellet design process.

It has been found that coated pellet damage during compression can be minimised by applying thick coatings.[37] According to Bodmeier,[34] thicker pellet coatings result in additional pellet expansion, due to the elastic polymer characteristics, increasing the ability of the pellets to deform with increasing coating level. In a study involving Eudragit L30D-55, it was found that softer pellets had more film rupture incidents, attributed to higher pellet deformation during compression. Minimal damage to coated pellets was found when the coated and core pellets had similar elastic and tensile properties. Pellet size also affects compaction properties and hence drug release. At the same coating level, smaller pellets were found to be more fragile than larger pellets, due to the reduced film thickness resulting from the larger surface area. However, since drug release is proportional to the pellet surface area, smaller pellets usually require higher coating levels to achieve a particular release profile.[34]

As already highlighted, MUPS tablets require an outer excipient phase, to protect the pellets and obtain tablets of adequate physical integrity. Ideal fillers act as cushions during compression, preventing direct contact between the pellets, facilitate rapid disintegration and do not affect drug release. Theoretically, a level of 29% v/v is required to fill the void-space between densely packed spheres. Besides their compaction properties, the excipients have to result in a uniform blend with the pellets. To avoid segregation of the pellet–excipient mixture, excipients of larger size or placebo pellets (*e.g.* containing GMS) could be used as outer phase.[25,34] The cushioning effect of an excipient generally depends on its particle size, volume and compaction properties. According to Gothoskar and Phale,[37] studies of 14 excipients showed that those with good plastic deformation afforded the best protection to pellet coatings. Examples are MCC (*e.g.* Avicel PH 200), high molecular weight PEG (*e.g.* 3350 or 6000), crospovidone, GMS, lactose or dicalcium phosphate.[25] The energy of compaction is predominantly absorbed by the matrix, if it has a lower yield pressure than the pellets and pellet coating, resulting in its preferential deformation.[35] Particle size has been highlighted as another important factor, with one study having found increased dissolution rate for particles larger than 20 μm.[37] In a MUPS study involving theophylline granules coated with Eudragit RS, tablets of sufficient hardness were obtained for up to 50–70% of pellets, while no stable compacts were obtained above 90% and fillers had little effect on drug release at less than 10% pellets. In another study, rapidly disintegrating MUPS tablets were prepared with pellets containing 75% theophylline and 25% inert excipients and Eudragit NE 30 D as the coating. They were mixed with inert granules prepared by wet granulation (1 : 1 MCC : corn starch), with particle size close to that of the pellets.[34]

LOGIMAX (AstraZeneca) is an example of a commercial MUPS tablet product, which contains ER pellets in an ER tablet matrix. According to Caldwell,[39] the ER pellets are prepared by spray layering of inert spheres with an API/HPMC layer and an EC : HPMC (80 : 20) barrier membrane.

8.4.4 Osmotic Drug Delivery Systems

Osmotic drug delivery systems utilise osmotic pressure as primary driving force for drug release, with release hence largely independent of pH and hydrodynamics.[40] Table 8.11 lists the key osmotic technologies for ER.

The push pull osmotic pump (PPOP) is the most established and well-known of the technologies in Table 8.11 and is hence discussed further in the next section.

Table 8.11   The key osmotic drug delivery technologies for extended drug release applications.[43]

Technology	Key attributes	Options assessment
Elemental osmotic pump (EOP)	- Single-layer tablet - Semi-permeable membrane, which is only permeable to aqueous media - Single drilled orifice	The original osmotic drug delivery technology, which is reported to require APIs with moderate solubility
Push pull osmotic pump (PPOP)	- Bilayer tablet - Semi-permeable membrane, which is only permeable to aqueous media - Single drilled orifice	Most established osmotic drug delivery technology, able to accommodate APIs with low to high solubility
Controlled porosity osmotic pump (CPOP)	- Single layer tablet - Semi-permeable membrane containing water-soluble additives, with micro- porous structure forming on hydration - Semi-permeable membrane is permeable to aqueous media and API - No drilled orifice	Coating controls liquid ingress as well as API egress, with technology hence requiring APIs that are soluble enough for diffusion-based release

8.4.4.1   *Push Pull Osmotic Pump (PPOP) Technology*

The PPOP technology was invented and commercialised in the 1980s by the Alza Corporation, mainly for CR of poorly soluble APIs. The key difference and advantage over other CR technologies is that it releases API in the solid *versus* liquid state. Due to technological changes and an increased focus on lipophilic biological targets, the proportion of poorly soluble APIs in development has increased significantly over the past decades. This has resulted in renewed interest in this technology, aided by the expiry of the main technology patents in the early 2000's. The PPOP technology is primarily associated with tablets, but a capsule format is also featured in this section.

8.4.4.1.1   PPOP Tablets.   The key components of a PPOP tablet are as follows:

  i. A bilayer tablet, with osmotic agent in layer 1 (*push layer*) and API in layer 2 (*pull layer*)
 ii. A rigid membrane coating, which is insoluble but permeable to aqueous media
iii. A small orifice in the coating (above layer 2), through which the API is released

The semi-permeable coating controls the ingress rate of liquid, which increases the osmotic pressure in the system. This causes material from the upper layer, including the API, to be transported through the delivery orifice

Table 8.12 Typical qualitative and quantitative compositional data for PPOP tablets.[42]

Ingredient function	Ingredient levels (per layer) (%)	Material examples
Pull layer		
API	1–50	–
Viscosity modifier	5–90	Polyox N10/N80
Filler/osmogen	5–20	NaCl, mannitol
Lubricant	0.5	Magnesium stearate
Push layer		
Swelling agent	70–90	Polyox coagulant
Osmogen	5–20	NaCl
Lubricant	0.5	Magnesium stearate
Dye	0.2	Blue Lake
Membrane coating		
Insoluble polymer	60–95	CA 398-10
Pore former	5–40	PEG

and become exposed to the surrounding fluid.[2] The coating is generally a mixture of water-insoluble cellulose acetate (CA) and water-soluble PEG. Their ratio determines the porosity and tortuosity of the membrane, and hence the liquid permeation through it, with acetone:water mixtures (*e.g.* 95:5 ratio) commonly used as the solvent system. According to Colorcon,[41] PEG levels in the coating should not exceed 30%, to maintain adequate coating integrity. Coating levels are typically 8–20% of the core tablet weight, but this depends on the tablet size and release rate target. 50 µm is the recommended minimum coating thickness, to maintain its mechanical integrity under the hydrostatic pressure of the core.[41,42] Verma *et al.*[43] even quoted typical membrane thicknesses of 200–300 µm, to ensure adequate strength. Table 8.12 shows typical qualitative and quantitative compositions of PPOP tablets.[42]

A library of rheology tests can be used to fine-tune the API-containing layer, to balance the viscosities of the two layers and thus mitigate the risk of so-called push-layer *break-through* or *work-around*, both of which are associated with high residual drug levels (Figure 8.20).[42]

The manufacturing process for PPOP tablets typically involves blending and granulation, bilayer compression, barrier coating, tablet drilling, elegance coating and printing. The last two stages are required since barrier coatings are generally transparent and product differentiation *via* embossing should be avoided.[41] Direct compression (DC) is the favoured full-scale process for layer 1 (push layer), since its main ingredient PEO does not lend itself well to roller compaction (RC) or wet granulation, due to its compression characteristics and rapid hydration respectively. For layer 2 (pull layer), either DC or RC are the favoured processes.[41,42]

The standard architecture of a PPOP tablet typically results in zero-order release, reaching 80% over 5–24 hours, following a characteristic lag-time

(a)   (b)   (c)

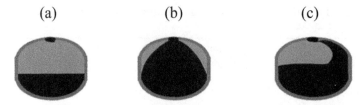

Figure 8.20   Different scenarios for drug release from PPOP tablets, namely (a) the ideal push–pull viscosity balance, (b) non-ideal push-layer "break-through" (generally as a result of low pull-layer viscosity and associated with incomplete drug release), and (c) non-ideal push-layer "work-around" (generally a result of high pull-layer viscosity and also associated with incomplete drug release).[41] Published with permission from Colorcon.

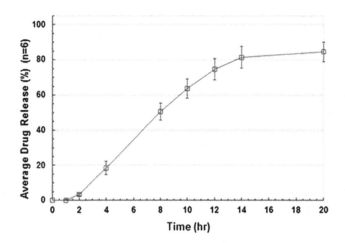

Figure 8.21   Dissolution profile of a PPOP tablet in 50 mM pH 6.8 phosphate buffer.[42]

that is a consequence of the release mechanism.[42] Figure 8.21 shows a typical dissolution profile for PPOP tablets, in this case for a formulation with only moderate lag-time.

As explained earlier, drug release from PPOP tablets occurs *via* a pre-formed small orifice. While it is not possible to advise on its optimal size, it can be said that small orifices risk unpredictable drug delivery, due to insufficient hydrostatic pressure relief, while large ones risk facilitating drug release *via* diffusion (in addition to the osmotic mechanism). One study involving the API nifedipine found that drug release was largely unaffected by orifice diameters of 0.25–1.41 mm, but became quite fast at 2.0 mm.[43] Although these values should be taken with caution, they emphasise the need to study the effect of orifice size during development.

PPOP tablets are probably most well-known for being the technology behind Pfizer's PROCARDIA XL (nifedipine), which launched in 1989 and turned a patent-expired compound and US$ 200 million product into the company's first product with annual sales exceeding US$ 1 billion. The secrets of its success were (i) that Alza's technology reduced dosing frequency from three-times-daily to once-a-day, significantly improving patient convenience and adherence, and (ii) that no alternative technology at the time was able to achieve similar performance in a robust manner, due to the physico-chemical properties of the compound (especially its low solubility). This makes PPOP tablets one of only a small number of technologies that managed to extend the commercial life of a compound far beyond its API patent expiry. However, the technology does have some important downsides, especially (i) the relatively high costs, due to the many processing steps and associated throughput limitations and (ii) the formulation expertise required to robustly develop these formulations. However, for the right API or indication, these downsides should be offset by the technology's key performance advantages.

8.4.4.1.2 PPOP Capsules. From a commercial stand-point, PPOP capsules offer no obvious advantages over PPOP tablets. The only reason why PPOP capsules are discussed here is that there are some niche benefits of a particular variant. That system is a modular PPOP capsule, whose key components are shown in Figure 8.22.[44]

The body and cap are not standard capsule shells, but were prepared specifically for this purpose. Using a conventional pan coater, high density polyethylene (HDPE) moulds were coated with a mixture of cellulose acetate and PEG 3350, dissolved in acetone and water. Upon coating, each shell is manually removed from its mould, after which time a laser hole is drilled into the end of the body. Shaped tablets containing the API and PEO (pull layer) and others containing high molecular weight PEO, MCC and sodium chloride (push layer) are prepared using conventional compression technology and inserted manually into the capsule body. The body and cap lock together due to their ridges, which eliminates the need for banding.

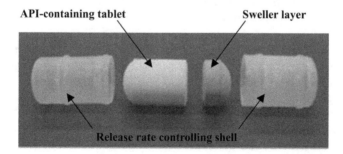

Figure 8.22 The components of a modular capsule system based on the PPOP technology.[44] Reprinted from *Journal of Controlled Release*, **152**, K. C. Waterman, G. S. Goeken, S. Konagurthu, M. D. Likar, B. C. MacDonald, N. Mahajan and V. Swaminathan, Osmotic capsules: A universal oral, controlled-release drug delivery dosage form, 264–269, Copyright 2011, with permission from Elsevier.

By using shells of different composition, *i.e.* different cellulose acetate to PEG ratios, and/or thickness, capsules can deliver different release rates, independent of the drug loading or API solubility. Waterman *et al.*[44] have quoted release of 80% label claim between 6 and 14 hours. The modular and API-sparing design, using prefabricated standard components, allows the rapid evaluation of CR drug delivery for compounds that are either in the very early stages of development or are life-cycle management candidates. This includes proof-of-concept clinical studies, if GMP requirements have been met.[44] However, one caveat is that this technology is unlikely to be available commercially.

8.4.5   Proprietary and Other Controlled Release Technologies

This section provides a brief overview of proprietary CR technologies, mainly to broaden the reader's horizon. Some of them have been around for some time and utilised in marketed products, while others are fairly new, with the support structure hence likely to differ significantly. General CR technologies can be found in Table 8.13, while gastroretentive (GR) technologies and so-called diagnostic tools are featured in Sections 8.4.5.1 and 8.4.5.2, respectively.

8.4.5.1   *Gastroretentive (GR) Technologies*

GR dosage forms are designed to be non-disintegrating and non-dissolving, to avoid them being readily ejected from the stomach, with key drivers for prolonged residence time as follows:[45]

- The API is locally active (in the stomach)
- The API has an absorption window (in the stomach or upper small intestine)
- The API is unstable in the intestinal or colonic environment
- The API exhibits low solubility at high pH, making the stomach more favourable for dissolution

Narrow absorption windows combined with extended release make GR a compelling alternative to frequent dosing, with key strategies for increasing gastric residence time as follows[45]

i. Delivery systems rapidly increasing in size upon ingestion—Examples are unfolding or swelling systems, to delay passage through the pyloric sphincter into the small intestine
ii. Density-controlled delivery systems—These are designed to either float or sink in gastric fluids and rely heavily on the filling state of the stomach
iii. Bioadhesive delivery systems—These are designed to adhere to mucosal surfaces, with the potential risk of adversely affecting the stomach lining and accumulating in the stomach

Table 8.13 Overview of general proprietary controlled release technologies.[59–61]

Technology	IP holder	Key attributes
Geomatrix™	Vectura group plc, UK (formerly Skyepharma, Switzerland)	- Multilayer tablet—with central layer containing API and hydrophilic polymer, whose drug release is controlled *via* available surface area and physical characteristics of the barrier layers - Use of established ingredients and unit operations - Claimed to be suitable for low and high solubility APIs, zero order and bi-phasic drug release, and the release of two or more APIs at different rates - Utilised in at least eight marketed products
Geoclock™	Vectura group plc, UK (formerly Skyepharma, Switzerland)	- Dry-coated tablet—with outer shell, prepared from mixture of hydrophobic wax and brittle material and at least one API-containing core - Outer layer composition governs release lag-time, which is largely independent of environmental pH or food constituents - Can be designed for IR or ER drug delivery (post lag) - Particularly suitable for regional drug delivery
TIMERx (including Geminex and SyncroDose)	Endo, Dublin, Ireland (formerly Penwest pharmaceuticals)	- Hydrophilic matrix tablet based on xanthan gum, locust bean gum and dextrose, with these polysaccharides forming a coating around the API-containing core and determining the API release - Geminex is a bilayer-based and SyncroDose a is GI site-specific/timed drug delivery platform utilising the TIMERx technology
Guardian™	Egalet Corp. (Wayne, Pennsylvania, USA)	- API-containing monolithic cores + outer shell - Combines hot-melt extrusion and injection moulding - Suitability for low and high API solubility has been claimed - Release governed by matrix composition and shape
Micropump®	Flamel technologies, S. A. (Venissieux, France)	- Multiparticulate-based CR platform—with core particles typically 200–500 μm in diameter - Drug release governed by particle coating(s) - Particles filled into capsules or compressed into tablets, with *ca.* 5000–10 000 particles per unit

Table 8.14   Overview of proprietary gastroretentive (GR) technologies.[62–65]

Technology	IP holder	Key attributes
Accordion Pill™	Intec pharma (Jerusalem, Israel)	- Drug-containing hydrogel sheet (*ca.* 50 mm × 25 mm × 0.7 mm) folded into size 00 gelatin capsule - Continuous process involving film casting, drying, lamination, ultrasonic welding and cutting - Single-layer design, used for soluble APIs, and multilayer design (with perforated outer layers), used for low solubility APIs. - 500–600 *versus* 300–400 mg maximum dose strength for soluble *versus* poorly soluble APIs - GR of up to 8–12 h for majority of subjects - Fed-state required, but not high fat or high calorie - Manufacturing process deemed main hurdle to commercial adoption, due to relatively low throughput (of no more than 20 000 units per day)
Acuform®	Depomed Inc. (Newark, California, USA)	- Tablet platform involving hydrophilic polymers (*e.g.* HPMC, PEO), characterised by significant volume expansion upon hydration - Generally large tablet, allowing high dose strength - Utilises standard unit operations and excipients - Licensed to at least six pharmaceutical companies and used in at least four marketed products - Requires high fat/high calorie feeding regime
RubiReten®	Rubicon Research Pvt Ltd (Mumbai, India)	- Utilises monolayer or bilayer tablet technology - Leverages solubility enhancers (*e.g.* Cremophor, Capmul, Gelucire, Lutrol, Captex, Vitamin E TPGS or Poloxamer) to increase API solubility - Matrix containing swelling agents (*e.g.* HPMC, PEO) and swelling enhancers (cross-linked PVP) - Requires high fat/high calorie feeding regime

The focus of this section is on size-increasing technologies (Table 8.14), since it is the author's experience that this strategy is best-placed to achieve the release targets.

The reliance on the state of the GI tract, which can vary significantly between patients and time periods, makes GR only commercially viable for compounds that can tolerate large concentration fluctuations.

8.4.5.2   Diagnostic Tools

Technologies used for diagnostics rather than as commercial dosage forms are often termed diagnostic tools (*e.g.* Enterion™ and IntelliCap®). Common applications are drug delivery to specific regions of the GI tract, to understand their contributions to drug absorption.[46]

8.4.5.2.1   Enterion™.   Enterion™ is a remote-controlled capsule-shaped device developed by Phaeton Research (Nottingham, UK) in collaboration with PA Consulting Group (Melbourn, Hertfordshire, UK).[46,47] It is 32 × 11 mm

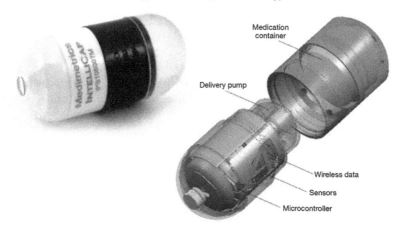

Figure 8.23 Picture and schematic illustration of the IntelliCap® system (dispensing type).[1] Published with permission from Medimetrics Personalized Drug Delivery B. V.

in size and has a 1 ml drug reservoir, which can be loaded with liquid or particulate formulations. A radioactive marker is placed in the device, for real-time visualisation of GI transit *via* gamma scintigraphy. When the capsule reaches the target location, its contents are ejected *via* application of an oscillating magnetic field.[47]

8.4.5.2.2 IntelliCap®. IntelliCap® (Medimetrics Personalized Drug Delivery, B. V.) is a CE-certified device for use in humans and larger mammals (*e.g.* beagle dogs).[48] Different versions exist, with the dispensing system being most relevant for CR applications and hence the focus of this section. It is a single-use capsule, 27 × 11 mm in size, with key features as shown in Figure 8.23.[49]

The capsule is transported along the GI tract by natural peristalsis, with temperature and pH measurement data allowing key locations to be determined, eliminating the need for *in vivo* imaging. The reason for this is that any passages through the pylorus and the ileocecal valve are accompanied by characteristic pH-changes.[48] Another differentiator of IntelliCap® is its ability to deliver any release profile, which makes it particularly interesting for CR development, allowing pharmacokinetic/pharmacodynamic (PK/PD) studies to be conducted without formulation development.[48] To date, IntelliCap® has been used exclusively as an R&D tool, to test different profiles *in vivo*. However, its developers have also expressed an interest in developing bespoke drug products based on this technology, in collaboration with pharmaceutical companies. This could open up new opportunities, either for (i) drug delivery profiles not achievable with current CR technology, or (ii) life-cycle management, with intellectual property (IP) extension through the device technology. This will only be possible if the devices are produced at sufficiently low unit cost and size—a challenge Medimetrics have claimed can be overcome.[1]

8.5 Evaluation of Controlled Release Dosage Form Performance

The most important performance attribute of CR formulations is their human PK profile, which is determined primarily by *in vivo* absorption rates, which, in turn, are governed by API release kinetics of the dosage form.[50] Since the latter can be determined by *in vitro* testing, it is possible to estimate CR performance *via* these means. However, accurate prediction of *in vivo* performance *via in vitro* testing is difficult, due to the complexity of human physiology. Performance evaluation *via in vivo* studies, preferably in humans, is hence generally required, with studies in pre-clinical species often a first step, due to their lower cost and faster cycle-times. A key exception is when an *in vitro–in vivo* correlation (IVIVC) has been established, since this allows dissolution to be utilised entirely *in lieu* of *in vivo* data. The next sections provide more background on the *in vitro* and *in vivo* evaluation of CR dosage forms.

8.5.1 *In vitro* Evaluation

Dissolution is generally the key *in vitro* test for CR formulations, since it is difficult to judge if drug release criteria are met without dissolution data.[50] *In vitro* drug release testing is most commonly performed using standard USP dissolution equipment (Section 8.5.1.1), but *in vitro* models closely resembling the GI environments (Section 8.5.1.2) are an increasingly popular alternative. In both cases, release profiles can be affected by many parameters, such as apparatus type, liquid media or agitation rate. Comparisons of dissolution data generated under different conditions or claims of *in vivo* performance based on dissolution data alone should hence be treated with caution.[50]

8.5.1.1 *USP Dissolution Testing*

USP dissolution testing serves a number of purposes, with the key ones listed below:[50]

 i. Control of product quality, especially with respect to release mechanism
 ii. As a potential alternative to *in vivo* testing for estimating formulation performance, typically through use of so-called biorelevant media (which exist as different recipes)
 iii. As a potential alternative to *in vivo* testing to justify formulation or manufacturing site changes

Table 8.15 summarises the key characteristics of apparatus 1 to 4 in the context of CR development.

For CR formulation design, it is important to have a stage-appropriate testing strategy. Since biorelevant media are more costly and challenging,

Table 8.15 Overview of USP dissolution apparatus.[15,50]

	USP apparatus	Key applications
1	Basket method	Primarily for multiparticulate formulations and/or dosage forms with tendencies to float or disintegrate slowly
2	Paddle method	Generally preferred for tablets—A sinker is often used to hold the tablet in place at the bottom of the vessel (with the sinker needing to be large enough to allow for tablet swelling)
3	Reciprocating cylinder/ modified disintegration method	Particularly useful for multiparticulate formulations, formulations containing poorly soluble drugs and/or for switching between different dissolution media
4	Flow-through cell method	Particularly useful for poorly soluble drugs and/or for switching between different dissolution media (assuming use of open-loop configuration)

it is advisable to use established media and methods for initial formulation screening, when fast turnaround, reliability and trending are most important. Although water has been a popular choice, established media, such as pH 6.8 phosphate buffer, are generally better alternatives. Following screening, evaluation of lead formulations in biorelevant media would be advisable. Single-stage dissolution testing may be sufficient, but two-stage testing (mimicking the transition from stomach to small intestine) should also be considered. An alternative would be testing in three media representing the gastric pH range, namely pH 1.2 0.01 N HCl, pH 4.5 acetate buffer and pH 6.8 phosphate buffer. Especially in the case of limited access to biorelevant media or complex and/or sensitive formulations, it is worth performing dissolution at different ionic strengths and pH values, to mimic fed and fasted states. In any case, it would be advisable to perform a range of *in vitro* tests ahead of any clinical evaluation, to confirm formulation robustness and suitability. This is particularly important when the clinical performance (*e.g.* PK) of other products needs to be matched. This should include dissolution testing of accelerated stability samples, since some CR formulations can be affected significantly by elevated temperature and/or RH. However, *in vitro* evaluation should not be limited to USP dissolution, but also consider *in vitro* models of human physiology (Section 8.5.1.2), especially for certain formulations or scenarios [*e.g.* relative bioavailability (RBA) or bioequivalence (BE) studies].

8.5.1.2 In vitro *Models of Human Physiology*

Reliable, rapid and low cost *in vitro* models of human physiology are challenging to design, owing to the complexity of human physiology, and can hence be considered the holy grail of CR formulation evaluation.

However, some very interesting options already exist, which are introduced in this section.

8.5.1.2.1 Dynamic Gastric Model (DGM).

The DGM is an *in vitro* system simulating the human stomach, invented and designed at the Institute of Food Research (Norwich, UK) on the basis of physiological parameters collected from human studies. The DGM can handle real food items and replicate the stomach's complex array of mixing modalities, dynamic biochemical release and temporal emptying patterns (including flow, shear and hydration). The DGM is modular, based on three stages, which are described below:

- Stage 1—Simulates the main body of the stomach (*fundus*), with initial low-shear mixing
- Stage 2—Simulates the lower stomach (*antrum*), which is responsible for mixing and shearing
- Stage 3—Provides a simulation of the small intestine

One of the DGM's main limitations is its emphasis on the stomach (*i.e.* stages 1 and 2), since CR dosage forms release limited drug in the stomach. Stage 3 is handled outside the device, more analogous to conventional dissolution testing.

8.5.1.2.2 Total Intestinal Model (TIM).

The TIM systems consist of several compartments interconnected by valves, which regulate GI transit. Each compartment can be studied separately, measuring even small changes in bioactive components, and mimics both the temperature and peristaltic movements of the body. The acidity and electrolyte concentrations, the presence of swallowed saliva, the "secretion" of gastric acid and enzymes, pancreatic juice with enzymes and bile salts are all dynamically monitored and regulated.[51]

The TIM-1 system represents the GI tract from stomach through to small intestine, while TIM-2 simulates the large intestine. Both systems can be adapted to a wide range of populations, such as infants, young adults, seniors, patients with impaired GI conditions and animals (*e.g.* dogs, pigs).

8.5.2 Pre-clinical Species

Studies in pre-clinical species are a potential alternative to clinical studies or *in vitro* testing. While their data will never be as valuable as clinical data, pre-clinical studies allow *in vivo* data to be generated more quickly and simply and at lower cost.[50] There is a range of options for evaluating CR formulations in pre-clinical species, including in discovery, where the use of rodents is prevalent. However, the focus of this section is on non-rodent species, due to the desire to perform pre-clinical studies in species that can accommodate dosage forms designed for humans. Selection of animal models for CR dosage forms is not a simple task, due to their differences in GI anatomy and physiology. Over the years, dogs have been used extensively as

Table 8.16 Attributes of key preclinical species for *in vivo* PK studies.

Species	Pros (+)	Cons (−)
Beagle dogs	- Relatively easy to work with - Can accommodate relatively large dosage form sizes - Quite similar GI anatomy to humans	- Longer gastric residence time than humans in fed state - Relatively short transit time for small intestine, which may limit release rate differentiation - Upper limit to absolute dose, due to relatively low body-weight of the animals (~10 kg)
Mini-pigs	- Similar GI transit time to humans (except for stomach) - Generally no upper dose limit, due to body-weight of the animals (~30–40 kg) - Can accommodate relatively large dosage form size	- Much longer gastric residence time than humans, especially in the case of larger dosage forms
Monkeys (non-human primates)	- Similar small intestinal transit time to humans	- Upper limit to absolute dose, due to relatively low body-weight of the animals (~5–7 kg) - Relatively small dosage form size required

a model for oral drug absorption, because of their similar gastric anatomy to humans.[50] However, each species has key advantages and disadvantages, with Table 8.16 comparing them for oral CR formulation evaluation.

The information in Table 8.16 should be taken into consideration, on a case-by-case basis, when selecting a species for the *in vivo* evaluation of CR formulations.

8.5.3 Clinical Evaluation

Available *in vitro* methods and animal models are generally not able to perfectly mimic the human *in vivo* situation, considering the nature of the GI tract, the factors affecting its activity and the various mechanisms employed to achieve CR. Specifically, *in vivo* drug absorption from dosage forms is known to be dependent on many factors other than dissolution, such as transit time, permeability, solubility, luminal content, metabolism and chemical stability in the GI tract. Adequate clinical strategies, like those discussed in the next sections, are hence required.[50]

8.5.3.1 *Identifying CR Issues/Opportunities Without Formulation Development*

Successful CR formulation development generally requires (i) an adequate window for drug absorption, (ii) API properties amenable to CR development and (iii) meaningful dose and release profile targets. For new chemical entities (NCEs) not yet evaluated in humans, the dose and release profile are

Table 8.17 Technology/dosing options, and their assessment, for different PK/PD strategies.

PK/PD strategy	Technology/dosing options	Options assessment
Use of diagnostic tools	- Enterion™ (Section 8.4.5.2)	- Can deliver liquids or solids to different regions of the GI tract, to map absorption profile, without the need for prototype formulation development - Device can only release entire payload, as per remote trigger
	- IntelliCap® (Section 8.4.5.2)	- Can deliver liquids or solids to different regions of the GI tract, to map absorption profile, without the need for prototype formulation development - Device can provide any desired release profile, including those not feasible by available formulation technologies
Mimicking CR profile *via* IR formulation	- Frequent dosing of low dose IR formulation (*i.e.* every few hours)	- Quite cheap, simple and flexible option to ascertain whether CR drug delivery could yield desired outcome - Lower GI tract contribution to absorption not probed, since repeat administration results in dose to be absorbed primarily in the upper GI
Modular CR prototype formulation	- Osmotic capsule (Section 8.4.4.1)	- Rapid evaluation of CR drug delivery with flexibility and low API requirements, assuming ready access to range of modular components - Of particular interest for poorly soluble APIs, especially for bridging to an osmotic tablet for commercialisation - Technology is most likely to be made available commercially

particularly difficult to define *a priori*. Changes in dose strengths are thereby a particular concern for CR formulations, since they can often not be readily adjusted without affecting release performance (especially for monolithic dosage forms). When pre-clinical data indicate the need for CR, it may be advisable to quantify CR risks or opportunities clinically without costly and time-consuming product development and supply. Key options for such a strategy are summarised in Table 8.17.

Key downsides of deferred CR formulation development are that this may delay the market entry of the commercial formulation, especially if there are unexpected development challenges.

8.5.3.2   Early Investment in CR Development

Early investment strategies have the advantage that development challenges are identified early, which may represent the fastest path to market. Early investment could be limited to fit-for-purpose formulation development, for fast-tracked evaluation in clinical PK studies, or involve more lengthy development, to ensure that the formulations have met robustness acceptance criteria. There is no "right" or "wrong" strategy, as it depends on the compound and the risk *versus* opportunity profile. However, in order to meet specific *in vivo* performance criteria, parallel evaluation of different *in vitro* release rates is established practice for clinical evaluation of ER formulations.

For lifecycle management (LCM) projects, early investment in CR development is commonly adopted, since drug product development activities are typically on the critical path and the physico-chemical and biopharmaceutical characteristics of the compound are usually well-known. For NCE projects, the risk profile is quite different, especially early in their development, due to the API knowledge gaps at this stage, including target dose (Section 8.5.3.1). The author had first-hand experience of a CR formulation of an NCE that could no longer be used during Phase 1, due to the lower than predicted dose range in first-time-in-humans (FTIH) trials. The known unknowns, commonplace at this development stage, can hence be a considerable strategic risk.

8.5.3.3 Formulation Bridging Strategies

Clinical bridging studies are a common feature of product development, whether for IR or CR formulations. However, *in vitro* screening for CR formulation selection is generally more challenging, especially in the case of ER applications, since drug release is affected by many factors, the extent of which can vary significantly between formulation types. It is hence particularly important for CR formulations to pursue well-considered bridging strategies. The most common clinical bridging strategy involves PK studies, whether they are (i) bioavailability (BA) or relative bioavailability (RBA) studies, primarily designed to provide directional information, or (ii) formal BE studies, which have strict acceptance criteria and are typically required for registration. BE studies are generally required when developing generic versions of a product or when bridging formulations to pivotal studies, during late stage development or post approval.

Similar to early investment strategies (Section 8.5.3.2), it is generally advisable to evaluate multiple formulations in clinical studies, typically with different release profiles/mechanisms, and to use appropriate *in vitro* screening for clinical formulation selection. Clinical PK studies can range from standard cross-over designs to sequential study arms evaluating candidates chosen from a registered design space. The latter can allow dosing and data analysis cycles with total durations of only 1–2 weeks between study periods. This is particularly interesting when the *in vivo* performance of a reference product needs to be matched or when there are many formulation parameters with the potential to influence drug release.

To minimise or eliminate the need for clinical studies, the possibility exists to develop an IVIVC. If successful, an IVIVC generally allows formulation modifications to be justified on the basis of *in vitro* data alone. Caveats are that this generally only applies to the established dissolution range of the IVIVC, and if the mechanism of release has remained broadly unchanged. Furthermore, establishing an IVIVC is not trivial, due to strict criteria, and there seem to have been historically more failures than successes.

Acknowledgements

The significant contributions by Dr Abdenour Djemai (GlaxoSmithKline, UK) to the authoring of Section 8.4.3.1.6 (Mini-tablets/Mini-matrices) are hereby gratefully acknowledged.

References

1. C. Seiler, *Chem. Eng.*, 2014, **877/78**, 39.
2. A. Nokhodchi, S. Raja, P. Patel and K. Asare-Addo, *BioImpacts*, 2012, **2**(4), 175.
3. I. Wilding, *Consultation for Merck Sharp & Dohme on a Multiparticulate Controlled Release Program*, 2012.
4. H. Patel, D. R. Panchal, U. Patel, T. Brahmbhatt and M. Suthar, *J. Pharm. Sci. Biosci. Res.*, 2011, **1**(3), 143.
5. J. Kristensen and V. W. Hansen, *AAPS PharmSciTech*, 2006, **7**(1), 22.
6. L. Gu, C. V. Liew and P. W. S. Heng, *Drug Dev. Ind. Pharm.*, 2004, **30**(2), 111.
7. C. Vervaet, L. Baert and P. J. Remon, *Int. J. Pharm.*, 1995, **116**, 131.
8. N. V. Dhandapani, A. Shrestha, N. Shrestha, A. Thapa, G. Sandip and R. S. Bhattarai, *All Res. J. Biol.*, 2012, **3**, 10.
9. J. M. Newton, in *Pharmaceutical Dosage Forms: Tablets*, ed. L. L. Augsburger and S. W. Hoag, CRC Press, Boca Raton, Florida, 3rd edn, 2008, vol. 1, ch. 10, p. 337.
10. M. M. Crowley, F. Zhang, M. A. Repka, S. Thumma, S. B. Upadhye, S. K. Battu, J. W. McGinity and C. Martin, *Drug Dev. Ind. Pharm.*, 2007, **33**(9), 909.
11. M. Wilson, M. A. Williams, D. S. Jones and G. P. Andrews, *Ther. Delivery*, 2012, **3**(6), 787.
12. M. A. Repka, S. Majumdar, S. K. Battu, R. B. Srirangam and S. B. Upadhye, *Expert Opin. Drug Delivery*, 2008, **5**(12), 1357.
13. D. Treffer and S. Schrank, *Pharmaceutical Solid State Research Cluster (PSSRC)*, 2013, website: http://www.pssrc.org/news/89.
14. A. Körner, L. Piculell, F. Iselau, B. Wittgren and A. Larsson, *Molecules*, 2009, **14**, 2699.
15. The Dow Chemical Company, 2000, http://www.colorcon.com/literature/marketing/mr/Extended%20Release/METHOCEL/English/hydroph_matrix_broch.pdf.
16. S. B. Tiwari and A. R. Rajabi-Siahboomi, *Pharm. Technol. Eur.*, 2008, **20**(9), http://www.pharmtech.com/modulation-drug-release-hydrophilic-matrices.

17. C. Seiler, R. Elkes, M. S. Beauchamp, S. R. Pygall, K. Bradley, A. Djemai, D. A. Thompson and S. Fitzpatrick, *38th Controlled Release Society Annual Meeting and Exposition*, National Harbor, Maryland, USA, 30 JUL–03 AUG 2011.

18. R. T. C. Ju, P. R. Nixon, M. V. Patel and D. M. Tong, *J. Pharm. Sci.*, 1995, **84**, 1464.

19. K. Hughes, *Colorcon Modified Release Forum*, London, UK, 21–24 APR 2010.

20. R. Steendam, *Amylodextrin and Poly(DL-lactide) Oral Controlled Release Matrix Tablets. Concepts for Understanding Their Release Mechanisms*, University of Groningen, 2005, ch. 2.

21. S. Mascia, C. Seiler, S. Fitzpatrick and D. I. Wilson, *Int. J. Pharm.*, 2010, **389**, 1.

22. Ashland, *Slides provided as Part of On-site Training*, 2010.

23. D. Ma, A. Djemai, C. M. Gendron, H. Xi, M. Smith, J. Kogan and L. Li, *Drug Dev. Ind. Pharm.*, 2013, **39**(7), 1070.

24. M. Ghimire, *Colorcon Modified Release Forum*, London, UK, 21–24 APR 2010.

25. S. Abdul, A. V. Chandewar and S. B. Jaiswal, *J. Controlled Release*, 2010, **147**, 2.

26. F. Siepmann and J. Siepmann, *Int. J. Pharm.*, 2013, **457**, 437.

27. M. Marucci, G. Ragnarsson and A. Axelsson, *Int. J. Pharm.*, 2007, **336**, 67.

28. F. Siepmann, A. Hoffmann, B. Leclercq, B. Carlin and J. Siepmann, *J. Controlled Release*, 2007, **119**, 182.

29. M. Marucci, J. Hjärtstam, G. Ragnarsson, F. Iselau and A. Axelsson, *J. Controlled Release*, 2009, **136**, 206.

30. J. Siepmann, *Consultation for Merck Sharp & Dohme (MSD)*, 2010.

31. D. Sauer, M. Cerea, J. DiNunzio and J. McGinity, *Int. J. Pharm.*, 2013, **457**(2), 488.

32. B. Gandhi and J. Baheti, *Int. J. Pharm. Chem. Sci.*, 2013, **2**(3), 1620.

33. C. J. Young, K. A. Fegerly and A. Rajabi-Siahboomi, *Reprint of Poster Presented at AAPS Annual Meeting and Exposition*, San Antonio, USA, 29 OCT–02 NOV 2006.

34. R. Bodmeier, *Eur. J. Pharm. Biopharm.*, 1997, **43**, 1.

35. B. E. Jones, *Int. J. Pharm.*, 2001, **227**, 5.

36. F. Podzceck, in *Pharmaceutical Capsules*, ed. F. Podzceck and B. E. Jones, Pharmaceutical Press, 2nd edn, 2004, ch. 6, p. 119.

37. A. V. Gothoskar and M. D. Phale, *Pharm. Technol.*, 2011, **35**, 7.

38. K. Lehmann, H. U. Petereit and D. Dreher, *Pharm. Ind.*, 1993, **55**(10), 940.

39. W. B. Caldwell, *40th Controlled Release Society Annual Meeting and Exposition*, Honolulu, Hawaii, USA, 21–24 JUL 2013.

40. R. K. Verma and S. Garg, *Eur. J. Pharm. Biopharm.*, 2004, **57**, 513.

41. Colorcon, *Slides provided as Part of On-site Training*, 2011.

42. W. B. Caldwell, R. J. Wald, C. D. Craig, C. L. Hostetler, D. G. Koehler-King, M. Milewski, S. R. Pygall, J. C. Mann and C. Seiler, *40th Controlled Release Society Annual Meeting and Exposition*, Honolulu, Hawaii, USA, 21–24 JUL 2013.

43. R. K. Verma, D. M. Krishna and S. Garg, *J. Controlled Release*, 2002, **79**, 7.

44. K. C. Waterman, G. S. Goeken, S. Konagurthu, M. D. Likar, B. C. Mac-Donald, N. Mahajan and V. Swaminathan, *J. Controlled Release*, 2011, **152**, 264.

45. A. Streubel, J. Siepmann and R. Bodmeier, *Expert Opin. Drug Delivery*, 2006, **3**(2), 217.

46. I. Wilding, *Drug Delivery Technol.*, 2001, **1**(1).

47. D. V. Prior, A. L. Connor and I. R. Wilding, in *Modified-release Drug Delivery Technology*, ed. M. J. Rathbone, J. Hadgraft and M. S. Roberts, Marcel Dekker, Inc., New York, 2002, ch. 24, p. 273.

48. E. Söderlind, B. Abrahamsson, F. Erlandsson, C. Wanke, V. Iordanov and C. von Corswant, *J. Controlled Release*, 2015, **217**, 300.

49. M. Koziolek, M. Grimm, D. Becker, V. Iordanov, H. Zou, J. Shimizu, C. Wanke, G. Garbacz and W. Weitschies, *J. Pharm. Sci.*, 2015, **104**(9), 2855.

50. Y. Qiu and G. Zhang, in *Handbook of Pharmaceutical Controlled Release Technology*, ed. D. L. Wise, Marcel Dekker, Inc., New York, 2000, ch. 23, p. 465.

51. TNO, *TIM Gastrointestinal Systems*, 2013.

52. V. Dias, V. Ambudkar, P. Vernekar, R. Steffenino and A. Rajabi-Siahboomi, *36th Controlled Release Society Annual Meeting and Exposition*, Copenhagen, Denmark, 18–22 JUL 2009.

53. S. Shelukar, J. Ho, J. Zega, E. Roland, N. Yeh, D. Quiram, A. Nole, A. Katdare and S. Reynolds, *Powder Technol.*, 2000, **110**, 29.

54. H. D. Williams, K. P. Nott, D. A. Barrett, R. Ward, I. J. Hardy and C. D. Melia, *J. Pharm. Sci.*, 2011, **100**, 4823.

55. Ashland Specialty Ingredients, *From Matrix to Film Coating - Your Full-service Pharmaceutical Technology Resource*, 2017.

56. Colorcon, *POLYOX™ Application Data*, 2009.

57. Y. Qiu and P. I. Lee, in *Developing Solid Oral Dosage Forms - Pharmaceutical Theory & Practice*, ed. Y. Qiu, Y. Chen, G. Zhang, L. Yu and R. V. Mantri, Academic Press, 2nd edn, 2016, ch. 19, p. 519.

58. Ashland Specialty Ingredients, *Natrosol™ Hydroxyethylcellulose - Selection Guide for Paints and Coatings*, 2017.

59. B. Prajapati and H. Solanki, *Internet J. Third World Med.*, 2008, **8**(1), 1–7.

60. C. Martin, A. De Baerdemaeker, J. Poelaert, A. Madder, R. Hoogenboom and S. Ballet, *Mater. Today*, 2016, **19**(9), 491–502.

61. Pharmacircle, *Overview of Flamel's Micropump-technology*, 2016.

62. Intec Pharma, *Accordion Pill™ Technology Information Material Presented/provided*.

63. Depomed-website, *Acuform® Technology Summary*, 2016.

64. Pharmacircle, *Overview of Rubicon's RubiReten®-technology*, 2016.

65. Rubicon, *RubiReten® Technology Information Material provided*, 2010.

66. K. A. Rao, E. Yazaki, D. F. Evans and R. Carbon, *Br. J. Sports Med.*, 2004, **38**, 482.

67. Röhm GmbH & Co. KG (now Evonik), *Specifications and Test Methods for EUDRAGIT® RL 12,5 and EUDRAGIT® RL 100 ; EUDRAGIT® RS 12,5 and EUDRAGIT ® RS 100*, 1999.

68. Lubrizol Corporation, *Carbopol® Polymers for Controlled Release Matrix Tablets*, 2008.
69. K. C. Sung, P. R. Nixon, J. W. Skoug, T. R. Ju, P. Gao, E. M. Topp and M. V. Patel, *Int. J. Pharm.*, 1996, **142**, 53.
70. H. D. Williams, R. Ward, A. Culy, I. J. Hardy and C. D. Melia, *Int. J. Pharm.*, 2010, **401**, 51.
71. I. Caraballo, *Expert Opin. Drug Delivery*, 2010, 7(11), 129.
72. Kos Industry, 2016, website: http://www.kosindustry.com/how-to-choose-the-right-multi-tip-tablet-tooling.html.
73. Glatt GmbH, 2012, https://www.glatt.com/fileadmin/user_upload/content/pdf_downloads/GPCG_10_e.pdf.
74. Z. Lian, A. Bell, D. Tewari, C. Seiler, A. Djemai and T. Dürig, *40th Controlled Release Society Annual Meeting and Exposition*, Honolulu, Hawaii, USA, 21–24 JUL 2013.

CHAPTER 9

Less Common Dosage Forms

STEPHEN WICKS

School of Pharmaceutical Chemical and Environmental Sciences, University of Greenwich, UK
*E-mail: s.r.wicks@greenwich.ac.uk

9.1 Introduction

Taking drugs by mouth is convenient for most patients. In 2012, 7468 oral products were marketed, or approved for marketing.[1] Injectables accounted for approximately 3000 products; topicals 1200. Other commercially important but less common dosage forms, mentioned in this report, included ophthalmic, inhalation and transdermal drug delivery systems. This chapter will address the fundamental formulation principles involved in the production of these more commercially important dosage forms; is not intended to be an exhaustive treatment of the science underpinning these principles. The goal is to introduce the reader to the critical technical issues, thus providing a starting point for the realisation of dosage form products by the pharmaceutical scientist alone, or in partnership with specialist development service-providers.

9.2 Why Are Alternative Dosage Forms Important?

The three main drivers for the development of non-oral dosage forms are: overcoming low oral availability, achieving potency with greater selectivity and greater convenience to the patient.

Drug Discovery Series No. 64
Pharmaceutical Formulation: The Science and Technology of Dosage Forms
Edited by Geoffrey D. Tovey

The biophysical properties required for a drug molecule to be absorbed from the gastrointestinal tract have been codified in the Biopharmaceutics Classification System.[2] A drug must be sufficiently soluble and permeable, relative to the therapeutic dose, to be absorbed from the gut. Overlaid upon this is the requirement for the drug to remain chemically stable, resisting chemical and microbiological degradation, in the gut lumen and wall. Also, the drug molecule should not be extensively metabolised in its first pass through the liver or expelled from gut wall cells by p-glycoprotein and/or other efflux mechanisms. Drugs that cannot meet the biophysical requirements for oral delivery may still be effective therapeutic agents when delivered directly to the target organ.

Given orally, for systemic rather than local therapy, drugs will be widely distributed in the body and are therefore designed to be both potent and selective thus avoiding side-effects and toxicity. Drugs applied directly to the target organ can be used at a lower dose and systemic exposure can be reduced or eliminated. The failure to develop safe or effective orally acting drugs for the treatment of mild to moderate lung disease arises from the superior efficacy and selectivity of drugs when given by inhalation.

Non-oral dosage forms can also provide greater convenience to the patient. Transdermal patches enable the delivery of drugs for hormone replacement therapies conveniently over a period of several days. β_2 Adrenergic receptor agonists, given by inhaler, are a discrete, effective and rapidly-acting 'rescue therapy' for the treatment of acute asthma attacks.

Modern drug development programmes generally aim to achieve oral delivery by conventional tablets or capsules but this is not always possible. Many clinically and commercially important drugs, unsuitable for oral therapy, have been developed in less common dosage forms. To avoid discarding potentially valuable therapeutic agents, the pharmaceutical scientist must remain alert to the possibility that certain drug molecules may best be delivered by non-oral means.

9.3 Alternative Dosage Form Formulation Design and Control

9.3.1 Drug Delivery in and *via* the Skin

This section aims to explain the basic principles of the design of transdermal drug delivery systems. This is a widely-researched subject with a rich body of literature.

A limited number of drugs are commercially available in transdermal drug delivery systems capable of delivering drugs to the systemic circulation. The skin is a low-capacity portal and so drugs intended for targeting to the systemic circulation must be highly potent with a high affinity for skin tissue.

Creams, ointments and lotions are ubiquitous amongst pharmaceutical and cosmetic preparations for the local delivery of drugs to the surface layers of the skin for 'conditioning' or the treatment of disease. Non-steroidal anti-inflammatory drugs, normally orally available, are also administered in topical preparations for the local treatment of inflammation and pain in muscle and soft tissue.

It is difficult to predict, *a priori*, whether the biophysical properties of a drug will be likely to enable transport through the skin, and so the feasibility of the transdermal route is estimated by the permeability coefficient, which is obtained using experimental designs based on the mathematics of diffusion.

The transdermal drug diffusion rate, per unit area, through skin is proportional to the concentration gradient of the drug across the skin. The transport process was described mathematically by Fick in 1855:

$$F = -D\partial C/\partial x, \tag{9.1}$$

where F is the rate of drug transfer, per unit area of skin, C is the concentration of diffusing substance in the surface layer of the skin and x is the effective path length of drug diffusion. D, Fick's proportionality constant, is known as the diffusion coefficient which is a complex rate parameter. D can be determined either experimentally or with the use of mathematical models. Fick's equation is a partial differential equation which can be solved, analytically, for specific diffusion geometries and boundary conditions. More complex situations can be solved by computer using numerical methods.

Practically, the rate of diffusion can be determined using a diffusion cell, the commonest being the Franz cell (Figure 9.1).

A test membrane is clamped between the flanges of the Franz cell donor and receptor half-cells. The donor compartment contains either a simple drug solution or a more complex formulation vehicle; the receptor compartment contains a solvent system capable of dissolving the diffusing drug. The liquid or semisolid solvents in the half-cells should not alter the physical or chemical structure of the test membrane.

The diffusion experiment commences by charging a concentrated drug solution into the donor compartment; the membrane and receptor solution being already in place and drug free. Samples are then removed from the receptor compartment, at various times, and the drug mass diffusing through the membrane is determined using a suitably sensitive and selective analytical method. The experiment is maintained under so-called 'sink conditions'. This means that the drug concentration gradient across the membrane remains effectively constant. To achieve this, the donor compartment drug concentration should remain practically unchanged and no significant accumulation of drug in the receptor compartment should be allowed.

Excised human skin is frequently used as the test membrane in Franz Cell experiments. The calculation of the diffusion coefficient reduces the complex structure of skin to that of an isotropic membrane. This approach therefore produces an aggregate flux rate for the full thickness skin controlled by rate-determining anatomical layers in the skin, mainly the stratum corneum.

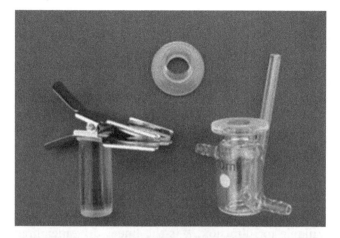

Figure 9.1 The Franz cell. The top image shows the component parts: the receptor half-cell and sampling port (lower right); the flange clip and support (lower left) and the donor half-cell. The lower image shows the assembled cell. The membrane is compressed between the donor and receptor cell flanges and clipped. Reproduced with the permission of Soham Scientific.

The mass of drug appearing in the receptor compartment can be plotted against elapsed time and the analytical solution to Fick's equation can be used to obtain a numerical estimate of the diffusion coefficient. The simplified solution of Fick's equation for the boundary conditions described above is given in eqn (9.2).

$$Q_t = \frac{DC_1}{l}\left(t - \frac{l^2}{6D}\right) \tag{9.2}$$

where Q_t is the mass of drug diffusing through the membrane, per unit area, at time, t. C_1 is the concentration within the surface layer of the membrane on the donor side, D is the diffusion coefficient and l is the membrane thickness. Eqn (9.2) has to be further modified for experimental use because it is usually not possible to measure, or describe, the concentration of drug in the surface layer of the membrane. Surface concentration is approximated by the drug–membrane sorption constant which is a quasi-partition coefficient. This is obtained in a preliminary experiment to quantify the ratio of drug concentration in the membrane and the donor compartment solution or formulation vehicle at equilibrium. A surface area term is also added resulting in eqn (9.3):

$$F_t = \frac{DAKC_{\text{Donor}}}{l}\left(t - \frac{l^2}{6D}\right) \tag{9.3}$$

where F_t is the mass of drug diffusing as a function of time, t. A is the surface area available for diffusion, K is the linear drug-membrane sorption constant, and C_{Donor} is the drug concentration in the vehicle contained in the donor half-cell. Experiments conducted under these conditions produce plots like those shown in Figure 9.2 for the diffusion of clebopride in rat skin. Eqn (9.3) describes the shape of the plots, mathematically.

Figure 9.2 and eqn (9.3) describe a pattern of drug permeation where a non-steady-state lag-time occurs due to the initial diffusion of drug in the membrane. A finite time elapses before drug appears in the receptor compartment. As diffusion proceeds, a steady-state transport rate is established where the mass of drug diffusing becomes linear with respect to time. The lag-time is obtained by extrapolating the linear steady-state portion of the curve to an intercept on the time axis. It follows from eqn (9.3), that:

$$L = \frac{l^2}{6D} \tag{9.4}$$

when L is the intercept, *i.e.* lag-time, on the time axis, and is a mathematical statement of the time of onset of systemic drug effects.

The slope of the steady-state portion of the curve corresponds to:

$$\frac{DAKC_{\text{Donor}}}{l} \tag{9.5}$$

which is also obtained from eqn (9.3). This is a mathematical statement of the transmembrane flux.

Eqn (9.4) and (9.5) are useful in visualising the key principles of transdermal formulation design. Considering eqn (9.5), the formulator is in engineering control of two variables: the drug concentration presented to the skin in the formulation vehicle or patch, and the area of skin across which diffusion takes place. Drug must be presented to the skin in a sufficiently high concentration to establish a concentration gradient that promotes the diffusion of drug through the skin at a clinically relevant rate.

Permeation will only be maintained, at a constant rate after the lag-time, if the concentration gradient across the skin remains essentially

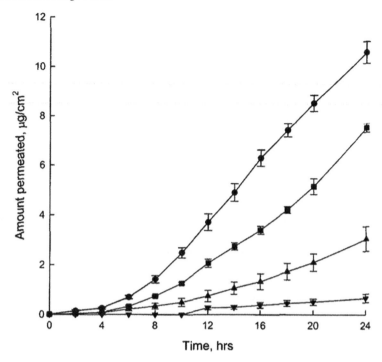

Figure 9.2   Permeation of clebopride through excised rat dorsal skins from various vehicles.[3] The plots show the characteristic lag-time followed by steady-state drug permeation rate. Reproduced from *Archives of Pharmaceutical Research*, Effects of vehicles and enhancers on transdermal delivery of clebopride, **30**, 2007, 1155, Y. S. Rhee, J. Y. Huh. C. W. Park, T. Y. Nam, K. R. Yoon, S. C. Chi and E. S. Park. © The Pharmaceutical Society of Korea 2007, with permission of Springer.

constant throughout the intended duration of dosing. The drug concentration at the skin surface must not substantially decrease during the dosing period if a constant infusion rate is to be maintained. This means that, unlike oral dosage forms, unused drug must remain in the delivery system after dosing. The Estraderm MX 100® patch, containing the hormone oestradiol hemihydrate, is applied to the skin with the aim of delivering an 'infusion' of drug over a 3–4 day period. The patch contains 3.0 mg of oestradiol hemihydrate and releases 100 μg of drug per day. 90% of the drug in the patch therefore remains unused at the end of the 3 day application period. Clinically, care must be taken to safely dispose of delivery systems after use. Commercially, this method of drug delivery may be considered uneconomic for expensive drugs where much of the payload is redundant.

From eqn (9.5), the drug infusion rate is also proportional to the patch area. Practical limitations on patch size must be considered in relation to the clinical indication. Generally, patches intended to deliver drugs systemically are quite small for reasons of convenience, comfort and discretion. Patches delivering non-steroidal anti-inflammatory drugs to larger areas of soft tissue and muscle are supplied in larger, adhesive sheets that can be cut to the size of the affected area.

The formulator is not in control of three variables in eqn (9.5): the diffusion coefficient; the sorption constant and skin thickness. The two former variables depend upon the physicochemical properties of the drug molecule and skin structure; the latter is determined anatomically.

The product of the sorption constant and diffusion coefficient is often used to describe the rate of drug permeation. This is referred to as the permeability coefficient, P.

$$P = K \times D \tag{9.6}$$

Drugs intended for delivery through the skin must have a high permeability coefficient, which reflects a high affinity for skin tissue and a suitable molecular volume, relative to skin pathways, to be able to penetrate the various anatomical layers of the skin. From a pharmacokinetic standpoint, the rate of diffusion through the skin must provide a high drug input rate, relative to drug clearance, ensuring that clinically effective blood or tissue levels result. Transdermal delivery feasibility is therefore a serendipitous finding rather than a matter of engineering design.

As of 2008, 19 transdermal drug delivery systems had been developed for commercial use, based on 18 drugs.[4] The physicochemical properties of the drugs cited are diverse and the list is limited. Whilst there is much commercial enthusiasm to exploit the transdermal route, few drug molecules have proven suitable for delivery this way.

Recognising the ethical challenge of obtaining donor human skin, research effort has been expended to identify membranes, for use in Franz cell experiments, that simulate the permeability characteristics of viable human skin.[5] Excised full-thickness human cadaver skin remains the membrane of choice, however animal skins and synthetic polymeric membranes have been proposed as valid alternatives. Additionally, several predictive mathematical models have been developed, based on the mathematics of diffusion and experimental provenance, allowing an estimation of the permeability coefficient on a theoretical basis.[6]

Formulators and researchers have sought ways to increase drug permeability through, and in, skin tissue to expand the use of the dermal route. Many chemical penetration enhancers have been investigated, some with highly vesicant properties. Enhancers in current use in commercial formulations, tend to be simple, non-irritant, non-vesicant molecules, e.g. ethanol and isopropyl palmitate. Iontophoresis has also proven to be commercially successful.[4] Here the chemical concentration gradient is augmented by a continuous low-voltage current that promotes the transport of drug molecules. Charged drugs are promoted by electrophoresis and uncharged drugs by the electroosmotic flow of water.

If a transdermal patch is the intended dosage form, the design must be robust to assure containment of the drug or drug vehicle, using high-quality materials and fabrication techniques. Intimate contact with the skin must be assured using a robust and effective adhesive.

Two types of patch design are common: the matrix patch and the reservoir patch. The matrix patch comprises three layers: an impermeable backing membrane onto which the drug, incorporated into an adhesive, is bonded and a protective release liner that is removed before applying the patch to the skin.

The reservoir patch comprises four layers: an impermeable backing membrane onto which a liquid or semi-solid drug reservoir is bonded that may additionally contain liquid penetration enhancers; a drug-free adhesive layer and release liner. In most cases, release from the patch is instantaneous and the rate of permeability is controlled by resistant structures in the skin. In very exceptional cases, high drug permeation rates through the skin will require the use of a polymeric rate-controlling membrane which is added as an extra layer to the multi-laminate transdermal patch between drug-containing components and the skin surface.

9.4 Ophthalmic Drug Delivery

Drugs intended for the treatment of chronic or acute eye disease can be instilled directly into the target organ. As discussed later, this provides opportunities to use active pharmaceutical ingredients, not suited for oral administration, to treat eye disease.

The antibiotic chloramphenicol can cause blood dyscrasias and aplastic anaemia when given orally and is therefore restricted for use in serious infections. Ophthalmic chloramphenicol is a safe and widely used broad-spectrum ocular antibiotic.[7] Similarly, timolol, a non-selective β-blocker, active on both β_1 and β_2 adrenergic receptors, is used in certain forms of glaucoma. Timolol is hydrophilic and subject to high first-pass liver metabolism.[8,9] Whilst inferior to other drugs in this class for the oral treatment of cardiovascular disease, it is effective at lowering intraocular pressure when instilled into the eye. In both cases, these drugs have become important, widely-used and safe drugs for the treatment of acute and chronic eye disease when given in an appropriate ophthalmic dosage form.

This section will deal with ophthalmic products designed to treat the more common infections and diseases of the anterior eye using non-oral dosage forms for application to the conjunctiva. Diseases such as age-related macular degeneration and diabetic retinopathy require more invasive intraocular drug delivery approaches which will not be discussed here.

9.4.1 Formulation Design and Controls for Ophthalmic Drug Delivery

Drug delivery directly to the eye is achieved by means of sterile liquids (solutions and suspensions), semi-solids (ointment) or solid objects that can be placed on the surface of the conjunctiva, or instilled into the conjunctival sac.

The simple eye-drop accounts for about 70% of the ophthalmic drug product market.[10] The ability to formulate a simple eye-drop depends upon the solubility of the drug in water relative to the dose or effective concentration. Unlike other solubility problems in pharmaceutical formulation, options to improve solubility are limited by the biocompatibility of conventional, pharmaceutically acceptable solubilising excipients, *e.g.* surfactants and water-miscible cosolvents. pH can be modified with the aid of dilute, low-capacity buffers, which can be used to increase the solubility of ionisable drugs. This type of buffer can be formulated outside of the normal conjunctival pH range because they can be readily overwhelmed by the buffering effect of tears, following instillation, thus avoiding pain and irritation. Derivatised cyclodextrins are an important new biocompatible technology able to solubilise hydrophobic drugs intended for ophthalmic dosing. Their use will be discussed in Section 9.4.2.

Eye-drops are *inefficient* drug delivery vehicles. Much of the administered dose is lost due to blinking or flooding by tears and typically only 7 µl (20%) of the instilled dose is absorbed. Many eye-drop formulations, therefore, include polymeric excipients that increase the viscosity of the formulation vehicle with the aim of increasing the time retained in the conjunctival sac. Direct instillation onto the surface of the eye results in diffusion through the cornea and conjunctiva. The kinetics of drug permeation will be determined by diffusion and are therefore the same as those discussed in Section 9.3.1. Drugs must be selected with suitable diffusion coefficients and sorption constants and be applied to the eye in a suitably high concentration to be effective. The residence time is critical in ensuring that the drug has sufficient time to partition into the surface layers of the anterior eye.

If the drug solubility in water is low, physically stable aqueous suspensions can be used as an alternative. Suspended particles are typically retained in the conjunctival sac for longer periods than drug administered in solution. Suspensions are dispersions of finely divided insoluble drug particles in water, or another suitably formulated vehicle, involving a suspending system comprising a viscosity-raising polymer and a surface-active dispersing agent, typically a non-ionic surfactant. Drug particle size is therefore a critical determinant of the duration of drug action, physical toleration (irritation due to 'grittiness') and dose uniformity. Dose uniformity in multiple dose preparations depends upon the quality of the suspension with respect to deflocculation and, hence, its resistance to excessively rapid settling.[10]

Ophthalmic ointments are the third conventional formulation widely used to treat the eye. Here, drug is dispersed in a semisolid base, typically mixtures of paraffin fractions, which melt at physiological temperatures (34–37 °C). Ointments improve the residence time of the drug in the conjunctival sac and can, under certain circumstances, sustain the release of drug.

Excipients used in the formulation of the three conventional ophthalmic drug products must be chosen to avoid irritation, inflammation and, where possible, interference with vision.[11] Eye preparations are produced using materials and methods designed to assure sterility and to avoid microbial contamination. Ultimately, eye preparations must be manufactured to comply with a test for sterility and particulate contamination. Antimicrobial preservatives are added to formulations supplied in multiple-dose containers. The type and concentration of the antimicrobial preservative must be justified and its efficacy demonstrated in a pharmacopoeial challenge test. Single-dose presentations are indicated when the use of an antimicrobial preservative is not possible, *e.g.* drug delivery to the severely injured eye. Single-dose presentations must be sterile but need not be preserved.

9.4.2 Derivatised Cyclodextrins and Ophthalmic Dosage Forms

Within the last 25 years, two cyclodextrin derivatives have become commercially important: hydroxypropyl β-cyclodextrin and sulphobutylether β-cyclodextrin. These materials are freely soluble in water, unlike the underivatised parent cyclodextrins, and can increase the apparent solubility of low-water-solubility drugs in aqueous solution, *e.g.* ophthalmic and parenteral formulations. This technology provides the basis to produce safe and effective pharmaceutical formulations without recourse to irritant surfactants or water-miscible cosolvents.[12]

Solubility is increased by an inclusion–complexation mechanism, which is a dynamic equilibrium between the entire drug, or one or more of its hydrophobic functional groups, and the hydrophobic interior of the cyclic cyclodextrin molecule when both are dissolved in aqueous solution. With the hydrophobic moiety 'encapsulated', the hydrophilic exterior of the cyclodextrin molecule facilitates the solubilisation of the drug in the aqueous vehicle. It has also been demonstrated that sequestration of a chemically labile moiety of the drug can also improve chemical stability and extend shelf-life. Derivatised cyclodextrins are safe and well tolerated.

The critical solubilisation concentration is important economically and can be determined by preparing aqueous solutions of increasing cyclodextrin concentration and measuring any proportionate increase in apparent drug solubility.[13]

The apparent drug solubility of a drug in the presence of a cyclodextrin forming a 1:1 inclusion complex can be described by:[13]

$$S_{\text{total}} = S_0 + \frac{KS_0[\text{CD}_{\text{total}}]}{KS_0 + 1} \tag{9.7}$$

where S_{total} refers to the apparent drug solubility; S_0 the intrinsic solubility of the drug, *i.e.* in the absence of the cyclodextrin and K is the 1:1 equilibrium constant for drug–cyclodextrin inclusion complexation. Whilst the 1:1 interaction is the simplest case; higher order interactions are possible. Eqn (9.7) serves to illustrate the general principles of the use of derivatised cyclodextrins as solubility enhancers; more complex mathematical relationships can be obtained for higher order interactions if needed.

Eqn (9.7), describes an initial linear increase of apparent drug solubility with increasing cyclodextrin concentration, $[CD_{total}]$. The increase in solubility, in the presence of the cyclodextrin, is determined by the magnitude of the intrinsic solubility in the aqueous solution in the absence of the derivatised cyclodextrin (S_0), K, a measure of drug–cyclodextrin affinity and the total cyclodextrin concentration in the solution. The greater the value of the stability constant K, the more effective the cyclodextrin will be at improving the solubility of the target drug. If the intrinsic solubility, S_0, can also be increased, then apparent solubility will increase proportionately, *e.g.* increasing solubility by ionisation effects or the addition of water-miscible cosolvents, provided this can be tolerated. Manoeuvres designed to increase S_0 may however reduce K by disturbing the inclusion equilibrium. It is therefore important to balance any detrimental effect of intrinsic solubility enhancement on the drug during cyclodextrin formulation optimisation exercises.[14]

9.5 Drugs Given by Inhalation

Drugs given for the treatment of airways disease, *e.g.* asthma and chronic obstructive pulmonary disease, can be delivered directly to the affected lung by inhalation. This has two very important advantages in common with the ophthalmic cases (see above). Firstly, the mass of drug delivered to provide a clinically effective unit dose is significantly smaller than that needed to be given by oral administration. Secondly, side effects are reduced because the systemic exposure is lower. Salbutamol, a short-acting β_2-agonist, has been given in tablet form at a unit dose of 2 mg; the corresponding unit dose from an inhalation device is 200 µg.

Inhalation drug delivery is a complex area of pharmaceutical science, involving both pharmaceutical scientist and engineer technical inputs. This section aims to introduce key technical issues from the standpoint of the generalist formulator.

The most widely used drugs given by inhalation are: β_2-agonists, steroids and anticholinergics for the treatment of asthma and chronic obstructive pulmonary disease (COPD). The inhaled route is intuitive, clinically, and the response to inhaled drugs is both rapid and complete.

Inhaled forms of the non-selective β-agonists adrenaline and isoprenaline became available in the 1950s. The isoprenaline pressurised metered dose inhaler (pMDI) became established as the standard of care and sales of inhalers rose by 600% between 1959 and 1965.[15] Unfortunately, asthma mortality also increased and the safety of inhaled β-agonist therapy was fiercely debated in the 1960s. When salbutamol, a β_2-agonist drug with greater bronchial smooth muscle selectivity over cardiac effects, was introduced in a pMDI formulation, it progressively replaced the pioneering isoprenaline pMDI. A reduction in side-effects and mortality became apparent. Whilst true that the direct administration of intrinsically non-selective drugs from non-oral dosage forms can improve selectivity and reduce side-effects, there are performance limits. Formulators, in partnership with medicinal chemists, should therefore seek continuous potency and selectivity improvement opportunities to produce superlative medicines for delivery by organ-targeting dosage forms.

Improved drug selectivity, using the inhaled route, is further illustrated in steroid asthma therapy. Cortisone, extracted from the adrenal cortex by Edward Kendal in 1936, was also introduced into asthma therapy in the 1950s. Research into improved steroid molecules continued, and the synthetic steroids, prednisolone and hydrocortisone, were introduced shortly thereafter. Long-term oral steroid therapy was associated with a range of serious side-effects and led to interest in safer modes of administration. In the early 1970s, the inhaled steroid beclomethasone dipropionate was introduced, revolutionising the treatment of asthma and COPD.[15]

In parallel with developments in inhaled β-agonist and steroid therapy, the importance of the cholinergic mechanism was recognised in COPD and acute asthma.[16] Whilst the use of so-called tertiary compounds, *e.g.* atropine, proved to be unsatisfactory, poorly absorbed quaternary compounds, *e.g.* ipratropium and oxitropium bromides, proved to be both safe and effective when administered by inhalation.

In the late 1980s, the discovery and development of salmeterol heralded the introduction of the first long-acting β$_2$-agonist. The later combination dosage form of salmeterol and fluticasone, a more potent inhaled steroid, is now an important product in the treatment of airways disease. The discovery and development of tiotropium bromide resulted in the production of a therapeutically and commercially important long-acting bronchodilator inhalation therapy based on the cholinergic mechanism.[16]

The administration of drugs *via* the lung, not intended for the treatment of airways disease, has been the focus of much research in the last 40 years. The lung has been proposed as a portal for the systemic delivery of proteins, peptides and other poorly orally absorbed drugs as an alternative to injections. Notwithstanding the academic interest, few therapeutically and commercially important drug products have been produced.[17] The disappointingly brief appearance and almost immediate withdrawal of inhaled insulin signposts the difficulties of the use of the lung as a general portal for systemic drug delivery.

Many anti-infective drugs, used to treat lung infections, can be delivered efficiently *via* the oral route producing clinically significant concentrations in the lung. In this case, drug delivery by inhalation is unnecessary.

9.5.1 Inhalation Drug Delivery Devices

Formulations for inhalation are generally administered either by a nebuliser or inhaler device. The two most common inhalers are the pMDI and the dry powder inhaler (DPI). Recently, there has been a resurgence of interest in non-pressurised metered dose inhalers, also known as 'soft-mist' inhalers. The soft-mist device is a variant of the nebuliser.

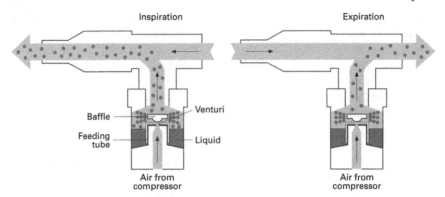

Figure 9.3   Conventional nebuliser design. The images show the input of compressed air into a venturi tube. Oversized droplets impact onto a baffle which is bypassed by the inhalable mist. The mist is either inhaled (left image), or lost to the environment during exhalation. Adapted by permission from BMJ Publishing Group Ltd,[18] *Thorax*, C. O'Callaghan and P. W. Barry, **52**, S31, 1997.

9.5.1.1   The Nebuliser

The nebuliser comprises four parts: an energy source; a disposable nebulisation chamber; a nebuliser liquid (the formulated product) and a disposable mouthpiece or mask. The nebuliser liquid is placed in the nebulisation chamber and aerosolised. The resulting vapour cloud, like that produced by an electronic cigarette, is inhaled *via* a tube leading to the mouthpiece or mask. This is a passive means of inhalation delivery designed to deliver drug over an extended period to a patient breathing normally (Figure 9.3). It is an intensive form of inhalation drug delivery suitable for patients incapable of, or unable to use hand-held inhalers, *e.g.* children and critically ill adults.

A compressed air energy source is the most widely used, clinically, however nebulisers using ultrasonic waves to produce inhalable mists are also widely used (Figure 9.4).

Nebuliser liquids are solutions, suspensions or emulsions presented in single or multiple-dose containers. Drug products may be ready-to-use liquids or concentrates, requiring dilution prior to administration. Nebuliser liquids are generally formulated at between pH 3 and 10. Single-dose preparations are sterile and preservative-free, however multiple-dose preparations require a safe and effective antimicrobial preservative system. This is not necessary if the drug is intrinsically antimicrobial or the container is engineered such that microbial contamination is prevented.

The fine-particle mass of suspension nebuliser liquids must be assessed as part of the control strategy for the product using a suitable pharmacopoeial test.

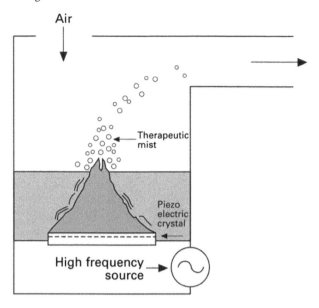

Figure 9.4 Diagram of the components of the ultrasonic nebuliser. Adapted by permission from BMJ Publishing Group Ltd,[18] *Thorax*, C. O'Callaghan and P. W. Barry, **52**, S31, 1997.

9.5.1.2 Pressurised Metered Dose Inhaler (pMDI)

The pMDI is the longest-serving inhaler since its introduction in 1950s. The basic formulation design remains essentially unchanged, however the sub-components of the device have been the subject of continuous improvement.

The pMDI comprises an aluminium canister, containing a liquid formulation, with a specialised actuator cap and a metering valve assembly. The canister is fitted into plastic mouthpiece actuator where the valve stem is inserted into a spray nozzle assembly (Figure 9.5).

The formulation of pMDIs requires a close collaboration between pharmaceutical scientists and engineers. The performance of the inhaler will depend not only upon the pressurised liquid formulation, filled into the canister, but also the device components, *i.e.* the materials used to form the canister, the metering valve and the design of the spray orifice.[20]

The base of the liquid formulation is a propellant, which exists in the liquid state whilst under pressure in the capped aluminium canister. A metered dose is emitted by placing the mouthpiece into the mouth, inhaling and drawing air through the mouthpiece actuator whilst depressing the valve stem. As the metering chamber depressurises, the propellant vaporises and expands, ejecting the contents of the metering chamber into the inspired airstream.

Unlike common household space aerosol sprays, that emit a constant stream of product while the actuator button is depressed, an aerosol fitted

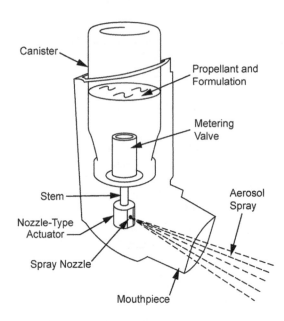

Figure 9.5   Diagram of the principal components of the pressurised metered dose
inhaler (pMDI). J. Kesavan, D. R. Schepers, J. R. Bottiger, M. D. King
and A. R. McFarland, Aerosolization of bacterial spores with pressurized metered dose inhalers, *Aerosol Science and Technology*, 2013, **47**,
1108–1117.[19] Reprinted by permission of the American Association for
Aerosol Research, http://www.aaar.org.

with a metering valve emits a fixed volume of spray, hence dose of drug,
regardless of the duration of valve depression. When the system is at rest, the
metering chamber is a 25–100 µl liquid lock that is open to the contents of the
aluminium canister, and seals the formulation system from the environment.
Actuation firstly seals the metering chamber from the canister contents and
then opens the metering chamber to the environment. The metering chamber
contents are emitted in a short puff and when actuation ceases, the system
returns to rest. The metering chamber will re-prime if the canister is inverted,
i.e. with the empty metering chamber in contact with the canister liquid.

On 26th August 1987, the Montreal Protocol on Substances that Deplete
the Ozone Layer was agreed and entered force on the 26th August 1989. The
Protocol controlled substances that deplete stratospheric ozone and set a
timetable for the cessation of the production and consumption of chlorofluo-
rocarbons (CFCs), including those used in pharmaceutical pMDIs. The International Pharmaceutical Aerosol Consortium (IPAC) was created to evaluate
the pharmaceutical acceptability of alternative propellants: HFA-134a (tetra-
fluoroethane) and HFA-227ea (heptafluoropropane) which emerged from this
process as safe and commercially useful. These propellants are now in use in
pMDI formulations and although 'greenhouse gases' pose no threat to stratospheric ozone due to the absence of chlorine from their chemical structures.

As with aqueous formulations, the solubility of drug in liquefied propellant will determine whether it is possible to produce a solution or suspension. In addition to the usual requirements of dosing precision and accuracy, throughout the life of the pMDI product, the emitted dose of solid drug particles must be inhalable *i.e.* with a mass median aerodynamic diameter below 8 μm. Solution pMDIs must be engineered to emit droplets such that, when the propellant flashes, solid particles in this size range result. In the case of suspension pMDIs, the drug particle size is reduced, prior to compounding, by fluid-energy milling or another suitable high-shear process.

The reformulation of beclomethasone dipropionate pMDIs as solutions in HFAs instead of as suspensions in CFCs, revealed dosing disparities.[21] Drug particles emitted from solution formulations had a significantly lower mass median aerodynamic diameter and the improved aerosol efficiency halved the clinical dose. Whilst solution formulations were comparatively rare in CFC-propellant-based systems, the physicochemical nature of the HFAs permitted the exploitation of high-performance solution aerosols. Further investigation of solution pMDI performance, using a design of experiments approach, revealed that the determinants of inhaler performance were: the content of any non-volatile excipients (used as pressure diluents); the actuator orifice diameter, the volume of the metering valve and the HFA propellant content.[22] These various factors interact to determine the initial emitted droplet size and hence final solid particle size following propellant flash.

The first formulations to appear after the CFC ban comprised fluid-energy milled drug stabilised with surfactant. As stated, the physiochemical nature of CFCs led to the widespread use of suspensions; solutions proving difficult to achieve. This was also a logical starting point for formulation development in the new, post-CFC era. Surfactants, primarily used to stabilise particles in suspension, also acted as metering valve lubricants. Where used, surfactants were typically those developed for use in CFC-based formulations with a proven safety record, *e.g.* oleic acid, sorbitan trioleate and lecithin. Difficulties solubilising surfactants in HFAs led to the use of cosolvents, *e.g.* ethanol.[23] Materials science studies with salbutamol revealed that the material properties of the drug are critically important: the formulation performance differed depending upon whether it was based on the free base or sulphate salt of the drug.[24]

The use of ethanol has been criticised and different strategies have evolved to reduce or eliminate its use in pMDI formulations. Successful formulations have been produced by adding magnesium stearate, a common tableting lubricant, to HFA-134a-based suspensions in ethanol-free formulations.[25] It has also been possible to formulate drug suspensions in HFA-134a without the addition of any excipients at all by coating the surface of the aluminium canister wall with specialist materials to prevent particle adherence.[26]

9.5.1.3   Dry Powder Inhalers

Whilst the pMDI was the original and subsequently preferred inhalation dosage form, dry powder inhalers (DPIs) are well-established alternatives. They operate on the principle that an accurate unit dose of drug powder can be filled into a suitable device from which the patient can inhale. Unlike the pMDI, the dose is released from the device by the inspiratory effort of the patient instead of by the action of a propellant. A major clinical compliance advantage of the DPI is the avoidance of the so-called 'hand–lung coordination' problem associated with pMDIs. If inspiration is not continuous whilst the pMDI is being actuated because the patient, surprised by the release of the dose, stops breathing in, the dose may be lost to accidental ingestion or oropharyngeal deposition. This is not a problem with the DPI as the drug powder is drawn from the inhaler on demand.

There are three basic DPI designs: pre-metered single-dose devices; pre-metered multi-dose devices and reservoir devices.[27]

Pre-metered single-dose devices are generally based on specially designed, part-filled hard gelatin capsules containing the drug. These inhalers require the patient to load the hard gelatin capsule into a simple plastic device prior to inhalation. The capsules are manufactured and controlled in a factory environment and supplied to the patient together with the inhaler. Prior to inspiration, the capsule is variously punctured by needles or shattered by blades in the device to release the powder formulation which can then be inhaled. Depending upon the design, pierced, predominantly intact capsules may be rotated laterally or axially, in an asymmetric airstream, to facilitate release of the powder formulation. Examples of these inhalers include the Rotahaler™ device designed by GlaxoSmithKline, and the Handihaler™ by Boehringer Ingelheim (Figure 9.6).

Reservoir devices comprise a plastic inhaler containing the powder formulation. Volumetric filling of a cluster of small holes is achieved by twisting the base of the device. As the base rotates excess powder is removed by scrapers, which also lightly compresses the powder into plugs. As the patient inspires, the small powder plugs are released into the airflow through the device and dispersed by the turbulent airflow created by strakes in the mouthpiece (Figure 9.7). A major advantage of this design is that the use of carrier excipients, *e.g.* lactose, may be avoided and the drug powder, engineered to be inhalable, can be used alone.

Finally, more complex devices have evolved where strips of up to 60-unit doses of powder are loaded into a pre-metered multiple-dose device (Figure 9.8).

The inhaler is designed to release a unit dose of powder for inhalation into a holding chamber. The dose is emitted when the patient inspires, drawing air through the device. Closure of the device, or the operation of a cocking mechanism, prepares the device for its next dose. These are extremely

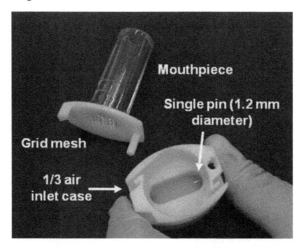

Figure 9.6 Simple dry powder inhaler for use with drug supplied in pre-metered hard gelatin inhalation capsules. The lower image shows the pin used to pierce the capsule allowing release of the drug powder formulation. Also shown is the spinning chamber with eccentric air inlets. The upper image shows the grid mesh used to facilitate drug-carrier separation. Reprinted from *The AAPS Journal*, The Delivery of High-dose Dry Powder Antibiotics by a Low-cost Generic Inhaler,**19**, 2017, 191–202, T. Parumasivam, S. S. Leung, P. Tang. C. Mauro, W. Britton and H. K. Chan.[28] © American Association of Pharmaceutical Scientists 2016 with permission of Springer.

complex devices constructed of many small precision-engineered plastic components.

Whilst many complex formulation design strategies can be found in the literature, many dry powder inhaler formulations generally depend upon two approaches: fluid-energy milling of drug particles or spray drying the drug to produce spherical, free-flowing engineered particles. The engineered drug particles may be mixed with a carrier particle, usually a suitable inhalation grade of lactose, or used without the carrier.

Drug particles need to be engineered in the respirable range, *i.e.* the sub 5–8 μm range. Fluid-energy milling can produce respirable particles, however, it is a process that results in small, cohesive particles with highly energetic surfaces. Cohesive particles are difficult to handle in manufacturing processes, *e.g.* factory-based metering into capsules and cartridges, and do not disperse well on inhalation. Fluid-energy-milled drug particles are therefore invariably mixed with larger lactose monohydrate carrier particles with geometric diameters in the range of 50–200 μm.

Paradoxically, the formulator needs to exploit strong interparticle forces between the drug and lactose monohydrate carrier, to produce a stable ordered mix that can be processed during manufacture, whilst ensuring that drug particles can be released from the carrier during inspiration by creating

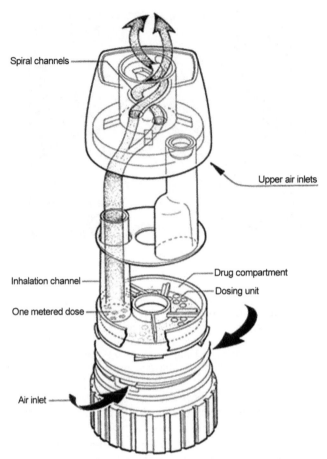

Spiral channels

Upper air inlets

Drug compartment

Inhalation channel

Dosing unit

One metered dose

Air inlet

Figure 9.7   The AstraZeneca Turbuhaler® multidose reservoir dry powder inhaler. The image shows the position of the drug reservoir and the rotating dosing disc. The black arrows show device airflows during inspiration and the vortex created in the mouthpiece to facilitate drug–carrier separation. Reproduced from G. Persson, E. Gruvstad and E. Stahl, A new multiple dose powder inhaler, (Turbuhaler), compared with a pressurized inhaler in a study of terbutaline in asthmatics, *European Respiratory Journal*, 1988, **1**, 681–684[29] with permission from the © ERS 1988.

turbulent airflows (with strakes) and dispersion by collision with impaction surface features (meshes or baffles)[31] (Figure 9.9). To form an ordered powder mix, the content of drug cannot be greater than 30% w/w and is normally formulated at 5% w/v or less.

Unsurprisingly, research has focussed on the modulation of interparticle interactions, between energetic drug particles and lactose carriers. The formulation of ternary powder systems, with the addition of a so-called

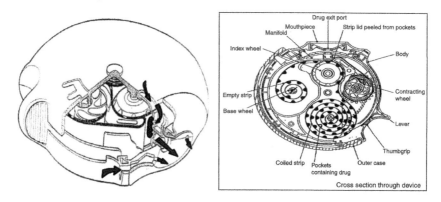

Figure 9.8 The GlaxoSmithKline Diskus™ multidose dry powder inhaler. The arrows in the image on the left show airflows through the device on inspiration. The image on the right is a cross section through the device to show the principal component parts. Reproduced from H. Chrystyn, The Diskus™: a review of its position among dry powder inhaler devices, *International Journal of Clinical Practice*, John Wiley and Sons,[30] © 2006 The Author.

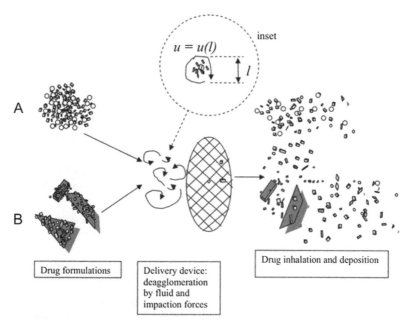

Figure 9.9 Depiction of the stages of drug powder dispersion in a dry powder inhaler. Formulations can be either drug alone (Case A) or an ordered mixture of drug and lactose (Case B). Dispersion, and segregation in Case B, occurs due to turbulence created by the device and inertial impact on baffles or meshes. Reproduced from *Medical Engineering & Physics*, **34**, 2012, N. Islam and M. J. Cleary, Developing an efficient and reliable dry powder inhaler for pulmonary drug delivery–a review for multidisciplinary researchers, 409–427,[32] Copyright 2012, with permission from Elsevier.

force-control agent to the drug–lactose mix, has found commercial application. Tablet lubricants, such as magnesium stearate and leucine, have been shown to improve the performance of drug–lactose dry powder inhaler systems. It has been suggested that drug–carrier interactions can be modified by force-control agents selectively blocking energetic sites on either the drug particle or the carrier or both.[31] Use of atomic force microscopy techniques has been shown to assist the formulator in quantifying interparticle forces in such powder formulations.[33]

There are many reports in the literature of particle engineering solutions that produce efficient powder formulations for dry powder inhalers. By far the most practical has been the use of the spray drying. The spray drying of solutions of drug, with or without excipients, produces spherical particles with excellent flow properties without the high-energy surface features associated with drug–lactose monohydrate carrier systems.[23]

9.5.1.3.1 The Evaluation and Equivalence of Dry Powder Inhalers. Dry powder inhalers are required to comply with pharmaceutical quality standards, typical of many standard dosage forms. There is, however, a requirement to characterise the emissions from inhalation dosage forms, particularly with respect to particle size.

Aerodynamic particle sizing is the most clinically relevant sizing method and size distributions can be characterised using a variety of cascade impactor or liquid impinger instruments described in many national pharmacopoeias. The Andersen Cascade Impactor, one of the more common impactor instruments, is operated at a fixed airflow rate of 28.3 l min^{-1}. The inhalation dosage form is actuated into the airflow, using a 'throat' of prescribed dimensions and fractions are collected on different collection stages. The mass median aerodynamic diameter is obtained from the distribution of impacted drug mass on different collection stages with different particle size cut offs; the fine particle dose is the mass of drug on stages 2–7 and the filter which corresponds to particles with a mass median aerodynamic diameter of 5.8 μm or less (Figure 9.10)

Whilst providing a useful method to evaluate the consistency of manufacturing, this is of doubtful value as a clinically relevant predictor of *in vivo* performance. Conventional aerodynamic particle-sizing methods require operation at a fixed flow rate, which contrasts with the variable flow rate profiles resulting from inspiratory manoeuvres by the patient.

Inhalation simulators now provide a means of sampling dry powder inhalers using computer controlled profiles mimicking patient inspiration. The simulator actuates the device with a realistic variable flow rate profile and stores the emitted dose in a holding chamber. The sample is then drawn out of the holding chamber using an aerodynamic particle-sizing instrument operating at a fixed flow rate. Whilst the relationship between flow-dependent and flow-independent dose delivery and clinical effect remains unclear, the use of inhalation simulators to assess the performance of dry powder

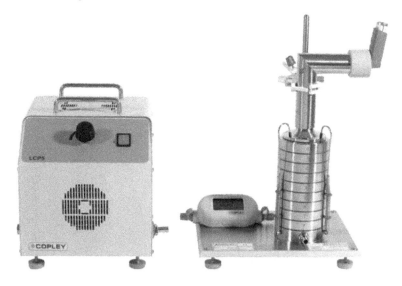

Figure 9.10   The Andersen Cascade Impactor (left). The basic configuration of the British Pharmacopoeia Apparatus D with modifications for use with dry powder inhalers (right). The dry powder inhaler is attached to the impactor using a standard throat and attachment port. The impactor draws air through the device by means of a vacuum pump connected to a rate-controlled air flow. With the permission of Copley Scientific.

inhalers might be more clinically relevant than fixed flow rate cascade impactors and liquid impingers.[34–36]

9.5.1.4   Soft-mist Inhalers

The soft-mist inhaler evolved from experimentation with pocket-sized devices capable of delivering a small, fixed volume of drug solution in an aerosol. In this regard, these systems can be viewed as hand-held, metered-dose nebulisers.

Early prototypes exploited piezoelectric crystal atomisation, extrusion through micronised holes or electrohydrodynamics. Each of these technologies required battery power, adding considerable expense to the device compared with other inhalers, and offered no ready means of use if the battery failed.

Applying precision fabrication technology from the microelectronics industry, devices have been constructed that force drug solutions through a two-channel nozzle using mechanical power (Figure 9.11). Commercial devices are now available with the ability to deliver up to 120 doses. Metered volumes are of the order of 15 μl and the emitted droplet size is in the range 1–5.8 μm.[37]

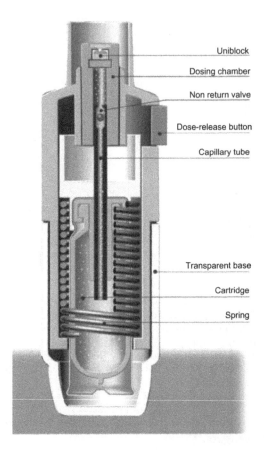

Uniblock
Dosing chamber
Non return valve
Dose-release button
Capillary tube
Transparent base
Cartridge
Spring

Figure 9.11 Schematic drawing of the components of a soft-mist inhaler. The compressed spring provides the mechanical force; the 'uniblock' contains the fine nozzles conducting the two liquid streams to impinge and atomise. Reproduced from T. R. MacGregor, R. ZuWallack, V. Rubano, M. A. Castles, H. Dewberry, M. Ghafouri and C. C. Wood, Efficiency of Ipratropium Bromide and Albuterol Deposition in the Lung Delivered *via* a Soft Mist Inhaler or Chlorofluorocarbon Metered - Dose Inhaler, *Clinical and Translational Science*, John Wiley and Sons.[37] © 2016 ASCPT.

References

1. D. Toscano, *Recent Advancements in Drug Delivery: Novel Formulations & Technologies Offer Improved Treatment Options | Articles | Drug Development and Delivery Back Issues*, Drug Development and Delivery, 2012, [cited 2017 Mar 24], available from: http://www.drug-dev.com/Main/Back-Issues/Recent-Advancements-in-Drug-Delivery-Novel-Formula-378.aspx.

2. R. Löbenberg and G. L. Amidon, Modern bioavailability, bioequivalence and biopharmaceutics classification system. New scientific approaches

to international regulatory standards, *Eur. J. Pharm. Biopharm.*, 2000, **50**(1), 3–12.

3. Y. S. Rhee, J. Y. Huh, C. W. Park, T. Y. Nam, K. R. Yoon, S. C. Chi and E. S. Park, Effects of vehicles and enhancers on transdermal delivery of clebopride, *Arch. Pharmacal Res.*, 2007, **30**(9), 1155.

4. M. R. Prausnitz and R. Langer, Transdermal drug delivery, *Nat. Biotechnol.*, 2008, **26**(11), 1261–1268.

5. G. P. Moss, D. R. Gullick and S. C. Wilkinson, *Predictive Methods in Percutaneous Absorption*, Springer, Berlin, 2015.

6. S. Mitragotri, Y. G. Anissimov, A. L. Bunge, H. F. Frasch, R. H. Guy, J. Hadgraft, G. B. Kasting, M. E. Lane and M. S. Roberts, Mathematical models of skin permeability: an overview, *Int. J. Pharm.*, 2011, **418**(1), 115–129.

7. R. J. Buckley, C. M. Kirkness, J. J. Kanski, A. E. Ridgway, A. B. Tullo and P. G. Watson, Is it time to stop using chloramphenicol on the eye? Safe in patients with no history of blood dyscrasia, *BMJ [Br. Med. J.]*, 1995, **311**(7002), 450.

8. P. Larochelle, S. W. Tobe and Y. Lacourcière, β-Blockers in hypertension: studies and meta-analyses over the years, *Can. J. Cardiol.*, 2014, **30**(5), S16–S22.

9. H. Patel, L. V. Wilches and J. Guerrero, Timolol-induced interstitial lung disease, *Respir. Med. Case Rep.*, 2015, **15**, 30–32.

10. A. Patel, K. Cholkar, V. Agrahari and A. K. Mitra, Ocular drug delivery systems: an overview, *World J. Pharmacol.*, 2013, **2**(2), 47.

11. J. C. Lang, Ocular drug delivery conventional ocular formulations, *Adv. Drug Delivery Rev.*, 1995, **16**(1), 39–43.

12. T. Loftsson and E. Stefánsson, Cyclodextrins in eye drop formulations: enhanced topical delivery of corticosteroids to the eye, *Acta Ophthalmologica*, 2002, **80**(2), 144–150.

13. V. J. Stella and Q. He, Cyclodextrins, *Toxicol. Pathol.*, 2008, **36**(1), 30–42.

14. M. E. Brewster and T. Loftsson, Cyclodextrins as pharmaceutical solubilizers, *Adv. Drug Delivery Rev.*, 2007, **59**(7), 645–666.

15. G. Crompton, A brief history of inhaled asthma therapy over the last fifty years, *Prim. Care Respir. J.*, 2006, **15**(6), 326–331.

16. F. P. Maesen, J. J. Smeets, M. A. Costongs, F. D. Wald and P. J. Cornelissen, Ba 679 Br, a new long-acting antimuscarinic bronchodilator: a pilot dose-escalation study in COPD, *Eur. Respir. J.*, 1993, **6**(7), 1031–1036.

17. D. C. Cipolla and I. Gonda, Formulation technology to repurpose drugs for inhalation delivery, *Drug Discovery Today: Ther. Strategies*, 2012, **8**(3), 123–130.

18. C. O'Callaghan and P. W. Barry, The science of nebulised drug delivery, *Thorax*, 1997, **52**(suppl. 2), S31.

19. J. Kesavan, D. R. Schepers, J. R. Bottiger, M. D. King and A. R. McFarland, Aerosolization of bacterial spores with pressurized metered dose inhalers, *Aerosol Sci. Technol.*, 2013, **47**(10), 1108–1117.

20. S. P. Newman, Principles of metered-dose inhaler design, *Respir. Care*, 2005, **50**(9), 1177–1190.
21. C. Leach, Effect of formulation parameters on hydrofluoroalkane-beclomethasone dipropionate drug deposition in humans, *J. Allergy Clin. Immunol.*, 1999, **104**(6), S250–S252.
22. D. A. Lewis, D. Ganderton, B. J. Meakin and G. Brambilla, Theory and practice with solution systems, *Respir. Drug Delivery IX*, 2004, **1**, 109–116.
23. G. Pilcer and K. Amighi, Formulation strategy and use of excipients in pulmonary drug delivery, *Int. J. Pharm.*, 2010, **392**(1), 1–9.
24. T. Z. Tzou, R. R. Pachuta, R. B. Coy and R. K. Schultz, Drug form selection in albuterol-containing metered-dose inhaler formulations and its impact on chemical and physical stability, *J. Pharm. Sci.*, 1997, **86**(12), 1352–1357.
25. R. Muller-Walz and C. Niederlander, inventors, Medical aerosol formulations, United States patent application US 10/473,874, 2002 Mar 11.
26. D. Lewis, D. Ganderton, B. Meakin, P. Ventura, G. Brambilla, R. Garzia, inventors, Chiesi Farmaceutici SpA, assignee. Pressurised metered dose inhalers (MDI), United States patent US 7,223,381, 2007 May 29.
27. I. Ashurst, A. Malton, D. Prime and B. Sumby, Latest advances in the development of dry powder inhalers, *Pharm. Sci. Technol. Today*, 2000, 3(7), 246–256.
28. T. Parumasivam, S. S. Leung, P. Tang, C. Mauro, W. Britton and H. K. Chan, The delivery of high-dose dry powder antibiotics by a low-cost generic inhaler, *AAPS J.*, 2017, **19**(1), 191–202.
29. G. Persson, E. Gruvstad and E. Stahl, A new multiple dose powder inhaler, (Turbuhaler), compared with a pressurized inhaler in a study of terbutaline in asthmatics, *Eur. Respir. J.*, 1988, **1**(8), 681–684.
30. H. Chrystyn, The Diskus™: a review of its position among dry powder inhaler devices, *Int. J. Clin. Pract.*, 2007, **61**(6), 1022–1036.
31. J. G. Weers and D. P. Miller, Formulation design of dry powders for inhalation, *J. Pharm. Sci.*, 2015, **104**(10), 3259–3288.
32. N. Islam and M. J. Cleary, Developing an efficient and reliable dry powder inhaler for pulmonary drug delivery–a review for multidisciplinary researchers, *Med. Eng. Phys.*, 2012, **34**(4), 409–427.
33. M. Tobyn, J. N. Staniforth, D. Morton, Q. Harmer and M. E. Newton, Active and intelligent inhaler device development, *Int. J. Pharm.*, 2004, **277**(1), 31–37.
34. P. K. Burnell, A. Malton, K. Reavill and M. H. Ball, Design, validation and initial testing of the Electronic Lung™ device, *J. Aerosol Sci.*, 1998, **29**(8), 1011–1025.
35. P. K. Burnell, T. Small, S. Doig, B. Johal, R. Jenkins and G. J. Gibson, *Ex vivo* product performance of Diskus™ and Turbuhaler™ inhalers using inhalation profiles from patients with severe chronic obstructive pulmonary disease, *Respir. Med.*, 2001, **95**(5), 324–330.

36. S. Stegemann, S. Kopp, G. Borchard, V. P. Shah, S. Senel, R. Dubey, N. Urbanetz, M. Cittero, A. Schoubben, C. Hippchen and D. Cade, Developing and advancing dry powder inhalation towards enhanced therapeutics, *Eur. J. Pharm. Sci.*, 2013, **48**(1), 181–194.

37. T. R. MacGregor, R. ZuWallack, V. Rubano, M. A. Castles, H. Dewberry, M. Ghafouri and C. C. Wood, Efficiency of ipratropium bromide and albuterol deposition in the lung delivered *via* a soft mist inhaler or chlorofluorocarbon metered-dose inhaler, *Clin. Transl. Sci.*, 2016, **9**(2), 105–113.

CHAPTER 10

Paediatric Pharmaceutics— The Science of Formulating Medicines for Children

MINE ORLU*, SMITA SALUNKE AND CATHERINE TULEU

UCL School of Pharmacy, Department of Pharmaceutics, 29–39 Brunswick Square, London WC1N 1AX, UK
*E-mail: m.orlu@ucl.ac.uk, s.salunke@ucl.ac.uk, c.tuleu@ucl.ac.uk

10.1 Introduction

10.1.1 Children and Their Specific Needs

Children are a heterogeneous population that includes new-borns (term or pre-term), infants, toddlers, pre-schoolers, school-age children and adolescents.[1] The stages of developmental physiological changes throughout childhood complicate pharmacotherapy. A complete consensus does not exist about the age ranges that define infancy, childhood and adolescence. The term 'child' has been used broadly to refer to individual ages 0 to 18 years. Biologically, a child (plural: children) is generally a human between the stages of birth and puberty. The guideline on clinical investigation of medicinal products in the paediatric population uses the age groups in relation to developmental stages. It reflects biological changes—the changes after birth; the early growth spurt; gradual growth from 2 to 12 years; the pubertal and adolescent growth spurt and development towards adult maturity.[2]

Drug Discovery Series No. 64
Pharmaceutical Formulation: The Science and Technology of Dosage Forms
Edited by Geoffrey D. Tovey

The subsets of the paediatric population widely differ in their therapeutic requirements due to their developmental and behavioural stage. From birth into adulthood, children change and develop physically, cognitively, socially and emotionally. Physical growth during childhood is apparent to the eye, but less obvious is the ongoing maturation of organ function. The physiological make-up of children differs not only from that of adults but also within their own age group. During the first few weeks and months of their life, changes occur in saliva production, body composition (*e.g.*, body water and fat content, protein binding characteristics), organ weight and maturity (*e.g.* renal maturation, hepatic maturation).[3] This can affect the absorption, distribution, metabolism and excretion of drugs and excipients and in turn can cause toxicity.[4] Additionally, there is extensive inter-individual variation; children of the same age may vary with respect to weight, height, body surface area and maturity.[3] There will always be an overlap in developmental stages. Understanding the physiological development differences and changes during the earliest period of life is important in paediatric drug testing.[5] One area that needs special attention is neonatal (in first month life) deaths which are falling more slowly than under-five deaths and accounted for nearly half (2.6 million) of all deaths in children under five in 2015. Preterm birth complications and birth asphyxia and trauma are now the leading causes of deaths in children younger than five years of age worldwide, highlighting the slower progress in reducing neonatal conditions compared with communicable diseases in childhood. Hence, when designing a paediatric drug product it is important to take into consideration the specific age category.

10.1.2   Children and Their Medicines

Lack of authorised medicines and consequent off-label use of adult's medicines is a significant problem in the paediatric population. In neonates, the situation is particularly challenging due to the vulnerability of new-borns and even lower patient numbers. If children are not young adults then why are they prescribed adult medicines on an "off-label" basis? Authorised medicines that are not available on the market do not bring any benefit to a child. The percentage of authorised and dose-capable medicines with a suitable dosage forms increases with age. The American Academy of Pediatrics has argued that the shortage of paediatric research creates an ethical dilemma for physicians, who "must frequently either not treat children with potentially beneficial medications or treat them with medications based on adult studies or anecdotal empirical experience in children". Research with adults cannot simply be generalized or extrapolated to infants, children and adolescents and hence research-involving children is essential if children are to share fully in the benefits derived from advances in medical science. Several challenges, including the relatively small numbers of children with serious medical problems, the need for developmentally appropriate outcome measures for children of different ages, the complexities of parental involvement and family decision making, and the adaptations required in research procedures and settings to accommodate children's physical, cognitive and emotional development, make the research in paediatrics more challenging than research in adults. Specific clinical investigations in paediatric populations are normally required due to age-related differences in the drug handling or drug effects, which may lead to different dose requirements to achieve efficacy or to avoid adverse effects. The development of medicines tailored for children's needs implies that a specific drug may be needed to be available in various dosage forms and/or strengths. Thus several medicinal drug products may be needed in order to treat a broad patient population from birth into adulthood. The dose capability and suitability of dosage form are considered for any authorised paediatric medicine. (*e.g.* for acetaminophen, two strengths of chewable tablets, a low-strength "swallowable" tablet, a syrup and drops in a different concentration for infants). Furthermore, compared with adults, children generally represent a smaller market for commercial sponsors of research. The commercial value of various preventive, diagnostic and therapeutic options for children, especially for rare diseases, may not be enough to offset the costs of developing them. On the one hand, there are several formulation, clinical and regulatory requirements for developing paediatric formulations, while on the other hand, the widespread use of off-label drugs does not incentivize companies to finance paediatric research on drugs that are already approved for use by adults. Challenges in carrying out paediatric research include the rarity of many childhood diseases, heterogeneity of the population and issues regarding consent. Efforts are needed to obtain good evidence with as few subjects as possible and to prevent unnecessary clinical trials. Approaches such as extrapolation and modelling and simulation are increasingly becoming part of paediatric medicine development to optimise available data from other populations and reduce the number of children needed in clinical studies, however, clinical research with children is essential for paediatric drug development in the majority of cases. Much progress has been made on understanding how diseases differ in children and adults, but more concerted effort is needed towards understanding the patient.

In general, several features distinguish pharmacotherapy in children from that in adults and explain why medicines must be studied in research with children to ensure their safe and effective use.[3]

These features include

- Lack of age-appropriate formulations that allow the accurate, safe and palatable administration of medicines to children of a wide range of developmental characteristics such as weight, height, body surface area and maturity
- Age- and development-dependent changes in how medicines are distributed in and eliminated from the body (pharmacokinetics);
- Age- and development-dependent changes in the response to medicines (pharmacodynamics);

- Age- and development-dependent changes in the adverse effects of medicines, both short and long term; and
- Unique paediatric diseases that require development of unique paediatric medications.

10.1.2.1 Children and Regulations Around Their Medicines

Historically, paediatric drug development was mainly promoted and incentivized as a voluntary process. However, voluntary market forces alone have proven to be insufficient to stimulate research or address the lack of dosage forms for children. The unmet need for safe and better medicines for children was well recognized by various agencies governing pharmaceutical regulations across the globe and has resulted in a dramatic progress and growing interest in the development of age-appropriate formulations to better serve the needs of the paediatric population.[6] Legislative and regulatory reforms were initially led by the United States Food and Drug Administration (FDA) to increase the information in the drug label on use of medicines in children. The FDA Amendments Act (FDAAA) in 2007 was an important landmark, which included reauthorisation of the 2002 Best Pharmaceuticals for Children Act (BPCA) and the 2003 Pediatric Research Equity Act (PREA). The BPCA grants 6 months market exclusivity as an incentive to conduct necessary paediatric studies (voluntarily), while PREA codified the authority of the FDA to mandate studies for certain drugs and biological products. The FDA Safety and Innovation Act (FDASIA) in 2012 made both the BPCA and PREA permanent. Subject to PREA, sponsors are required to provide information related to the development of paediatric formulations as part of a Pediatric Study Plan (PSP) submitted at the end of Phase 2 research. The European Union (EU) adopted its own comprehensive reforms when Regulation (EC) No. 1901/2006 or the "Paediatric Regulation" came into force in January 2007. The paediatric regulation aims to improve the health of children of Europe by a system of obligations and rewards facilitating the development and availability of appropriately authorized medicines for children between birth and 18 years; by improving the information on the actual use of medicines in children; by ensuring that medicines for use in children are of high quality and ethically researched.[7] The regulation requires companies to develop a Paediatric Investigation Plan (PIP) at an early phase in the development of a new medicine, new route of administration or new indication or for any variations to patented authorised medicines (unless a waiver is granted). The PIP describes the plan for paediatric development of medicines, including the pharmaceutical design of the preparation(s) to be developed for each of the target age group.[8-10] The PIP is assessed and subjected to agreement upon by a scientific Paediatric Committee (PDCO) of the European Medicines Agency (EMA). The EMA/PDCO PIP decisions are binding at the time of marketing authorisation and industry can only apply for marketing authorisation of the (adult) medicine when the EMA has confirmed that the PIP was followed or a deferral was obtained. In contrast with the US PSP, the EU PIP is agreed at the end of Phase 1, though deferrals can be agreed for the initiation or completion of initial proposals if justified. Both legislations provide frameworks together with the incentives and rewards and ensure that new medicines are adapted to children's needs and that the paediatric population is not neglected despite the forces of the market. However, a more harmonised approach across these jurisdictions would be beneficial.[11] The International Conference on Harmonisation of Technical Requirements for Registration of Pharmaceuticals for Human Use (ICH) (which brings together authorities and industries in the EU, USA and Japan) also adopted a guideline, ICH E11, addressing the conduct of clinical trials in the paediatric population.[12] Notably, this guideline categorises the paediatric population into five distinct age groups, including "children" aged 2–11 years and adolescents aged 12 years and above (for the purposes of this research, this includes persons aged 12–17 years old). While these groups reflect clinical applications, the EMA further subdivided children into "pre-school children" aged 2–5 years and "school-children" aged 6–11 years in relation to formulation development considerations.[2] These remain the principal regulatory reforms and there has been comparatively little progress in other countries. Acknowledging that the majority of children in less developed countries live less healthy lives as compared with those in more developed countries, the limited availability of appropriate medicines for children is key concern to the World Health Organization (WHO). It has spearheaded important campaigns promoting awareness and accelerating action to address three challenges associated with paediatric medicines, namely availability, accessibility and affordability. These were aptly entitled 'Making Medicines Child Size' and the 'Better Medicines for Children Project', and notable outcomes of these initiatives include the WHO Model List of Essential Medicines for Children and a 'points to consider' document on the formulation of paediatric medicines.[13] The objective was to inform regulatory authorities and manufacturers on issues that require special attention in pharmaceutical formulation. In 2010, the WHO published a Model Formulary for Children built on the EML that provides prescribing guidance on use of the essential medicines. As a result, for the first time, medical practitioners worldwide have access to standardized information on the recommended use, dosage, adverse effects and contraindications of these medicines for use in children.[14] Recommendations to improve children's access to better medicines had also been made by other Australian professional and government advisory groups since the late 1990s, but with little resulting action.

The changing regulatory landscape has generated a need for research to create better and safer medicines for children and to advance the current platforms and technologies that are already used in this patient population.[15] Pharmaceutical sponsors, regulatory agencies and allied stakeholders have reached an influential period in the new era of developing paediatric medicines. Given the lengthy drug development process, it is somewhat premature to measure the overall global effects of these legislations and initiatives. Nevertheless, proof of concept and progress to date has been encouraging, including improved drug labelling, completion of PIPs with new paediatric indications and formulations and emerging research into the previously neglected areas of neonatology and off-patent medicines.[11] Some argue that economic barriers and lack of adequate incentives continue to impede the necessary focus on unmet clinical need, and instead, development of paediatric medicines seems to shadow drug development in adults.[16,17] While these reforms continue to serve as platforms steering research and development, distinctive opportunities and challenges in the field also emerge.

With the mission of better medicines for children finally on the global agenda, the challenges are now to collaboratively further shape the paediatric drug development agenda and effectively use the existing data to address these challenges in formulating medicines for children and to bridge the adult–children medicine gap. Researchers and academics are putting in all the efforts to respond to many unanswered questions about medicines for children, through research and international collaboration both at country or regional levels and at the global level. Key developments include the range of pioneering paediatric drug development initiatives, such as formation of the International Alliance for Better Medicines for Children (IABMC) in 2006; establishment of the Paediatric Task Force by The International Federation of Pharmaceutical Manufacturers and Associations (IFPMA) in 2008; and establishment of European Paediatric formulation Initiative (EuPFI)[†] in 2007. The European Union funded the Global Research in Paediatrics (GRIP) project that brings together over twenty collaborating organizations and more than a thousand researchers to harmonize paediatric research tools and share research strategies. However, despite the expansion of research in the development of paediatric medicinal products, there are still unmet needs and challenges for medicines for children. These research efforts have resulted in some progress in medicines with a larger market (*e.g.* anti-infectives, antibacterials, medicines for the respiratory and central nervous system) but not in all areas of priority paediatric health need.[18] Younger and more vulnerable age groups, where the need for better evidence is even greater, have been less well studied and many of the off-patent medicines remain unevaluated. Hurdles such as regulatory capacity, affordability and patient and caregiver acceptance still hinder access to safe and appropriate medicines for children. In addition, research is still needed to define appropriate dosages and formulations for other priority medicines for children.

[†]The European Paediatric Formulation Initiative (EuPFI) is a consortium founded in 2007 and working in a pre-competitive way on paediatric drug formulations. Members are from academia, hospital pharmacies, pharmaceutical industry (Innovators, Generics, Contract Research Organizations (CRO), Specials and Excipient Manufacturers) with the European Medicine Agency (EMA) as an observer. The main objective of the members is to resolve scientific, regulatory and technological issues associated with paediatric formulation development.

10.2 Paediatric Drug Development: Key Attributes

10.2.1 Dosage Form Design

Appropriate dosage form design is essential for any type of drug product development to help ensure safety, efficacy and quality. When designing a paediatric drug product it is also important to consider age-related physiological and behavioural growth and their influence on the pharmacokinetics and pharmacodynamics and medicine use. The variance across the paediatric population is also an important factor determining the appropriateness of the formulation. The dosage form design should be tailored according to the specific needs of paediatric subsets ranging from neonates to adolescents. The EMA has issued a specific guideline related to the pharmaceutical development of medicines for paediatric use in 2013. This key reference describes the regulatory expectations for a paediatric medicinal product design including the end-user acceptability to optimize therapeutic outcomes.

The design of paediatric dosage form is driven by the key points to consider listed below.[19]

1. Efficacy
   a. Adequate bioavailability to ensure pharmacotherapeutic effect
   b. Disease to be treated (chronic or acute condition)
   c. Dose flexibility (enabling dosing to different age-subsets, acceptable dose size)
2. Safety
   a. Dosing accuracy (minimal risk of dosing error, no requirement for manipulation prior to use)
   b. Excipients (determination of qualitative and quantitative composition considering patient's tolerability)
   c. Stability (shelf-life and in-use stability)
3. Patient access
   a. Manufacturability (availability of robust process, ease of production, transport and storage, commercial viability)
   b. Cost (affordability for patient/healthcare provider)
4. Patient acceptability and adherence
   a. Patient age subset
   b. Patient ability (suits patient capability)
   c. Patient willingness (meets patient preferences)
   d. Administration-related requirements (easy and convenient preparation of point of care, acceptable for care-givers and healthcare professionals)
   e. Compliance (minimal effect on life style)

Oral drug delivery is the most widely preferred route of administration of paediatric medicine.[20] Historically liquid formulations have been reflected as the choice of formulation as the main barrier is the swallowability of intact conventional solid oral dosage forms for younger children. However the issues related to the need of use of specific excipients (*e.g.* co-solvents, preservatives, sweeteners, flavours) and packaging/administration devices may be the barrier to the use of liquid formulations to treat childhood conditions.

Progress in the development of novel drug delivery systems has enabled solid oral dosage forms as age-appropriate formulations for paediatric use. The World Health Organization suggests the use of flexible solid oral dosage forms as the preferred way of administering medicines to children. The flexible solid oral dosage forms include dispersibles, orodispersibles and multiparticulates. The design of age-appropriate formulations considers the aspects of the Quality Target Product Profile relating to patient and caregiver's needs, capabilities and preferences. Dispersibles are presented as solid dosage forms that are dispersed or dissolved in a liquid to form a solution or suspension prior to administration. Oral liquid dosage forms are normally considered acceptable from birth, taking into consideration appropriateness of volume, composition and palatability.[7] Dispersible and soluble tablets should disintegrate within three minutes in a small amount of water, to yield a homogenous dispersion or solution. Orodispersible formulations include tablets (ODT) and oral thin oral dispersible films (ODFs) that rapidly disintegrate in saliva, usually within seconds. These formulations are well suited for drugs with high aqueous solubility; however, their applicability in practice may be restricted by limited drug loading. Multiparticulates describe powders, granules, pellets and minitablets that are presented as multiple, discrete unit dosage forms. The flexibility of the multiparticulates is due to the possibility of administration by sprinkling on soft food.[21] The points to consider for paediatric dosage form design are explained using the recent development of lopinavir/ritonavir sprinkle formulation as follows.

The recognition of the challenge related to the traditionally available antiretrovirals in liquid and conventional solid formulations led to the development of paediatric lopinavir/ritonavir in a new formulation. Lopinavir/ritonavir are produced in pellets by melt-extrusion technology and are enclosed in capsules. The dosage form has enabled dose flexibility *via* the possibility of sprinkling the oral pellets on a compatible soft food prior to administration to infants. The medicinal product can also be taken as a whole capsule by older children. This new design has also addressed the demand for a heat-stable and easy to transport/store formulation. The pellets are also functionalized by taste masking. The palatability is one of the key requirements for acceptability of orally administered dosage forms. The acceptability may be perceived as a pre-condition to long term adherence. In this respect, the new sprinkle pellet formulation shows promise for higher patient acceptability and adherence. A multi-disciplinary approach may be required (collaboration with experts on pre-clinical, packaging and devices as well as behavioural science) to obtain further understanding of the overall acceptability and longer term adherence to paediatric medicinal products.

10.2.2   Excipients

Excipients play a fundamental role in medicines. They are included in a dosage form to convert a pharmacologically active compound/drug substance into a pharmaceutical product that can be administered to or taken by the patient and that is acceptable to them. Although not pharmacologically active, they can enhance product performance by ensuring the stability of the active substance, or protecting against microbial contamination during use (*e.g.*, parabens, benzoic acid). Some excipients (*e.g.* polyethylene glycol, sodium pyrophosphate, mannitol) can in fact accelerate the passage of orally administered active substances through the intestinal tract, thus adversely influencing the gastrointestinal absorption of the active principle.[22-24] There are many instances in which excipients have been shown to have a significant effect on the bioavailability of the drug.[25] They can contribute to reactions leading to degradation or to interactions between the drug and the excipient.[26,27] To further complicate the issue, these effects can be drug-, dose-, formulation- and/or subject-dependent. For instance by modifying absorption for parenteral products, excipients can change exposure patterns and thus influence both safety and efficacy outcomes.[28]

Advancements in functionality of excipients have now rendered the traditional view of excipients as *"simple inert pharmaceutical fillers"* obsolete. Today excipients, which have a critical effect on the quality and bioavailability of some drug products and novel dosage forms, do not anymore fit within the traditional definition as *"an inert substance used as diluent or vehicle for a drug"*.[29] The evolution of the excipient definition from *"the inert substance used as a medium for giving a medicament"* to *"any constituent of a medicinal product other than the active substance"* is summarised in Figure 10.1.

10.2.2.1   Issues of Excipients in Paediatrics

An objective of development of medicines for the relevant paediatric subsets is providing formulations that have sufficient bioavailability, acceptable palatability, acceptable dose uniformity and stability. Developing such age-appropriate formulations is more complex and may involve a broader range of excipients than for adult dosage forms.[35] There are many aspects to be considered when selecting an appropriate excipient, such as influence of excipient on the overall quality, stability and effectiveness of drug product, compatibility with drug, route of administration, dosage form, their quantities in relation to the target age group, treatment duration and severity of disease, patient acceptability and safety profile.[2] The current literature indicates that certain excipients acceptable in adult formulations (*e.g.* benzyl alcohol, ethanol, propylene glycol, ethanol, parabens) are associated with

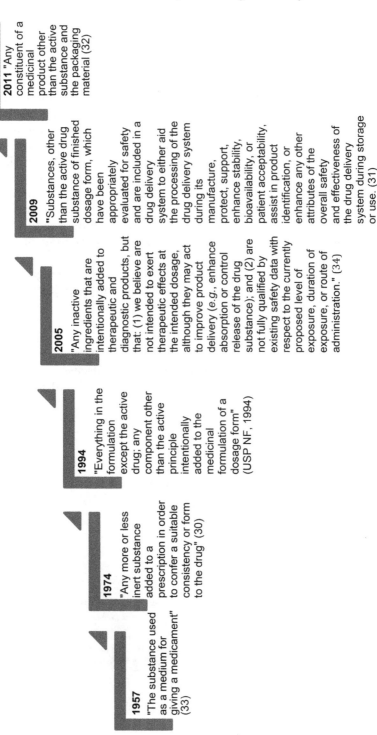

Figure 10.1 Evolution of the definition of excipients.[30–34]

elevated toxicological risks and safety issues when used in children, even in proportionally lower concentrations.[36] Nevertheless, excipients with a potential cause for concern may be essential to the development of a specific dosage form. Hence, the screening and careful selection of excipients in a paediatric medicinal product is one of the key elements of pharmaceutical development[7] and the excipients chosen, their concentration, and the attributes relevant to their function in the drug product need to be justified in terms of safety for the targeted age group, treatment, route of administration, duration, allergies, and severity of disease in their PIP application.[37] The EMA recommends that selection of a particular excipient and excipient quantity should be justified based on overall risk to benefit evaluation of the product itself for its intended use and target age group.[7] For example, an excipient, which raises a minor safety concern, may still be allowed in exceptional cases taking into account the seriousness of the clinical indication or the advantages offered by a particular pharmaceutical form, route of administration, *etc.*

A combination of clinical, formulation and regulatory challenges (Figure 10.2) have to be addressed in the process of selecting and justifying the excipients for paediatric preparations.

10.2.2.1.1 Clinical Issues.   The five-year report to the European Commission (EC) on the public health effects of the Paediatric Regulation indicated that safety of excipients is one of three major topics discussed by the Paediatric Committee Formulation Working Group (PDCO FWG) members.[38]

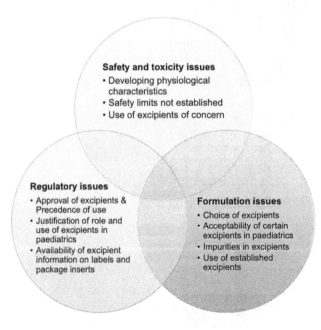

Figure 10.2   Issues of excipients in paediatrics.

The WHO 'Points to Consider' document,[20] the EMA reflection paper,[2] and the EMA Guideline on Pharmaceutical Development of Medicines for Paediatric Use, list known concerns about the use of excipients in paediatric patients. There are a number of reviews on the risks and benefits of excipients in compounded formulations,[39–41] which are mainly used in children due to unavailability of medicines for children. Table 10.1 summarises the adverse effects of commonly used excipients in paediatrics. There are the theoretical arguments on why the use of the excipients in children is matter of concern. These include: the developing physiological characteristics of children, inappropriate labelling of excipients on paediatric medicines and non-established safety limits of excipients in paediatrics.

Developing Physiological Characteristics of Children. Physiological differences between children and adults may affect the ways in which any xenobiotic works in the body. Agents that are effective in adults are not always effective in children. Infants have slower gastric emptying time, but faster intramuscular (IM) absorption, limited protein binding and immature enzymes. Their livers are immature and may not metabolise excipients as rapidly as expected; their kidneys are also small and immature. The immaturity of an infant's physiology (*e.g.* glomerular filtration rate, nervous system *etc.*) may contribute to elimination and functional sensitivities of chemical exposure.[51,52] The differential hepatic and renal clearance mechanisms, coupled with the immaturity of the blood–brain barrier in new-borns may lead to possible accumulation of excipients, which can lead to toxicity, such as central nervous system depression, renal failure, metabolic acidosis and seizures, as seen with propylene glycol, benzyl alcohol and benzoic acid.[53–55] Furthermore, children have larger liver:body and brain:body weight ratios and higher blood–brain barrier permeability, and small infants often have a two to three times longer half-life for elimination of medicines than adults, requiring lower doses of medicines. Consequently, even when a medicine has a known effect in adults, a linear dose per kg correlation often does not hold true with regards to small children. Dose-related adverse effects of excipients are of particular concern in preterm low birth-weight infant because of the known immaturity of hepatic and renal function in this population. For instance, dose related reversible central nervous system (CNS) effects have been reported in children after receiving intravenous injection for long term therapy in which propylene glycol was a cosolvent.[56–58] Furthermore, the growth is not a linear process; age-associated changes in body composition and organ functions are dynamic and can be discordant during the first decade of life.[3] Compared with adults, neonates and infants can be anticipated to have the greatest differences in pharmacokinetics and susceptibility to excipient toxicity—the youngest being the most likely to exhibit aberrant responses. It is difficult to generalize about age-dependent deficiencies in the metabolism of excipients because different enzymatic pathways seem to exhibit dissimilar maturational patterns.[3] It is dependent on the timing of the exposure during developmental life-stages, the kinetic and dynamic

Table 10.1 Reported adverse effects caused by excipients especially in children.

Lactose	Diarrhoea, malabsorption, vomiting, flatulence (in patients with lactose intolerance), jaundice, hypoglycaemia, CNS symptoms, cataracts (in patients with galactosaemia)	42–44
Sweeteners and flavouring agents:		
Aspartame	Headache, *grand mal* seizures, memory loss, gastrointestinal symptoms, dermatological symptoms (large quantities). Potentially toxic metabolites methanol, aspartic acid and phenylalanine. Phenylalanine is harmful in patients with phenylketonuria. Aspartic acid is neurotoxic and epileptogenic. Lastly, aspartame has been blamed for causing hyper-activity in children; the US the acceptable daily intake is 50 mg kg^{-1} day^{-1}	36, 45–47
Fructose	Hypoglycaemia (in patients with fructose intolerance)	44
Menthol	Hypersensitivity reactions, systemic allergic reactions. In infants cause isolated cases of spasm of the larynx. A few cases of nervous or digestive system disturbance have been associated with excessive inhalation or oral exposure to menthol	44 and 45
Peppermint oil	Atrial fibrillation, muscle pain, cooling or burning sensations	44
Saccharin Saccharin sodium	Irritability, hypertonia, insomnia, opisthotonus and strabismus, cross-sensitivity with sulfonamides; the most frequently described adverse reactions are dermatological and represented by urticaria, pruritus, dermatitis and photosensitivity. Other systemic reac-tions have been however reported: irritability, insomnia, opisthotonos and strabismus in children assuming saccharin-containing feed formulas. Approved for children over 3 years of age. Banned in Canada, allowed in the USA and Europe. The American Medical Associa-tion recommended limiting the use of this synthetic sweetener in food and pharmaceutical products intended for the paediatric population, the average acceptable intake is 0.6–0.9 mg kg^{-1} day^{-1} for the general population and 0.6–2.3 mg kg^{-1} day^{-1} for diabetic patients	36, 46, 48
Sodium cyclamate	Incidence of bladder cancer increased in rats. Use is restricted in many countries, Banned in the USA and Canada, allowed in Europe	48
Sorbitol, mannitol, xylitol	Large amounts: osmotic diarrhoea; "fructose intolerance" 0.15 g kg^{-1} day^{-1} is well tolerated in males and 0.3 g in females. The medicinal intake of sorbitol in paediatric population has been associated with disorders of the gastrointestinal tract, above all diarrhoea and malabsorption. A maximum intake limit neither for the paediatric population nor for adults has been defined, however, it has been suggested that a 20 g daily intake should possibly represent a reasonable limit for an average weight adult	36 and 46

Sucrose	Tooth decay, carcinogenicity, increased degradation of active drug, allergic reactions (very rare); diabetes mellitus or rare hereditary problems of fructose intolerance, glucose–galactose malabsorption, or sucrose–isomaltase insufficiency represent risk factors for sucrose adverse effects	44, 45, 48
Colouring agents:		
Azo dyes	Anaphylactic reactions, angioedema, asthma, urticaria, hyperkinesis, cross-sensitivity with acetylsalicylic acid, sodium benzoate and indomethacin (tartrazine FD&C yellow 5 = E102, sunset yellow FD&C 6 = E110)	46
Quinoline dyes	Contact dermatitis	46
Triphenylmethane dyes	Bronchoconstriction (brilliant blue FCF: FD&C blue 1 = E133), erythema multiforme-like skin rash (fast green FCF: FD&C green 3), anaphylaxis, angioedema (fluorescein: FD&C yellow 7)	46
Xanthine dyes	Photosensitizer (eosin: FD&C red 22), carcinogenicity (erythrosine: FD&C 3 = E127)	46
Preservatives and antibacterial agents:		
Benzalkonium chloride	Dose-related bronchoconstriction, cough, burning sensation, occasionally facial flushing, pruritus	47 and 48
Benzoic acids and benzoates	Displacement of bile from albumin binding sites in premature neonates, 'gasping syndrome'	48 and 49
Benzyl alcohol	A number of neonatal deaths and severe respiratory and metabolic complications (32–105 mg kg^{-1} day^{-1}), bronchitis, haemoptysis, hypersensitivity reactions (rare)	47
Boric acid	Is not used internally owing to its toxicity: death from ingestion of less than 5 g in young children	45
Parabens	Skin sensitization and cross-sensitization with each other concern has been expressed over the use of methylparaben in infants' parenteral products because bilirubin binding may be affected, which is potentially hazardous in hyperbilirubinaemic neonates. The WHO has set an estimated total acceptable daily intake for methyl-, ethyl-, and propyl-parabens at no more than 10 mg kg^{-1}	45, 48, 49
Sodium benzoate	Non-immunological contact urticaria, anaphylaxis. It has been recommended that sodium benzoate injection should not be used in neonates	45
Sodium borate	Damaged skin, severe toxicity (vomiting, diarrhoea, erythema, CNS depression, kidney damage). Lethal oral intake 5 g in children	45
Surfactants and solubilising agents:		
Ethanol	Accumulation of acetaldehyde. In the USA, the maximum quantity of alcohol included in over the counter (OTC) medicines is: 10% v/v for use by individuals of 12 years of age and older, 5% v/v for children aged 6–12 years of age, and 0.5% v/v for children under 6 years of age. In Europe there are no limits set	45, 46, 48

(continued)

Table 10.1 (continued)

Ethylene glycol	Renal failure (in 1937, children treated with sulphanilamide elixir developed renal failure traceable to the ethylene glycol which had been used as a solvent). The WHO has set an estimated acceptable daily intake of polyethylene glycols at no more than 10 mg kg^{-1}	45 and 46
Glycerol	>40% in volume: mucositis, diarrhoea, electrolyte disturbances	46
Polysorbate	Hypersensitivity, serious adverse effects, E-Ferol syndrome: thrombocytopenia, renal dysfunction, hepatomegaly, cholestasis, ascites, hypotension and metabolic acidosis, including some deaths, in low-birth-weight infants. The WHO has set an estimated acceptable daily intake at no more than 25 mg kg^{-1}	45 and 48
Propylene glycol	One-third as intoxicating as ethanol, effects on central nervous system, ototoxicity, cardiac arrhythmias, seizures, osmotic laxative effects, contact dermatitis lactic acidosis (especially in neonates and children less than 4 years of age) acceptable daily intake up to 25 mg kg^{-1}. Not recommended for children under 4 years of age (limited alcohol dehydrogenase) half-life 17 h in neonates (5 h in adults)	45, 47, 48
Polyethylene glycol	Metabolic acidosis in neonates and infants <6 months. Children between 1 and 6 years: 6.5 g to treat constipation	45
Miscellaneous groups, e.g. antioxidants, lubricants, etc.:		
Liquid paraffin	Lipoid pneumonia caused by aspiration or use of ophthalmic preparations. Should not be used in very young children	45
Potassium metabisulphite	Bronchospasm, anaphylaxis (especially in those with a history of asthma or atopic allergy)	45
Povidone	Anaphylactic reaction	48
Sulphites	Wheezing, dyspnoea, chest tightness (in patients with known reactive airway disease). Anaphylaxis, hives, itching	47 and 50
Thymol	Respiratory arrest, nasal congestion oedema (reported in new-born). Not for children under 5 years of age	45

characteristics of the specific excipient, and the exposure situation.[59-61] The toxicity of some common excipients, like lactose, may differ across the various paediatric sub-groups and between paediatric and adult patient groups.[62] More than one system can be susceptible and different pathology may occur depending on the dose and timing of exposure of excipients. Also depending on the dose and timing of exposure during gestation, effects may be severe and immediately obvious, or subtle and delayed. Certain excipients may lead to life-threatening toxicity in paediatric patients when multiple doses of medications with the same preservative are employed (*e.g.* benzyl alcohol and benzoic acid).[63]

Safety Limits Not Established for Paediatrics. The literature available on excipient use in the paediatric population reveals that the harm caused due to the excipients is often associated with use of higher amounts of excipients than the Acceptable Daily Intake (ADI)[‡] for adults. For instance, neonates receiving propylene glycol in doses exceeding 2000 mg kg^{-1} day^{-1} exhibited significantly higher degrees of hyperosmolality than their counterparts receiving more than 200 mg kg^{-1} day^{-1}.[64] High doses of propylene glycol have been associated with cardiovascular, hepatic, respiratory adverse events and with toxic effects on the CNS in new-borns and infants.[65] In a UK-based study Whittaker *et al.* described that during their hospital stay, 38 infants were exposed to over 20 excipients including ethanol, propylene glycol and high concentrations of sorbitol.[41] By calculating age-corrected exposure, the authors showed that in several neonates weekly exposure to excipients exceeded the limit that was considered safe in adults.

The underlying issue is that the accepted daily and cumulative intake of excipients has usually not been established for paediatrics and the applicability of the adults ADI to infants and children is questionable. The Joint Expert Committee on Food Additives (JECFA) of the Food and Agriculture Organisation of the United Nations (FAO) and the WHO set ADI for substances used as food additives. No limit of acceptable exposure has been defined for substances used as excipients in medicinal product formulations, neither for the adult nor for the paediatric population. The issue of sensitivity of children compared with adults has been largely ignored. Children are more likely than adults to exceed the ADI or tolerable daily intake (TDI),[§] due their low body weights. The concern is even greater for children from six months to 12 years. Poly-pharmacy increases the probability of common excipients exceeding safe threshold levels, potentially putting patients at an increased risk of developing adverse effects. The need for the development of a child- and neonatal-specific ADI has been highlighted in the literature.[66]

[‡]The ADI is "an estimate of the amount of a food additive, expressed as μg or mg per kg body weight, that can be ingested daily over a lifetime by humans without appreciable health risk."[94]

[§]A TDI is an estimate of the amount of a substance in air, food or drinking water that can be taken in daily over a lifetime without appreciable health risk. TDIs are calculated on the basis of laboratory toxicity data to which uncertainty factors are applied.

The permissible daily intake (PDE) for excipients has been determined for a few excipients as part of revision of "Guideline on Excipients in the Label and Package Leaflet of Medicinal Products for Human Use".[67] For instance, a recent refection paper from the EMA has suggested a PDE for propyl paraben. However a PDE has not been set for children under 2 years old because of the uncertainty about the metabolizing capacity at this very early age, and the absence of animal data corresponding to this age group.[68]

In such scenarios, safety assessment of excipients in paediatrics has to be considered on case-by-case basis by systematically assessing the available literature. It would be useful for formulators to have the list of an acceptable range/level or consolidated information on safety and toxicity studies on excipients, to establish the acceptable level for the most common excipients used in paediatric formulation products.

10.2.2.1.2   Formulation Issues.   A key consideration for paediatric dosage forms is understanding the limitations in the type of excipient that can be used and also the amounts and concentrations that can be administered. For instance, injectable products require a unique formulation strategy. The formulated product must be sterile, pyrogen-free, and, in the case of solution, free of particulate matter. No colouring agent may be added solely for the purpose of colouring the parenteral preparation. The formulation should preferably be isotonic, and sterility requirements demand that an excipient is able to withstand terminal sterilization or aseptic processing. These factors limit the choice of excipients available.[28] For formulation of oral liquids, several excipients may be needed as solvents, bulking agents, viscosity modifiers, wetting agents *etc.* to make a solution or suspension suitable for volumetric dosing. This may result in a higher potential for drug–excipient and for excipient–excipient incompatibilities and, thus, adds to the complexity of preformulation studies. Also, excipients may contain (or develop over time) trace amounts of their own degradation products that may negatively affect the stability of the API, the colour and/or the level of taste masking in the formulation. Examples are aldehydes and peroxides. Modern concepts of design of experiment (DoE) and quality by design (QbD) need to be applied to understand the robustness of such formulations and to establish the critical quality attributes of excipients to be used for routine manufacturing of the paediatric product.[35] Hence, from a formulator's perspective, one of the challenges in working with excipients may relate to limited choice. There is no reference list available of excipients generally considered safe for use in paediatric formulations.

10.2.2.1.3   Acceptability of Certain Excipients in Paediatrics.   There are limitations on choice and concentration of certain groups of excipients for paediatric patients. The selection of colourants, sweeteners and preservatives is based on several acceptance criteria. These include regulatory acceptance, toxicity, function (such as mouthfeel, viscosity and taste), disease state (acute *versus* chronic and the disease itself), administration (dose strength, volume and frequency), patient population, market potential and dosage-form characteristics.

Sweeteners. A key stumbling block to administering medicine orally to children is 'taste', with over 90% of paediatricians reporting that a drug's taste and palatability were the biggest barriers to completing treatment. Taste (and aftertaste) are particularly crucial for compliance in children.[69] Therefore, natural (*e.g.* sucrose, dextrose, fructose and lactose) or artificial (*e.g.* saccharin, cyclamate and aspartame) sweeteners and flavouring agents are frequently used to improve the palatability of medications and ensure good compliance. The choice of natural *versus* artificial sweeteners (*e.g.* syrup *versus* sugar-free (SF) preparations) is critical. Artificial sweeteners, although typically well tolerated, may have adverse reactions when used in children.[70] Hence sweeteners and their levels have to be judiciously chosen. The decision in choosing sweeteners has to be balanced with supportive information and not overly constraining. Trade-offs have to be identified and carefully considered by all stakeholders (*e.g.*, clinical, regulatory, pharmaceutical development and marketing). For example, paediatric drug products may need more than one type of sweetener and taste modifier to effectively mask the bitterness of the API that is strong in intensity and long in duration. Nutritive sweeteners and sugar alcohols alone do not provide relative sweetness. High-intensity sweeteners do not provide bulk, build viscosity or provide beneficial mouthfeel effects and as such do not work in formulations by themselves.[71] Thus, they are often used in combination with each other. As long as there is evidence of absence of adverse effects, multiple sweeteners may be acceptable, however, pharmaceutical companies have to provide thorough justification or clarification on the need for and concentration of the sweeteners or reduce the number of sweeteners.

Colourants. Colourants are dyes, pigments or other substances that can impart colour when added or applied to foods, drugs, cosmetics, medical devices or the human body. Selection of the appropriate colourant and its purpose in a specific pharmaceutical dosage form plays an important role in manufacturing of pharmaceutical dosage forms. In selecting a colourant for a given application, prime consideration is given to the type of formulation in which the colourant is to be incorporated. Colour also influences the taste and flavour perception and may affect patient compliance.[72] Tablet colour has been linked with taste, where pink is considered to be sweeter than red, and yellow is considered to be salty irrespective of its actual ingredients.[73] Colour preferences among children have been shown to be stereotypically gender dependent, and they seem to prefer brightly coloured medicines.[74]

The number of colouring agents that are acceptable for use in medicines is limited but their wide use in the food industry has indicated that a number of colouring agents in current use have been associated with reports of hypersensitivity and hyperkinetic activity, especially among children.[75]

The safety of azo dyes remains a big issue.[76] Some of these dyes are no longer used in food, but the restrictions do not extend to many medicines designed for children. For instance, Allura Red AC is not recommended for children. It is banned in selected countries like Denmark, Belgium, France, Switzerland and Sweden. The use of azo dyes for paediatric medicines is discouraged. The 2007 "Guideline on Excipients in the Dossier for Application for Marketing Authorisation of a Medicinal Product" clearly indicates that azo dyes (and other synthetic colouring agents) should not be used in (new applications for) paediatric drug products.[77]

Several regulations are available on the aspects of colourants including their procedures for use, provisionally and permanently certified and uncertified colour additives and use levels and restrictions for each colouring additive.[78-80] Restrictions or bans on the use of some colouring agents have been imposed in some countries, while the same colours may be permitted for use in a different country. As a result the same colour may have a different regulatory status in different territories of the world.[81] With the differences in colourant regulations worldwide and the need for various performance attributes based on the dosage form, there are numerous considerations that must be assessed.

Preservatives. Antimicrobial preservatives are normally added to prevent microbial proliferation arising under in-use conditions. The use of preservatives is currently one of the most controversial issues in paediatric drug development. The use of preservatives is discouraged in general, especially when considering the suitability of related formulations to the paediatric population. Two general issues are linked to the use of these preservatives, one of which is the choice of materials. Plastic containers and dispensing devices pose problems such as permeation of preservatives through the container or interaction with the plastic materials. A second issue is the high incidence of local side effects attributed to preservatives. The discussion is controversial, and published preclinical and clinical studies are not always consistent. It seems to be clear that short-term use of preparations containing preservatives at low concentrations is well tolerated, but preservatives can cause serious inflammatory effects with long-term use.[82] The adverse effects may include chemical irritation, hyperactivity and allergic reactions. Hence evidence of safety of preservatives used is required, together with thorough justification for the choice of the preservative. Typically, the use of the older preservatives (*e.g.* imidurea, bronopol, hexachlorpene) in new products has been largely discontinued because of safety considerations.[83] There is a limited number of approved preservatives available for multi-use oral or topical products, and options are even more limited for dosage forms such as parenteral. For instance, benzyl alcohol is not recommended for use in parenteral products due to fatal toxic syndrome in low weight neonates.[84] The long-chain alkyl alcohols, cetyl and stearyl alcohol used as preservatives in topical products can lead to contact allergies and irritant reactions.[85]

2-Phenylethanol can be mildly irritant to skin, eye and mucous membranes. A large number of clinical and experimental studies have revealed that preservatives in topical ophthalmic medications have been demonstrated to produce effects from inflammation/hypersensitivity to permanent cytotoxic effects involving all structures of the eye.[86] Benzalkonium chloride and other quaternary ammonium preservatives have direct toxic effects on the cells and damage the cornea.[87]

Alternatively, it is known that a combination of preservatives can have a synergistic effect on antimicrobial efficacy, allowing smaller amounts to be used, in total and per excipient and this approach might be considered if they are known to be safe *e.g.* benzalkonium chloride (BKC) is ineffective against some strains of *Pseudomonas aeruginosa*, *Mycobacterium* and *Trichophyton* but combinations with benzyl alcohol, 2-phenylethanol or 3-phenylpropanol enhances anti-*Pseudomonad* activity, probably by increasing the permeability of the cells to the antimicrobial agents.[88]

There is a regulatory expectation that the reason for preservative inclusion, proof of efficacy, safety information, control methods in the finished product and details of labelling in the finished product should all be addressed by the applicant.[89] The EMA has recommended that the levels of preservatives within a formulation should be maintained at the minimum concentration consistent with antimicrobial effectiveness in each individual preparation.[68] Pharma companies are encouraged to formulate preservative-free products. Preservative-free approaches are still in their infancy and much more research and analysis of existing information is required before they can be considered on an equal footing with preserved approaches.

10.2.2.1.4   Regulatory Approval of Excipients and Precedence of Use.

There is no general approval process for excipients and they are approved together with a drug (as a drug product) under particular settings (*e.g.*, indication, route, dose-levels). The excipients are scrutinized through cross-references to pharma/food/cosmetic compendia, reference in an Abbreviated New Drug Application (ANDA) or NDA for a particular function in a drug product and permitted list of colours and flavours in EU food legislation. The precedence of use of marketed excipients is assessed by reference to the FDA's Inactive Ingredients Database (IID), the Japanese Pharmaceutical Excipients Dictionary (JPED), and drug catalogues such as Dictionnaire Vidal or Rote Liste. However, if there is no precedence of use in a drug product, then the excipient is considered as new excipient and the manufacturer has to develop the safety information appropriate to their intended use. The FDA has issued guidance concerning the safety testing required for novel excipients.[34] The IPEC Europe Safety Committee also has published a guide for the qualification of excipient ingredients by excipient suppliers and pharmaceutical users.[31] The additional safety data is required to introduce a novel excipient to a pharmaceutical product. The resources and time associated with this requirement makes formulation scientists hesitant to try new excipients. Hence the biggest challenge for formulators is the limited and scattered information on known and approved excipients available for use in paediatrics.

Justification of Role and Use of Excipients in Paediatrics. Excipients may have avoided detailed regulatory attention because it was not always perceived that they have a purpose but now marketing authorisation (MA) applicants are required to state and justify the role an excipient has to play. Recent legislative changes require that companies provide the supportive data and complete justification on use of excipients in paediatric formulations proposed in PIP. However, insufficient justification of the chosen excipients related to age and daily dose of excipient(s) and insufficient discussion on the feasibility of replacing excipients with potential safety issues are concerns the regulators often encounter in PIPs.[90]

From the regulators point of view, it is not yet clear to what extent a precautionary approach to the excipient composition should be envisaged in the PIPs. For example, it is not clear whether to accept or ask companies to replace the excipients that may cause problems in children with less common deficiencies, *e.g.* hereditary fructose/galactose intolerance? or lactose (which may cause problems in some children with lactose intolerance).[91] A structured risk analysis framework[11] assessing the available information may allow an informed discussion among regulators, industry and academia to come up with a transparent and consistent approach to this dilemma for future applications.

Availability of Excipient Information on Labels and Package Inserts. With regard to labelling of the medicinal products, Article 54 (1) (c) of Council Directive 2001/83/EEC requires that excipients known to have a recognized action or effect need to be declared on the labelling of all other medicinal products. According to Article 59 (1) (a) a full statement of the active substance and excipients should be included in the package leaflet. Also all excipients, which are present in the product, should be listed in the Summary of Manufacturing Product Characteristics (SmPCs), even those present in small amounts. Recently, the EMA has undertaken the task of updating the information in the package leaflet to update the thresholds and toxicological profile and to adjust them in relation to different age groups. A concept paper on the need for revision of the guideline was released in 2012. However regulatory authorities do not yet adequately regulate or enforce its guidelines on the requirements for quantitative information on excipients on package inserts or labels. Although it is acceptable that safeguard is granted to the intellectual property of drug developers (namely, quantitative details), information on excipients should be sufficient to allow precautions to be taken when needed. The need for drug users (health care professionals, patients, caregivers) to obtain adequate information on the drug product excipient composition is commonly acknowledged.[92,93] Information on excipient content could prove helpful in a clinical setting where no alternatives are available.

In a regulatory context, it is important to consider all the existing guidance documents that support the development of paediatric formulations known to be safe and effective for neonates, infants and children of all ages. The pharmaceutical companies are struggling to find the existing information on safety and toxicity of excipients in paediatrics as it is scattered around various sources. In general, there is a tendency to apply the precautionary principle as justification for excluding excipients from medicines given to paediatric population. However excluding excipients is not always appropriate. The Safety and Toxicity of Excipients for Paediatrics (STEP) database project was hence developed by EuPFI consortium in collaboration with USPFI,[¶] to bridge the gap in resources for safety and toxicity of excipients for paediatrics and address the challenges in information gathering and evaluation.[96,97] Similar discussions were being carried out in Pediatric Formulation Initiative (PFI) in the USA to address safety issues and problems associated with the lack of adequate paediatric[‖] formulations.

10.2.3   Administration

The non-acceptability of medicine can have major implications including medicine errors, under- or over-dosing and poor adherence and therefore suboptimal therapy. Patient acceptability is defined by the EMA as the overall ability and willingness of the patient to use and their caregiver to administer the medicine as intended. Higher acceptability renders the medicine less prone to any type of modification prior to administration. All major components of formulation design can influence patient acceptability. The key design aspects include composition (qualitative and quantitative), route of administration, dosage form, dosing frequency, packaging, administration device and user's instructions. The understanding of paediatric patient acceptability to formulations has not been fully established yet. There is lack of standardization of the measurement of acceptability and data interpretation, nevertheless further research is expected to be conducted to define the dosage form attributes and their perception to determine the patient acceptability.

Acceptability is a term different from palatability or swallowability for orally administered formulations. Palatability is defined as the overall appreciation of a medicinal product in relation to its smell, taste, aftertaste and

---

[¶]The United States Pediatric Formulation Initiative (US-PFI) is a project of the Eunice Kennedy Shriver National Institute of Child Health and Human Development (NICHD). The PFI was established in 2005 to address the issue of the lack of appropriate formulations in children and to use this activity as a means to improve paediatric formulations.

[‖]The European spelling "paediatric" is used throughout the chapter unless specifically referring to the USA. In cases of US references the US spelling "pediatric" has been adopted.

texture. The sensory evaluation of the dosage form influences the patient's ability and willingness to take the medicine. The gold standard to assess the sensory attributes of dosage forms is human panel study. Providing there is evidence for the correlation between *in vivo* animal or *in vitro* characterization studies and human panel data, predictive methodologies can also be applied to understand the patient acceptability. Taste has been the mostly studied among the other sensory attributes. The taste assessment can be performed by applying *in vitro* and *in vivo* methods. The *in vitro* tool, e-tongue has been studied to evaluate the taste of medicinal formulations, though there are limitations of the method depending on the physicochemical properties of the drug molecule. The *in vivo* animal model (Brief Access Taste Aversion) is promising as a predictive method to assess the perceived aversive taste of drug formulations.

Acceptability is also not a synonymous term for medicine adherence (or compliance) which is generally defined as the extent to which patients take medications as agreed with their healthcare providers. Acceptability can be seen being the first stage of adherence due to its effect on the agreement of the child to take the medicine, it does not result in the optimum adherence, as controlled by multiple factors ranging from the clinical condition to the treatment setting. The age subset of the paediatric population also has an effect on the compliance with the medicine. Adolescents may show a different adherence profile due to their autonomy and self-management of their medicine compared with younger children.

10.3   Patient Centric Pharmaceutical Drug Product Design and Future Visions

The objectives of the Paediatric Regulation in Europe (2007)[95] was to stimulate the development of paediatric medicines but also to provide more information on their use, as a response to the lack of evidence and approval of medicines for children. In fact similar initiatives started in the 80s in the USA.

The tools in place in the European Union encompass the PDCO, the European Network for Paediatric Research (Enpr-EMA), Paediatric Use Marketing Authorizations (PUMA) and importantly PIPs. Although a holistic approach in paediatric drug development is required, with concomitant advances in clever clinical trial designs, modelling/simulation approaches, refining endpoints and biomarkers, PIPs are crucial as they offer a framework for developing clinically and age-relevant paediatric dosage forms so that children of all ages and their caregivers have access to safe and accurate medicines.

There has been a lack of evidence to guide the design of age-appropriate and acceptable dosage form, which has resulted in a longstanding knowledge gap in paediatric formulation development. A list of criteria for screening PIPs with regard to paediatric-specific quality issues and referring them to the PDCO Formulation Working Group for discussion has been published (http://www.ema.europa.eu/docs/en_GB/document_library/Other/2014/01/WC500159380.pdf). This provides a structured framework for pharmaceutical design options against pre-determined criteria relating to efficacy, safety and patient access, this latter being particularly complex due to the diverse paediatric population.

There is a drive now to carefully consider and balance the quality target product profile against not only technical challenges and development feasibility but also the varied needs and abilities of children as well as their carers. Patient centricity can be defined as '*Putting the patient first in an open and sustained engagement of the patient to respectfully and compassionately achieve the best experience and outcome for that person and their family*'. No doubt the binding elements of the Paediatric Regulation has steered research with and for children and their families to refine end-user requirements in order to guide dosage form design and formulation selection.

In a decade the Paediatric Regulation has certainly had a positive effect on paediatric drug development, yet the years to come will reveal the true extent of this effect as we catch up with long deferrals for completion of paediatric studies requested by pharmaceutical companies and gather real life outcomes from post marketing studies.

References

1. ICH. ICH, *Topic E 11 Clinical Investigation of Medicinal Products in the Paediatric Population, Note for Guidance on Clinical Investigation of Medicinal Products in the Paediatric Population (CPMP/ICH/2711/99)*, 2001.
2. EMA, *European Medicines Agency Committee for Medicinal Products for Human Use (CHMP), Reflection Paper: Formulations of Choice for the Paediatric Population*, 2006.
3. G. L. Kearns, S. M. Abdel-Rahman, S. W. Alander, D. L. Blowey, J. S. Leeder and R. E. Kauffman, Developmental pharmacology–drug disposition, action, and therapy in infants and children, *N. Engl. J. Med.*, 2003, **349**(12), 1157–1167.
4. M. Molteni, E. Clementi and G. Zuccotti, Introduction to the new "perspectives in paediatric pharmacology" series, *Pharmacol. Res.*, 2011, **63**(5), 361.
5. FDA, *General Clinical Pharmacology Considerations for Pediatric Studies for Drugs and Biological Products, Guidance for Industry*, US Department of Health and Human Services Food and Drug Administration, Center for Drug Evaluation and Research (CDER), 2014, https://wwwfdagov/downloads/drugs/guidances/ucm425885pdf.
6. F. L. Lopez, T. B. Ernest, C. Tuleu and M. O. Gul, Formulation approaches to pediatric oral drug delivery: benefits and limitations of current platforms, *Expert Opin. Drug Delivery*, 2015, **12**(11), 1727–1740.
7. EMA, *Guideline on the Pharmaceutical Development of Medicines for Paediatric Use*, European Medicines Agency Committee for Medicinal Products for Human Use and Paediatric Committee, 2013.

8. K. Wer, in *The Global Framework of Paediatric Drug Development, European Union Paediatric Regulation: Theory and Practice, Guide to Paediatric Drug Development and Clinical Research*, ed. K. Rose and J. N. Van den Anker, Basel, Karger, 2010, pp. 1–20, DOI: 10.1159/000315565.

9. F. T. P. Rocchi, The development of medicines for children. Part of a series on Pediatric Pharmacology guest, in *Pharmacol. Res.*, ed. G. Zuccotti, E. Clementi and M. Molteni, 2011, vol. 64, (3), pp. 169–175.

10. D. A. van Riet-Nales, Child Friendly Medicines: Availability, Pharmaceutical Design, Usability and Patient Outcomes, Thesis Utrecht University, 2014.

11. M. A. Turner, J. C. Duncan, U. Shah, T. Metsvaht, H. Varendi and G. Nellis, *et al.*, Risk assessment of neonatal excipient exposure: Lessons from food safety and other areas, *Adv. Drug Delivery Rev.*, 2014, **73**, 89–101.

12. EMA. ICH, *Topic E 11 Clinical Investigation of Medicinal Products in the Paediatric Population*, 2001.

13. WHO, *Forty Sixth Report: Development of Pediatric Medicines: Points to Consider in Formulation. WHO Technical Report Series*, World Health Organization, 2012, p. 790.

14. WHO, *Model Formulary for Children WHO Library Cataloguing-in-publication Data*, World Health Organization, 2010.

15. S. Abdulla and I. Sagara, Dispersible formulation of artemether/lumefantrine: specifically developed for infants and young children, *Malar. J.*, 2009, **8**(suppl. 1), S7.

16. C. Milne and J. B. Bruss, The economics of pediatric formulation development for off-patent drugs, *Clin. Ther.*, 2008, **30**(11), 2133–2145.

17. V. R. C. Ivanovska, L. van Dijk and A. K. Mantel-Teeuwisse, Pediatric drug formulations: a review of challenges and progress, *Pediatrics*, 2014, **34**(2), 361–372.

18. D. Van Riet-Nales, K. de Jager, A. Schobben, T. Egberts and C. Rademaker, The availability and age-appropriateness of medicines authorized for children in The Netherlands, *Br. J. Clin. Pharmacol.*, 2011, **72**(3), 465–473.

19. T. Sam, T. B. Ernest, J. Walsh and J. L. Williams, A benefit/risk approach towards selecting appropriate pharmaceutical dosage forms – an application for paediatric dosage form selection, *Int. J. Pharm.*, 2012, **435**(2), 115–123.

20. WHO, *Development of Paediatric Medicines: Points to Consider in Pharmaceutical Development. QAS/08.257/Rev.3*, World Health Organization (WHO), 2011.

21. S. Ranmal, Acceptable medicines for children: end-user insights to support dosage form design, Doctoral thesis, UCL (University College London), 2015.

22. D. Adkin, S. Davis, R. Sparrow, P. Huckle and I. Wilding, The effect of mannitol on the oral bioavailability of cimetidine, *J. Pharm. Sci.*, 1995, **84**(12), 1405–1409.

23. A. W. Basit, F. Podczeck, J. M. Newton, W. A. Waddington, P. J. Ell and L. F. Lacey, Influence of polyethylene glycol 400 on the gastrointestinal absorption of ranitidine, *Pharm. Res.*, 2002, **19**(9), 1368–1374.

24. A. Rostami-Hodjegan, M. R. Shiran, R. Ayesh, T. J. Grattan, I. Burnett and A. Darby-Dowman, *et al.*, A new rapidly absorbed paracetamol tablet containing sodium bicarbonate. I. A four-way crossover study to compare the concentration-time profile of paracetamol from the new paracetamol/sodium bicarbonate tablet and a conventional parac-etamol tablet in fed and fasted volunteers, *Drug Dev. Ind. Pharm.*, 2002, **28**(5), 523–531.

25. A. García-Arieta, Interactions between active pharmaceutical ingredients and excipients affecting bioavailability: impact on bioequivalence, *Eur. J. Pharm. Sci.*, 2014, **65**, 89–97.

26. P. Crowley and L. G. Martini, *Excipients in Pharmaceutical Products: Encyclopedia of Pharmaceutical Technology*, Marcel Dekker Inc, New York, 2002.

27. F. Nishath, M. Tirunagar, Q. Husna Kanwal, A. Nandagopal and V. R. Jangala, Drug-excipient interaction and its importance in dosage form development, *J. Appl. Pharm. Sci.*, 2011, **01**(06), 5.

28. S. Pramanick, D. Singodia and V. Chandel, Excipient selection in paren-teral formulation development, *Pharmatimes*, 2013, **45**(3), 65.

29. Webster, *Webster's II New College Dictionary*, Houghton Mifflin Company, Boston, 1995.

30. W. Saunders, *Dorland's Medical Dictionary*, W.B. Saunders, Philadelphia, PA, 25th edn, 1974.

31. IPEC, *The International Pharmaceutical Excipients Council, Qualification of Excipients for Use in Pharmaceuticals*, 2008.

32. EC, *Council and the European Parliament, the Falsified Medicines Directive (Directive 2011/62/EU)*, Official Journal of the European Union, 2011.

33. Faber & Faber, *The Nurse Dictionary*, Morton's, London, 24th edn, 1957.

34. FDA, *Guidance for Industry: Nonclinical Studies for the Safety Evaluation of Pharmaceutical Excipients*, 2005.

35. S. Maldonado and D. Schaufelberger, Pediatric formulations, *Am. Pharm. Rev.*, 2011, http://www.americanpharmaceuticalreview.com/Featured-Articles/37186-Pediatric-Formulations/.

36. V. Fabiano, C. Mameli and G. V. Zuccotti, Paediatric pharmacology: remember the excipients, *Pharmacol. Res.*, 2011, **63**(5), 362–365.

37. ICH, *Pharmaceutical Development Q8(R2)*, 2009.

38. EMA, *5-year Report to the European Commission General Report on the Experience Acquired as a Result of the Application of the Paediatric Regulation*, 2012.

39. A. Lowey and M. Jackson, *Handbook of Extemporaneous Formulation – a Guide to Pharmaceutical Compounding*, Pharmaceutical Press, London, 2010.

40. M. Nahata, Safety of 'inert' additives or excipients in paediatric medicines, *Arch. Dis. Child.*, 2009, **94**, 392–393.

41. A. Whittaker, A. E. Currie, M. A. Turner, D. J. Field, H. Mulla and H. C. Pandya, Toxic additives in medication for preterm infants, *Arch. Dis. Child.*, 2009, **94**(4), F236–F240.

42. T. He, M. G. Priebe, H. J. Harmsen, F. Stellaard, X. Sun, G. W. Welling and R. J. Vonk, Colonic fermentation may play a role in lactose intolerance in humans, *J. Nutr.*, 2006, **136**(1), 58–63.

43. A. K. Campbell, J. P. Waud and S. B. Matthews, The molecular basis of lactose intolerance, *Sci. Prog.*, 2005, **88**(3), 157–202.

44. A. Kumar, A. T. Aitas, A. G. Hunter and D. C. Beaman, Sweeteners, dyes, and other excipients in vitamin and mineral preparations, *Clin. Pediatr.*, 1996, 443–450.

45. R. C. Rowe, M. E. Quinn and P. J. Sheskey, *Handbook of Pharmaceutical Excipients*, Pharmaceutical Press, 6th edn, 2009.

46. S. Pawar and A. Kumar, Issues in the formulation of drugs for oral use in children: role of excipients, *Pediatr Drugs*, 2002, **4**(6), 371–379.

47. *Anonymous. "Inactive" Ingredients in Pharmaceutical Products: Update (subject Review)*, American Academy of Pediatrics Committee on Drugs Pediatrics, 1997, vol. 99, (2), pp. 268–278.

48. I. Costello, P. Long, I. Wong, C. Tuleu and V. Yeung, *Paediatric Drug Handling*, Pharmaceutical Press, Cornwall, 2007.

49. A. Kumar, R. Rawlings and D. Beaman, The mystery ingredients: sweeteners, flavorings, dyes, and preservatives in analgesic/antipyretic, antihistamine/decongestant, cough and cold, antidiarrheal, and liquid theophylline preparations, *Pediatrics*, 1993, **91**, 927–933.

50. A. Pagliaro, Administering drugs to infants, children, and adolescents, in *Problems in Pediatric Drug Therapy*, ed. L. A. Pagliaro and A. M. Pagliaro, American Pharmaceutical Association, Washington, 4th edn, 2002.

51. J. C. Larsen and G. Pascal, Workshop on the applicability of the ADI to infants and children: consensus summary, *Food Addit. Contam.*, 1998, **15**(suppl. 001), 1–9.

52. P. Peters, Developmental toxicology: adequacy of current methods, *Food Addit. Contam.*, 2009, **15**(s1), 55–62.

53. K. Allegaert, S. Vanhaesebrouck, A. Kulo, K. Cosaert, R. Verbesselt and A. Debeer, *et al.*, Prospective assessment of short-term propylene glycol tolerance in neonates, *Arch. Dis. Child.*, 2010, **95**(12), 1054–1058.

54. A. Kulo, A. Smits, G. Naulaers, J. de Hoon and K. Allegaert, Biochemical tolerance during low dose propylene glycol exposure in neonates: a formulation-controlled evaluation, *Daru, J. Pharm. Sci.*, 2012, **20**(1), 5.

55. N. Shehab, D. D. Streetman and S. M. Donn, Exposure to the pharmaceutical excipients benzyl alcohol and propylene glycol among critically ill neonates, *Pediatr. Crit. Care. Med.*, 2009, **10**(2), 256–259.

56. M. L. Glover and M. D. Reed, Propylene glycol: the safe diluent that continues to cause harm, *Pharmacotherapy*, 1996, **16**(4), 690–693.

57. Y. Lolin, D. A. Francis, R. J. Flanagan, P. Little and P. T. Lascelles, Cerebral depression due to propylene glycol in a patient with chronic epilepsy-the value of the plasma osmolal gap in diagnosis, *Postgrad. Med. J.*, 1988, **64**(754), 610–613.

58. M. J. Kelner and D. N. Bailey, Propylene glycol as a cause of lactic acidosis, *J. Anal. Toxicol.*, 1985, **9**(1), 40–42.

59. R. Hasegawa, M. Hirata-Koizumi, M. Dourson, A. Parker, A. Hirose and S. Nakai, *et al.*, Pediatric susceptibility to 18 industrial chemicals: a comparative analysis of newborn with young animals, *Regul. Toxicol. Pharmacol.*, 2007, **47**(3), 296–307.

60. S. S. Olin and B. R. Sonawane, Workshop to develop a framework for assessing risks to children from exposure to environmental agents, *Environ. Health Perspect.*, 2003, **111**(12), 1524–1526.

61. G. Ostergaard and I. Knudsen, The applicability of the ADI (acceptable daily intake) for food additives to infants and children, *Food Addit. Contam.*, 1998, **15**, 63–74.

62. S. Edge, A. Kibbe and K. Kussendrager, in *Handbook of Pharmaceutical Excipients*, ed. R. C. Rowe, P. J. Sheskey and M. E. Quinn, Pharmaceutical Press, London, 2005.

63. B. Glass and A. Haywood, Stability considerations in liquid dosage forms extemporaneously prepared from commercially available products, *J. Pharm. Pharm. Sci.*, 2006, **9**, 398–426.

64. M. MacDonald, P. Getson, A. Glasgow, M. Miller, R. Boeckx and E. Johnson, Propylene glycol: increased incidence of seizures in low birth weight infants, *Pediatrics*, 1987, **79**(4), 622–625.

65. K. Arulanantham and M. Genel, Central nervous system toxicity associated with ingestion of propylene glycol, *J. Pediatr.*, 1978, **93**, 515–516.

66. P. Nydert and M. Ali, *Methyl Paraben, Usage of Preservatives in Drugs for Premature Neonates*, 2007.

67. EMA, *Concept Paper on the Need for Revision of the Guideline on Excipients in the Label and Package Leaflet of Medicinal Products for Human Use (CPMP/463/00)*, 2012.

68. EMA, *Committee for Medicinal Products for Human Use (CHMP), Reflection Paper on the Use of Methyl- and Propylparaben as Excipients in Human Medicinal Products for Oral Use. EMA/CHMP/SWP/272921/2012*, 2015.

69. C. Milne, The pediatric studies initiative, in *New Drug Development: A Regulatory Overview*, ed. M. Mathieu, PAREXEL International, Waltham, Mass, 8th edn, 2005.

70. J. Walsh, A. Cram, K. Woertz, J. Breitkreutz, G. Winzenburg and R. Turner, *et al.*, Playing hide and seek with poorly tasting paediatric medicines: do not forget the excipients, *Adv. Drug Delivery Rev.*, 2014, **73**, 14–33.

71. S. Chattopadhyay, U. Raychaudhuri and R. Chakraborty, Artificial sweeteners – a review, *J. Food Sci. Technol.*, 2014, **51**(4), 611–621.

72. J. Delwiche, The impact of perceptual interactions on perceived flavor, *Food Qual. Prefer.*, 2004, **15**, 137–146.

73. R. Srivastava and A. More, Some aesthetic considerations for over the-counter (OTC) pharmaceutical products, *Int. J. Biotechnol.*, 2010, **11**, 267–283.

74. C. Smith, H. Sammons, A. Fakis and S. Conroy, A prospective study to assess the palatability of analgesic medicines in children, *J. Adv. Nurs.*, 2013, **69**(3), 655–663.

75. I. Pollock, E. Young, M. Stoneham, N. Slater, J. D. Wilkinson and J. O. Warner, Survey of colourings and preservatives in drugs, *Br. Med. J.*, 1989, **299**(6700), 649–651.

76. D. McCann, A. Barrett, A. Cooper, D. Crumpler, L. Dalen and K. Grim-shaw, *et al.*, Food additives and hyperactive behaviour in 3-year-old and 8/9-year-old children in the community: a randomised, double-blinded, placebo-controlled trial, *Lancet*, 2007, **370**(9598), 1560–1567.

77. EMA, *European Medicines Agency Committee for Medicinal Products for Human Use (CHMP): Guideline on Excipients in the Dossier for Application for Marketing Authorisation of a Medicinal Product. CHMP/QWP/396951/06*, 2007.

78. FDA, *Food and Drug Administration,CFR – Code of Federal Regulations Title 2, Subchapter A—General, Part 70 Color Additives*, 2015.

79. EC, *European Parliament and Council Directive 94/36/EC of 30 June 1994 on Colours for Use in Foodstuff*, Official Journal of the European Communities, 1994.

80. EC, *Opinions – Scientific Committee on Medicinal Products and Medical Devices*, 1998.

81. A. Kumar and A. K. Madan, Color additives: legislative perspective in the United States, Europe, Australia, and India, *Int. J. Pharm. Compd.*, 2014, **18**(4), 293–300.

82. E. Tu, Balancing antimicrobial efficacy and toxicity of currently available topical ophthalmic preservatives, *Saudi J. Ophthalmol.*, 2014, **28**(3), 182–187.

83. D. Elder and P. Crowley, Antimicrobial preservatives part one: choosing a preservative system, *Am. Pharm. Rev.*, 2012, http://www.americanpharmaceuticalreview.com/Featured-Articles/341194-Antimicrobial-Preservatives-Part-One-Choosing-a-Preservative-System/.

84. J. Gershanik, B. Boecler, H. Ensley, S. McCloskey and W. George, The gasping syndrome and benzyl alcohol poisoning, *N. Engl. J. Med.*, 1982, **307**(22), 1384–1388.

85. A. E. Aakhus and E. M. Warshaw, Allergic contact dermatitis from cetyl alcohol, *Dermatitis*, 2011, **22**(1), 56–57.

86. J. Hong and L. Bielory, Allergy to ophthalmic preservatives, *Curr. Opin. Allergy Clin. Immunol.*, 2009, **9**(5), 447–453.

87. L. Rosin and N. Bell, Preservative toxicity in glaucoma medication: clinical evaluation of benzalkonium chloride-free 0.5% timolol eye drops, *Clin. Ophthalmol.*, 2013, 7, 2131.

88. M. Brown and R. Richards, Effect of ethylenediaminetetraaceticacid on the resistance of Pseudomonas aeruginosa to antibacterial agents, *Nature*, 1965, **207**, 1391–1393.

89. BfArM, *Bescheid des Bundesinstitut für Arzneimittel und Medizinproduke für benzalkoniumchlorid-haltige Arzneimittel zur Anwendung in der Nase, A 37489/38186/03*, Bonn, 2003.

90. R. Quijano, E. Desfontaine, S. Arenas-López and S. Wang, Pediatric formulation issues identified in paediatric investigation plans, *Expert Rev. Clin. Pharmacol.*, 2014, 7(1), 25–30.

91. D. Van Riet-Nales, P. Kozarewicz, B. Aylward, R. de Vries, T. Egberts and C. Rademaker, *et al.*, Paediatric drug development and formulation design—a European perspective, *AAPS PharmSciTech*, 2016, 1–9.

92. J. Lass, K. Naelapaa, U. Shah, R. Kaar, H. Varendi and M. Turner, *et al.*, Hospitalised neonates in Estonia commonly receive potentially harmful excipients, *BMC Pediatr.*, 2012, **12**(1), 136.

93. C. Cordner, N. Caldwell and P. Elliot, What else is in our children's medicine?, *Arch. Dis. Child.*, 2012, **97**(5), e2–e3.

94. WHO, *Principles for the safety assessment of food additives and contaminants in food*. Environmental Health Criteria no. 70, World Health Organisation, Geneva, 1987.

95. European Union, Regulation (EC) No 1901/2006 of the European Parliament and of the Council of 12 December 2006 on medicinal products for paediatric use and amending Regulation (EEC) No 1768/92, Directive 2001/20/EC, Directive 2001/83/EC and Regulation (EC) No 726/2004, 2007, http://ec.europa.eu/health/files/eudralex/vol-1/reg_2006_1901/reg_2006_1901_en.pdf.

96. S. Salunke, B. Brandys, G. Giacoia and C. Tuleu, The STEP (Safety and Toxicity of Excipients for Paediatrics) database: part 2 – the pilot version, *Int. J. Pharm.*, 2013, **457**(1), 310–322.

97. S. Salunke, G. Giacoia and C. Tuleu, The STEP (Safety and Toxicity of Excipients for Paediatrics) database. Part 1 – A need assessment study, *Int. J. Pharm.*, 2012, 435(2), 101–111.

CHAPTER 11

The Formulation of Biological Molecules

TUDOR ARVINTE*[a,b], AMELIA CUDD[a], CAROLINE PALAIS[a] AND
EMILIE POIRIER[a]

[a]Therapeomic Inc., BioPark Rosental, Mattenstrasse 22, WRO-1055, 4002
Basel, Switzerland; [b]Department of Pharmaceutical Sciences, University of
Geneva – University of Lausanne, 30 Quai Ernest Ansermet, 1205 Geneva,
Switzerland
*E-mail: tudor.arvinte@unige.ch

11.1 Definitions: Biologics and Their Formulations

The terms *biological products, biologics, biological, biopharmaceuticals, biopharmaceutics* and *biotechnology* are widely used in popular, scientific,
industrial and financial literature. Originally, in the scientific and industrial
community, the terms *biotech drug, biopharmaceutics* and all other derivatives of these words were used for products manufactured using biotech-
nological methods. The best examples are the recombinant proteins, which
entered the market in 1982 with the launch of recombinant human insulin. The wider use of the biopharmaceutical and biotechnology terminology
was critically discussed by R. A. Rader.[1,2] The press and financial documents
have used the term "biotechnology" to include small-molecule drugs that
were chemically synthesized and not manufactured by means of biotech-
nological methods, *e.g.* using *Escherichia coli* or yeast. The risk of the misuse

Drug Discovery Series No. 64
Pharmaceutical Formulation: The Science and Technology of Dosage Forms
Edited by Geoffrey D. Tovey
© The Royal Society of Chemistry 2018
Published by the Royal Society of Chemistry, www.rsc.org

of biotechnology wordings was concluded to be serious for the industry: "… terms are so misused and abused that they are losing their meaning. If the industry does not use biopharmaceutical terminology consistently, it may well lose its identity."[1,2] Good definitions of biological products can be found in documents from regulatory bodies such as the U. S. Food and Drug Administration (FDA)[3] or the Royal Pharmaceutical Society (RPS).[4,5] However, it may be noted that even in the best of cases some confusion might arise. For example in the RPS document biopharmaceutics and biologics are correctly described in Chapter 2.1.3 as "… therapeutic agents often obtained from a variety of natural sources (bacteria, yeast or increasingly mammalian, including human, cells) and include 'living entities' such as cells and tissues…";[4] however, in the same document, biopharmaceutics is defined as "study of the physical and chemical properties of drugs and their dosage forms as related to the onset, duration and intensity of drug action".[5]

According to the FDA[3] biological products (biologics) are compounds "isolated from natural sources", or derived from molecules isolated from natural sources, developed and used as pharmaceutical products. They "include a wide range of products such as vaccines, blood and blood components, allergenics, somatic cells, gene therapy, tissues, and recombinant therapeutic proteins".[3,6] The technologies employed in their preparation are referred to as biotechnology methods. These methods are often used at the leading front of scientific talent and intensive effort, and the final products are usually complex molecules difficult to identify, characterize and prepare in a consistent way for administration or store as stable products. However, in many cases these complex treatments are the only ones available for an illness.[3,6]

"Formulation", or "to formulate" a compound, is the process of preparing a drug product from the drug substance (the active pharmaceutical molecule), making it suitable for use in clinical therapy. For small molecules which are generally delivered by the oral route the term "active pharmaceutical ingredient" (API) is often used. In traditional small-molecule pharmaceutical development, APIs are mixed with other chemical compounds, excipients, to produce a dosage form that is clinically suitable for the intended therapy (*e.g.* tablet, injection or cream). As well as ensuring that the small-molecule API is capable of being delivered in an active form to the patient, the formulation must also ensure that the small molecule is stable during the shelf-life of the product. The addition of the excipients must be determined to have a stabilizing effect on the API.

Some activities that have to be performed in the formulation of biopharmaceuticals are also required for small molecules, but many are quite different. One analogy[7] with our day to day experience with food is to compare the stability and "formulation" of small chemicals like salt and sugar (or aspirin and paracetamol) with the stability of dairy products such as eggs, yogurt or ice cream, dried fruits and meat, each of which poses a challenge in stabilizing their proteins in a "fresh" state.

11.2   Formulation of Small Molecules *vs.* Biopharmaceutics

Whereas small-molecule drugs have molecular weights below 1 kDa, proteins are much larger, for example, the molecular weight of human calcitonin is 3.4 kDa, hirudin is 6.9 kDa, human growth hormone 22 kDa, antibodies 150 kDa, and Factor VIII about 260 kDa. The proteins consist of polypeptide chains of amino acids (the primary structure), which may be in some cases glycosylated. The polypeptide chains adopt well-defined three-dimensional structures (the secondary structure). The tertiary structure refers to the structures of complex proteins that have more polypeptide chains covalently linked, each peptide chain having its own secondary structure. An additional layer of complexity that can arise in proteins is the "quaternary structure" which is formed by the association of independent tertiary structural units *via* non-covalent interactions (of electrostatic, hydrophilic or hydrophobic nature).

Biopharmaceuticals can degrade, resulting in changes in chemical structure and in physical structure. Potential chemical changes are multiple, such as deamidation (the loss of an amide group from the side chain of an asparagine or glutamine amino acid residue), oxidation (of cysteine, methionine, tryptophan or tyrosine amino acid residues), proteolysis (hydrolysis of the peptide backbone) and disulphide bond exchange. Deamidation and oxidation are common degradation pathways in proteins. Physical degradation includes changes in conformation (loss of secondary and/or tertiary/quaternary structure), adsorption to surfaces (non-covalent interaction with materials such as vial stoppers or transfusion tubing or concentration at *e.g.* the solution–air interface), precipitation (changes in the solution and/or protein properties which mean that the protein is no longer in solution) and aggregation (physical association of more than one protein molecule by either non-covalent or covalent bonding). Physical degradation is sometimes also induced by chemical degradation. Protein aggregation is one of the most challenging issues in the development of a recombinant protein drug since, besides being inactive, aggregated proteins are known to be able to induce immunogenic reactions, resulting in possible side effects, as well as loss of desired pharmacological activity.[8]

At present, most biological molecules are administered by injection, usually intravenous or intra-arterial, subcutaneous or in some cases intrathecal or intraocular. This means that, after isolation and/or purification and before administration, they must be put into a liquid state, the "formulation". Several corollaries follow: the formulation must be fluid enough to be used with a syringe; the formulation must deliver the active biological molecule consistently; the active biological molecule must be soluble in the formulation. Subsequent requirements are as follows: the active biological molecule must be physically and chemically stable in the formulation, and retain its biological activity, at least long enough to be prepared and administered; particulate matter is unacceptable in the liquid product; the formulation containing the active biological molecule may not contain ingredients which could harm the patient (toxicological considerations).

Thus, for biopharmaceuticals, the formulation activities are more complex and laborious when compared with those needed for small molecules.

Small changes in biopharmaceuticals properties during manufacturing, formulation and storage may result in loss of biological activity as well as an increase in the risk of the occurrence of unexpected side effects. This complexity-of-formulation impact on the product quality is caused by different factors such as (i) the perishable nature of the flexible biomolecules, (ii) the importance of maintaining chemical stability as well as conformation and aggregation stability, (iii) the complex interactions that can occur with formulation ingredients and primary packaging and clinical application material and (iv) the requirement for sterile manufacturing and storage, since the biopharmaceuticals are preponderantly administered by parenteral routes.

The combination of chemicals in which the active biological molecule is stored, diluted or mixed with an infusion solution, or administered without dilution to the patient is called its formulation. In general, water containing various combinations of sugars (trehalose, *e.g.*, Avastin®, Lucentis®), surfactants (polysorbate 20, *e.g.*, Avastin, Lucentis), salts (sodium phosphate, *e.g.*, Avastin, Lucentis) and/or amino acids (histidine, *e.g.*, Lucentis) are used. Other components may also be employed (zinc, *m*-cresol, glycerol, *e.g.*, Lantus®; acetic acid, phenol, sodium chloride, *e.g.*, Miacalcin®; citric acid, mannitol, polysorbate 80, *e.g.*, Humira®; sucrose, *e.g.*, Remicade®) in order to provide whatever is required by the physical and chemical properties of the biological for final product stability and clinical efficacy. In 2014 biopharmaceuticals represented eight out of the fifteen top-selling pharmaceuticals (Table 11.1); their formulations are described in Table 11.2. Detailed information on the formulation of each marketed biological is given in the "Full Prescribing Information" of the product, provided by the FDA.[6]

Table 11.1 The fifteen top-selling drugs in 2014. The biopharmaceuticals, six monoclonal antibodies and two peptides*, are highlighted in bold; total sales 62.0 $Billion. The sales of the seven small molecules represented 51.4 $Billion. Adapted from ref. 36.

Number	Drug	Company	Sales 2014 $ billions
1	**Humira**	**AbbVie**	**11.8**
2	**Lantus***	**Sanofi**	**10.3**
3	Sovaldi	Gilead Sciences	9.4
4	Abilify	Otsuka	9.3
5	**Enbrel**	**Amgen**	**8.7**
6	Crestor	AstraZeneca	8.5
7	**Remicade**	**Johnson & Johnson**	**8.1**
8	Nexium	AstraZeneca	7.7
9	**Rituxan**	**Roche**	**6.6**
10	**Avastin**	**Roche**	**6.1**
11	Lyrica	Pfizer	6.0
12	**Herceptin**	**Roche**	**5.6**
13	Spiriva	Boehinger Ingelheim	5.5
14	Januvia	Merck&Co	5.0
15	**Copaxone***	**Teva**	**4.8**

Table 11.2 Administration procedures and formulations of the eight top-selling biopharmaceutics in 2014 from Table 11.1: the information is from the package inserts available on-line at the FDA website. *s.c.*, subcutaneous administration; *i.v.*, intravenous administration; p75-TNF, human 75 kDa (p75) tumor necrosis factor (TNF).

Name		
Type of molecule		
Trade name	How supplied	
Company	Administration route	Formulation
Adalimumab	**Liquid** prefilled syringe; prefilled pen; single-use vial	**50 mg ml⁻¹**
Recombinant human IgG1	*s.c.* **administration**	6.2 mM citric acid; 1.0 mM sodium citrate; 5.5 mM sodium dihydrogen phosphate; 8.6 mM sodium phosphate dibasic; 105 mM NaCl; 66 mM mannitol; 0.1% (w/v) Tween 80
Humira		**pH 5.2**
AbbVie	**Liquid** (launched in 2016); prefilled syringe; prefilled pen; *s.c.* **administration**	**100 mg ml⁻¹**
		228 mM mannitol; 0.1% (w/v) Tween 80; **pH 5.2**
Insulin glargine	**Liquid** vial; cartridge	**3.64 mg ml⁻¹ or 100 units ml⁻¹**
Recombinant human peptide		30 µg ml⁻¹ Zn; 25 mM *m*-cresol; 20 mg ml⁻¹ glycerol 85%; the vial formulation contains also 0.002% (w/v) Tween 20
Lantus	*s.c.* **administration**	**pH 4**
Sanofi-Aventis		
Etanercept	**Liquid** prefilled syringe; autoinjector	**50 mg ml⁻¹**
Fusion protein of human p75-TNF receptor and human IgG1	*s.c.* **administration**	25 mM sodium phosphate; 25 mM L-arginine hydrochloride; 100 mM sodium chloride; 1% sucrose; **pH 6.3**

Drug	Form / administration	Formulation
Enbrel Amgen	**Lyophilized** multiple use vial *s.c.* **administration**	**25 mg ml^{-1}** 220 mM mannitol; 29 mM sucrose; 9.9 mM tromethamine; **pH 7.4**
Infliximab Chimeric murine/ human IgG1 **Remicade** Johnson & Johnson	**Lyophilized** vial *i.v.* **administration** after dilution with 0.9% NaCl	**10 mg ml^{-1}** 1.6 mM monobasic sodium phosphate monohydrate; 3.4 mM dibasic sodium phosphate dihydrate; 146 mM sucrose; 0.005% (w/v) Tween 80 **pH 7.2**
Rituximab Chimeric murine/ human IgG1 **Rituxan** Roche	**Liquid** single use vial *i.v.* **administration** after dilution with 0.9% NaCl or 5% dextrose	**10 mg ml^{-1}** 25 mM sodium citrate dihydrate; 154 mM NaCl; 0.007% (w/v) Tween 80 **pH 6.5**
Bevacizumab Recombinant humanized IgG1 **Avastin** Roche	**Liquid** single use vial *i.v.* **administration** after dilution with NaCl	**25 mg ml^{-1}** 42 mM monobasic sodium phosphate monohydrate; 8.5 mM dibasic sodium phosphate anhydrous; 158.6 mM α,α-trehalose dehydrate; 0.04% (w/v) Tween 20 **pH 6.2**
Trastuzumab Humanized IgG1 **Herceptin** Roche	**Lyophilized** vial *i.v.* **administration** after dilution with 0.9% NaCl; 5% dextrose prohibited	**21 mg ml^{-1}** 2.4 mM L-histidine HCl; 2.1 mM L-histidine; 52.9 mM α,α-trehalose dihydrate; 0.009% (w/v) Tween 20 **pH 6**
Glatiramer acetate Synthetic peptides **Copaxone** Teva	**Liquid** prefilled syringe *s.c.* **administration**	**20 mg ml^{-1}** 220 mM mannitol **pH range: 5.5–7.0**

11.3 The Importance of Formulation

A good formulation is a key factor for the success of a biological drug from beginning to end, throughout the research and drug product development stages.[9] Formulations for biological molecules must be based on an in-depth understanding of the physical and chemical properties of the molecule itself, which should be acquired as early as possible in development. Protein structure, conformation, chemical degradation and aggregation studies in aqueous solutions at different pH values, temperatures, using different buffers and in the presence of different ingredients (*e.g.* ions, sugars, detergents) provide experimental bases for the development of a successful formulation.

Molecule-specific studies can reveal conditions that stabilize a molecule. Other important requirements to be fulfilled by a good formulation include an easy administration procedure, optimal release of the active protein at the administration site, optimal activity of the molecule at the target site, minimum side effects and realistic scale-up with robust and cost-effective manufacture of the formulation.

Unfortunately, a frequent mistake in biopharmaceutical research and development in industry as well as in academia, is to consider that protein formulation activities are straightforward and a matter of routine. The complexity of physical and chemical properties of proteins previously mentioned, and their interrelationship, strongly influence the stability of the formulation, as well as the biological activity. This makes it difficult to formulate proteins using standard procedures, in a routine process, with standard excipients. Since the formulation of proteins is difficult and unpredictable, these activities often take many years, and not infrequently fail.

The failure of pre-clinical and clinical tests can also be related to formulation problems. Failure of a project due to a poor formulation is in general not known and not investigated, discussed or published because the team members do not want to be considered responsible for the failure of the project: it is easier and more convenient to "blame the molecule" as being not active or toxic. Over many years in the industry, our group came across such cases where the formulation was very probably the cause of failure. A few cases are also described in the literature.

One example that demonstrates well the importance of formulation for clinical outcome is the Johnson & Johnson (J&J) human erythropoietin (EPO) product, Eprex®. Originally, the product was launched in 1989 but then was reformulated in 1998. During reformulation J&J replaced albumin with polysorbate 80 (because of a concern about albumin transmitting variant Creutzfeld–Jacob disease). Following the reformulation, a new EPO-related adverse event began to be reported. Patients were identified who suffered from pure red cell aplasia (PRCA). This adverse event was linked to the use of polysorbate 80 in combination with uncoated rubber stoppers. The EPO example demonstrates how a formulation issue[10] (and not the underlying protein, which was not changed) may cause an adverse event and, in a

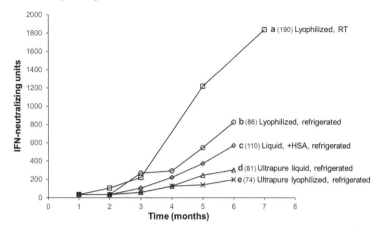

Figure 11.1 Neutralizing antibodies formed in patients showed that the immunogenicity of human interferon-α2A (IFN-α2A) is highly dependent on the formulation and storage conditions: (a) lyophilized powder stored at room temperature (RT), (squares); (b) lyophilized powder stored under refrigeration (circles); (c) human serum albumin (HSA)-containing liquid stored under refrigeration (diamonds); (d) ultrapure liquid formulation (HSA-free) stored under refrigeration (triangles) and (e) ultrapure lyophilized powder stored under refrigeration (crosses). Comparing (a) and (b): immunogenicity of the lyophilized formulation stored at room temperature (a) was higher than the immunogenicity of the same lyophilized formulation stored refrigerated (b). Comparing (b) and (e): the use of ultrapure IFN-α2A increased the stability of the lyophilized formulation stored refrigerated. Comparing (c) and (d): the use of ultrapure IFN-α2A increased also the stability of the liquid formulation stored refrigerated. Comparing (d) and (e): the lyophilized formulation was less immunogenic than the liquid formulation. IFN-neutralizing units are arbitrary units of neutralizing activity; the numbers in parenthesis are the number of patients that had neutralizing antibodies. Adapted from data in ref. 11 and 39.

pre-approval context, may lead to the abandonment of the entire project if the problem cannot be solved.

Another example is alpha interferon.[11] Depending on the formulation administered, patients develop neutralizing antibodies to the drug. Once a patient develops neutralizing antibodies the drug is not active any more. Such problems have led to removal of some formulations from the market. Ryff[11] showed that the type of formulation influences the formation of neutralizing antibodies (Figure 11.1). Comparison of curves (a and b) in Figure 11.1, shows that the immunogenicity of the lyophilized formulation stored at room temperature (curve a) was higher than the immunogenicity of the same lyophilized formulation stored refrigerated (curve b). Comparison of curves (b and e) and (c and d), Figure 11.1, shows that the use of ultrapure IFN-α2A increased the stability of both the lyophilized formulation and liquid

formulation stored refrigerated. The lyophilized formulation was less immunogenic than the liquid formulation (comparison of curves d and e). Ryff[11]
concluded that "Refrigerated lyophilisate and a new human serum albumin
(HSA)-free formulation of IFN-α2A, produced according to the latest process
specification, are less immunogenic than earlier products".

A recent example showing the importance of formulation for the success
of a marketed product is the case of Omontys® (peginesatide) marketed by
Affymax Inc., Cupertino, CA. Omontys was launched in March 2012 with
a multiuse vial (MUV) formulation which differed from the single-use vial
(SUV) formulation used for pre-market clinical trials and registration.[12]
Although the pre-market clinical trials rate of hypersensitivity reactions was
0.84 per 1000 with no fatalities, within the first year after launch, anaphylaxis and deaths occurred (hypersensitivity rate of 3.5 cases per 1000, with 7
deaths). The drug was withdrawn from the market in February 2013. Analysis
of subvisible particles using the orthogonal methods nanoparticle tracking
analysis and flow imaging, performed by the FDA and National Institute of
Health (NIH), showed that "...Standard physical and chemical testing did
not indicate any deviation from product specifications in either formulation.
However, an analysis of subvisible particulates using nanoparticle tracking
analysis and flow imaging revealed a significantly higher concentration of
subvisible particles in the multiuse vial presentation linked to the hypersensitivity cases... Although the constituents of the MUV formulation are all generally recognized as safe, formulation composition is widely understood as
having the potential to alter the properties of biological therapeutics, including the subvisible particulate (SVP) profile".[12] Besides showing the importance of formulation for the success of a biological drug, the Affymax case
also demonstrates "the utility of characterizing subvisible particulates not
captured by conventional light obscuration".[12]

Formulation scientists, besides working on stabilizing the protein drug,
should always adapt the formulation to the medical requirements and to
marketing requirements. For example, if the medical therapy requires either
intravenous administration in a hospital or a daily patient self-administration
by subcutaneous injection, the subcutaneous injection will be preferred for
marketing reasons. A lyophilized formulation requires reconstitution prior
to administration, and after reconstitution should be used within 24 hours
if it does not contain preservatives. For self-administration, reconstitution
of a lyophilized protein formulation is possible, but a stable liquid formulation is preferred by patients and marketing (in general stability should be for
1.5–2 years at 4 °C). Thus, formulation scientists are part of a wider project
team and the formulation choice is based on different factors such as clinical
aspects, commercial context and patient and medical staff requirements.

11.4   The Importance of Analytical Methods

A key factor in developing a good biopharmaceutical formulation is to use
many sensitive analytical methods to characterize the molecules in stability
testing. Therefore, the formulation team should possess knowledge of the

mechanisms of chemical and physical degradation of the specific biopharmaceutical and of the analytical techniques used to detect the degradation.

The analytical methods for biopharmaceutical formulations should aim to characterize the biomolecule in its complex formulation environment. For example, the study of protein aggregates in a 150 mg ml^{-1} antibody solution should be performed using methods that do not require dilution.[13] For methods such as size exclusion chromatography or gel electrophoresis, a dilution to around 1 mg ml^{-1} antibody concentration is required. However, the dilution step from 150 mg ml^{-1} to 1 mg ml^{-1} will change the aggregation states of the antibody; large structures will not be detected if they bind to the column or if solubilized, and the data, although important, will not provide a good characterization of the aggregates. Thus, analysis of the properties of biomolecules requires specific, "tailor-made" analytical methods. "These methods should be adapted to the necessities of the formulation and not to the needs of the analytical techniques".[13]

11.5 Potential for Particle Formation *In vivo*: Studies of Aggregates Formation After Mixing of Biopharmaceutics with Human Plasma and Human Blood

For intravenous administration, recent work from our group showed the importance of having a formulation and diluent that are compatible with the biological drug in human blood.[14,15] Use of one diluent over another may lead to particulate formation in the bloodstream when the drug is injected intravenously, despite successful release and stability testing for particulates according to the U.S. Pharmacopoeial Convention General Chapter <788> and the FDA Guidance on Immunogenicity Assessment for Therapeutic Protein Products. The importance of assessing the compatibility of a therapeutic monoclonal antibody with diluent and human plasma during product development was also emphasized in a recent publication from an FDA research group.[16]

11.5.1 Aggregation in Plasma

The prescribing information for Herceptin® states that for intravenous infusion 0.9% NaCl solutions should be used; the use of 5% dextrose is prohibited. No reasons for the prohibition of the use of 5% dextrose as an infusion solution are given in the Herceptin Package Insert, or in the literature. Since 5% dextrose is a commonly used infusion solution, we investigated Herceptin in dextrose using new orthogonal analytical methods (*i.e.* different and independent methods) and found that it induced Herceptin aggregation.[17] No aggregation was found when Herceptin was diluted in 0.9% NaCl. The Herceptin in 5% dextrose solution (1.2 mg ml^{-1} Herceptin) contained aggregates of about 200 nm in diameter.[14,18] Standard analytical methods, such as field flow fractionation, did not detect these aggregates.[17]

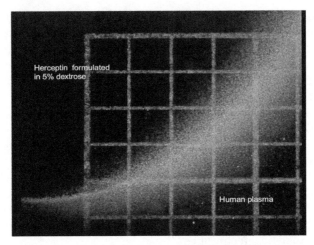

Figure 11.2   Light microscopy of the aggregates formed at the interface of a human plasma solution and a solution of Herceptin in 5% dextrose. The white 'curved band' is the interface where aggregation occurred. The Herceptin protein concentration prior to mixing with human plasma was 1.06 mg ml^{-1}; no particles were formed when human plasma was mixed with 5% dextrose (data from ref. 18). The dimensions of the squares are 250 μm × 250 μm. Herceptin in 5% dextrose is the wrong infusion solution that could be prepared by mistake in a hospital when a 5% dextrose infusion bag is used instead of an infusion bag containing 0.9% NaCl.

After mixing human plasma with Herceptin diluted in 5% dextrose (*ex vivo* experiments, using the same concentrations as those used in human therapy) we observed a strong, fast aggregation at the interface of the human plasma and the Herceptin solution (which appears as a 'crescent moon' in Figure 11.2).[18] The same was observed for Avastin diluted with 5% dextrose and added to human plasma, but not for Remicade diluted with 5% dextrose. Herceptin diluted in 0.9% NaCl (according to the prescribing information) mixed with human plasma did not result in particle formation.[18] As expected, Avastin and Remicade diluted, as indicated in the package inserts, with 0.9% NaCl solutions, also did not form particles when mixed with human plasma.[18] The particles formed when Herceptin and Avastin were diluted in 5% dextrose and subsequently mixed with human plasma were heterogeneous and ranged from about 0.5 μm up to 10 μm in diameter. The particles did not solubilize quickly: they were still present 2.5 hours after the initial mixing of human plasma with the antibody in 5% dextrose.[18] Aggregates and fibrils formed from the initial particles, in a few cases reaching sizes greater than 100 μm.[17] Electron micrographs revealed that when dextrose solutions of Herceptin and Avastin are mixed with human plasma, spherical structures are formed, with large globular structures ("berry-like" structures) evolving by apparent agglomeration of the smaller globular structures similar in appearance to lipoproteins. It is suggested that plasma lipoproteins may show an affinity to Herceptin and Avastin aggregates formulated in dextrose.[14] To explain the very fast formation (within

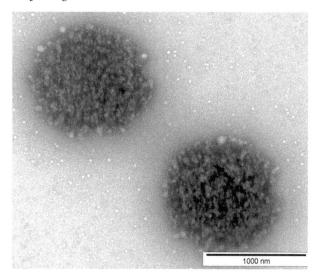

1000 nm

Figure 11.3 Transmission electron microscope picture of globular structures formed after mixing of human plasma with a solution of Herceptin in 5% dextrose. Adapted from ref. 14. The globular "berry like" structures appear to be formed by the conglomeration of lipoproteins. Herceptin in 5% dextrose forms antibody aggregates of about 200 nm in diameter (ref. 17). We proposed that the larger globular structures of about 1 μm shown in this figure are formed from lipoproteins and the 200 nm Herceptin aggregates.

seconds) of the "berry-like" structures observed by electron microscopy (Figure 11.3) we proposed an aggregation model that consists of the binding to lipoproteins of the Herceptin aggregates formed in 5% dextrose.[14]

11.5.2   Aggregation in Blood

Experiments similar to those published for the mixture of Herceptin diluted in dextrose and human plasma[17] were also performed for mixtures of biotech drug formulations with whole human blood. When Herceptin diluted in dextrose is mixed with human blood, a strong aggregation of human red blood cells (RBC) occurs: no aggregation was observed when whole blood was mixed with dextrose alone (Figures 11.4 and 11.5). One possible explanation for the aggregation is that the Herceptin–lipoprotein aggregates (the "berry-like" structures, Figure 11.3) form initially in 5% dextrose and subsequently stick to and aggregate the erythrocytes. Support for this model came from analytical centrifugation experiments (Figure 11.6). The sedimentation velocity of the RBC was slower in a mixture with Herceptin diluted in dextrose (at the same concentrations as applied in human therapy) and the RBC pellet height was larger. These changes are consistent with a coating of the RBC by the "berry-like" structures of Herceptin–lipoproteins (Figure 11.3): this coating increases the size of the individual RBC, which results in a slower sedimentation and a larger pellet (Figure 11.6).

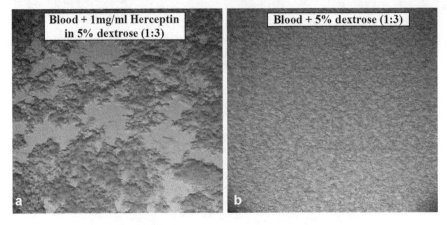

Figure 11.4 (a) Light microscope images showing the aggregation of red blood cells that occurred when Herceptin diluted in 5% dextrose was mixed with whole blood. Herceptin in 5% dextrose is the wrong infusion solution that could be prepared by mistake in a hospital when a 5% dextrose infusion bag is used instead of an infusion bag containing 0.9% NaCl. (b) No aggregation was observed when 5% dextrose solution was mixed with human blood. Fresh healthy-donor blood (anticoagulant citrate dextrose) was obtained from Zürich hospital (Blutspende Zürich, Switzerland). 5 μl of human blood were mixed in an Eppendorf tube with 15 μl of Herceptin diluted in 5% dextrose or of 5% dextrose alone: 1:3 (v/v) mixture. Light microscope investigations were performed within 2 min after mixing using a Leica microscope, objective 20×. Individual red blood cells (7 μm in diameter) can be observed in (b).

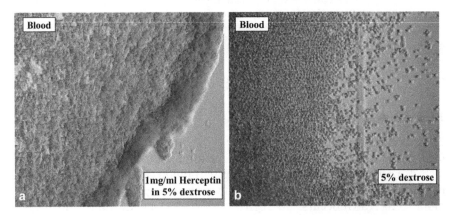

Figure 11.5 (a) Light microscope pictures showing the aggregation of red blood cells that occurred at the mixing interface when Herceptin in 5% dextrose was mixed with whole blood. Herceptin in 5% dextrose is the wrong infusion solution that could be prepared by mistake in a hospital when a 5% dextrose infusion bag is used instead of an infusion bag containing 0.9% NaCl. (b) No aggregation occurred at the mixing interface of a 5% dextrose solution and human blood. For these experiments 4 μl of Herceptin diluted in 5% dextrose or of 5% dextrose alone were added to 4 μl of human blood, 1:1 (v/v). Fresh blood from one healthy human donor (anticoagulant citrate dextrose) was obtained from Zürich hospital (Blutspende Zürich). Light microscope investigations were performed within 1 min after mixing using a Leica microscope, objective 20×. Individual red blood cells (7 μm in diameter) can be observed in (b).

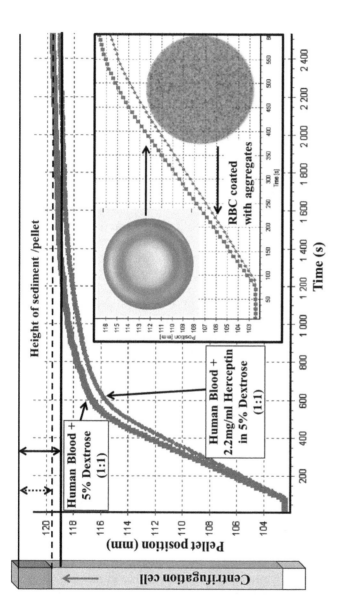

Figure 11.6 Front tracking analysis of the sedimentation of human blood mixed with 2.2 mg ml^{-1} Herceptin in 5% dextrose (1 : 1) (v/v) and human blood mixed with 2.2 mg ml^{-1} Herceptin in 5% dextrose 1 : 1 (v/v). Accelerated sedimentation analysis was performed using the analytical centrifuge LUMiSizer® 610 (L.U.M. GmbH, Berlin, Germany) which permits the characterization of particle sedimentation by measuring the intensity of the transmitted light as function of time. The pellet height during the sedimentation induced by centrifugation is the height from the bottom of the cuvette towards the steady-state of the sedimentation profile, as shown inside the graph. The pellet position is determined from near infrared transmission profiles, which were recorded with an interval of 10 seconds during 40 minutes of separation at a thermostat-regulated temperature of 25 °C. The centrifugal force applied was 1000 r.p.m. corresponding to 146 *g*. Fresh healthy donor blood (anticoagulant citrate dextrose) was obtained from Zürich hospital (Blutspende Zürich). The insert is a zoom of sedimentation curves in the first 600 seconds. The slower sedimentation of the red blood cells (RBC) and the larger height of the equilibrium pellet in the Herceptin–dextrose solution can be explained by the coating of the RBC with Herceptin–lipoprotein "berry-like" structures (see Figure 11.3). The coated RBC will have larger sizes, which will be responsible for the slower sedimentation and for the larger pellet height at the end of centrifugation. These coated RBC may induce side effects *in vivo* through obstruction of the blood capillaries and through reduced gas transport capabilities.

The results of experiments by an FDA group[16] confirmed our observations and indicated that the dextrose-mediated aggregation in human plasma of therapeutic monoclonal antibodies involved complement proteins C3 and C4 and Factor H, and that an acidic pH plays an important role in the formation of particles under these conditions.[16] The molecular-level aggregates analysed in the FDA paper[16] and the larger aggregates between 100 nm and 10 μm investigated in our work[14] show that complex interactions and different phenomena may occur when biopharmaceutical drugs enter the bloodstream or are administered by injection into a tissue.

During our studies over the last twenty years, we have observed that the aggregation of biopharmaceuticals in plasma is a very complex process. We noted the following: (i) Aggregation in plasma occurs for a wide range of pharmaceuticals, such as peptide drugs, small proteins, antibodies, large proteins, vaccines and virus-like particles. (ii) Plasma aggregation can also be induced by the formulation ingredients alone. (iii) The size of aggregates with different biopharmaceuticals varies. In some cases only small structures are formed (*e.g.* below 1–5 μm diameter) while in other cases large structures are also formed, *e.g.* more than 10 μm diameter. (iv) The aggregation is in some cases dependent on the healthy donor: in some healthy humans a biopharmaceutical will aggregate and in others, not. (v) Aggregation in human plasma from patients can be different from the aggregation in plasma from healthy donors. (vi) The aggregation in animal plasma can be different between species and different from that in humans. (vii) Plasma aggregation can depend on the manufacturing clone of the recombinant biopharmaceutical drug. (viii) Plasma aggregation may be dependent on the formulation of the biopharmaceutical drug. (ix) Plasma aggregation can sometimes occur at low concentrations of the biological drug (*e.g.* 1 mg ml^{-1}) and does not occur at higher concentrations (*e.g.* 20 mg ml^{-1}). (x) There are cases where plasma aggregation takes place in both infusion solutions, 5% dextrose as well as 0.9% NaCl.

As an example, the donor-to-donor variation was discussed in Arvinte *et al.*[14] for Ilaris®, where aggregation of an antibody drug candidate occurred in five out of ten healthy human volunteers; no aggregation occurred in any donor when another mAb (the mAb candidate that was finally marketed as Ilaris) was used.

Plasma aggregation may also occur in clinical settings and contribute to unexpected side effects, not only when a wrong infusion solution is used, but also during combination therapy, as discussed in Arvinte *et al.*[15] In this example, chemotherapy agents co-administered with Avastin include Paraplatin® (carboplatin), Platinol® (cisplatin) and Taxol® (paclitaxel). These drugs can be administered either with 5% dextrose or with 0.9% NaCl. For Platinol, it is recommended to pre-administer large volumes, *i.e.* 1 to 2 l, of 5% dextrose containing saline and mannitol to the patient. Administration of Avastin to these patients, even if given in 0.9% NaCl infusion solution, may result in aggregate formation since the patient already has large amounts of dextrose in the bloodstream. In this context, it may be noted that aggregates which form when human plasma is mixed with 5% dextrose–Avastin solutions could be one origin of the reported "arterial thrombolytic events, including cerebral infarction, transient ischemic attacks, myocardial infarction and angina that occurred at a higher incidence in patients receiving Avastin in combination with chemotherapy as compared with those receiving chemotherapy alone".[19]

The study of the aggregation phenomena that occur when biopharmaceuticals as well as small molecules are administered by injection *in vivo* is a new research and development field, which will grow and become an important part of pharmaceutical development and contribute to our understanding of the mechanisms and behaviour of biopharmaceutical drugs *in vivo*.

11.6 New Formulation Strategy: High-throughput Analysis and High-throughput Formulation

At the beginning of the 1990s, high-throughput screening (HTS) techniques which were emerging at that time in pharmaceutical research were introduced for protein formulation development by the team of T. Arvinte, at that time at Ciba-Geigy/Novartis. This approach later became known in the field as high-throughput formulation (HTF) and high-throughput analysis (HTA).[20–22] HTF and HTA methods were successful, leading to the development of (i) a stable liquid formulation of human calcitonin (hCT) for nasal application (marketed as Cibacalcin® nasal), (ii) a lyophilized formulation of recombinant hirudin stable at room temperature (marketed as Revasc®, launched 1997) and (iii) a lyophilized formulation of the monoclonal antibody Canakinumab stable under refrigeration (marketed as Ilaris®, launched 2009). Throughout the last 20 years, we have continued to develop methods suitable for the use of HTS in the characterization of a drug's physical and chemical stability, allowing faster characterization of drug substances and consequently faster formulation development. The high-throughput approaches permit the identification of conditions in which the molecules are stable. In the next step, it is important to understand the mechanisms responsible for these stabilizations. This enables an optimization of the formulation not based only on trial and error (the initial HTF screening experiments are by definition based on trial and error) but also on an understanding of the stabilization mechanisms at a molecular level.

11.6.1 HTF Methods

A number of variables must be balanced to provide a developable or marketable formulation for a biological. Stabilization of the molecule in formulation and for storage will be desired; retention of activity will be required. The influences of multiple factors such as protein concentration, solution pH, ions, sugars, detergents, incubation temperature, stress conditions (heating, agitation, pH, light exposure, high concentrations) have to be

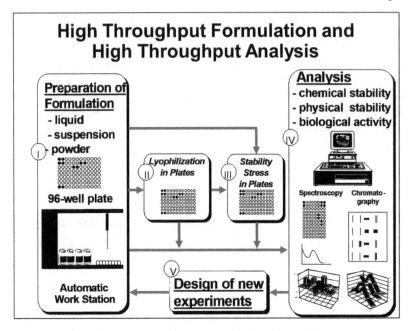

Figure 11.7 Scheme showing the platform for high-throughput formulation development for biologicals. Adapted from ref. 9. High-throughput formulations (HTF): first (I) various formulations containing the biological
under development are pipetted or prepared in multiwell plates. Then
the formulations are either analyzed directly (IV); (II) lyophilized
and stressed in plates (lyophilized formulation development); or
(III) stressed as liquids (stable liquid formulation development).
High-throughput analysis (HTA): (IV) analytical methods, such as
spectroscopy and chromatography are used to measure the biological
chemical stability, physical stability and biological activity. In an iterative way, based on the results from one experimental cycle (usually we
used one or two plates for one experimental cycle) new experiments
are planned and new formulations are prepared and analyzed.

tested. Formulation development should include forced degradation studies in order to find the best formulation as well as to evaluate degradation
pathways.

We developed a robust platform for rapid screening of many formulations,
illustrated graphically in Figure 11.7, Table 11.3 and Figure 11.8. In principle, the platform is comprised of sample preparation, sample analysis and
design of further experiments based on the results of the analysis (Figure
11.7). The platform is iterative. Thus, based on the results of multiwell studies, new experiments are designed, until an optimum formulation is found.
Use of multiwell plates enables investigation of numerous formulations
checked under several varying parameters in a short time. Analytical techniques may be those designed for the specific biological under development.

For example, in a proof-of-concept study,[22] a well known biological, salmon
calcitonin, was tested in 100 different formulations composed of 20 different

Table 11.3 Example of an HTA output grid from salmon calcitonin formulation screening. Reproduced from *Pharmaceutical Research*, A High Throughput Protein Formulation Platform: Case Study of Salmon Calcitonin, **26**, 2009, 118, M. A. H. Capelle, R. Gurny and T. Arvinte,[22] © Springer Science + Business Media, LLC 2008. With permission of Springer. The table shows the results of the tyrosine fluorescence, 1,8-ANS fluorescence, Nile Red fluorescence and turbidity assays seven days after preparation, for 100 formulations: 20 buffers at 10 mM ionic strengths and different pH values. Each assay was assigned a U, A, N, T which represents the UV turbidity assay, 1,8-ANS emission, Nile Red emission and tyrosine emission, respectively. The samples found to be unstable with the respective assays are indicated by bold italics; the remaining samples were found to be stable. The best formulation found was number 6, 10 mM Na–acetate buffer in the pH range from pH 3.5 to pH 5.5. It is interesting to note that the best formulation does not have an optimum pH and that the stabilization is not due to Na+ or acetate alone (there is a formulation with Na+ and one with acetate which are not good), but to the 10 mM Na–acetate buffer in a broad pH range.

	10 mM buffers	pH 2.5	3.0	3.5	4.0	4.5	5.0	5.5	6.0	6.5	7.0	7.5	8.0	8.5	9.0	9.5	10.0	10.5
1	Glycine–Hydrochloric acid	UT / AN	UT / AN	UT / AN														
2	Citric acid –Sodium citrate		UT / AN	UT / AN	UT / AN	UT / AN	UT / AN	UT / AN	UT / AN									
3	Citric acid–Sodium phosphate dibasic		UT / AN	UT / AN	UT / AN	UT / AN	UT / AN	UT / AN	UT / AN	UT / AN	UT / AN	UT / AN						
4	Citric acid–Potassium phosphate dibasic		UT / AN	UT / AN	UT / AN	UT / AN	UT / AN	UT / AN	UT / AN	UT / AN	UT / AN							
5	β,β′-Dimethylglutaric acid–Sodium hydroxideβ′		AN	UT / AN	UT / AN	UT / AN	UT / AN	UT / AN	UT / AN	UT / AN	UT / AN	UT / AN						
6	Acetic acid–Sodium acetate			AN	UT / AN	UT / AN	UT / AN	UT / AN		UT / AN	UT / AN	UT / AN						
7	Potassium phthalate–Sodium hydroxide				UT / AN	UT / AN	UT / AN	UT / AN	UT / AN									
8	Succinic acid–Sodium hydroxide				UT / AN	UT / AN	UT / AN	UT / AN	UT / AN									
9	Acetic acid–Ammonium acetate					AN	UT / AN	UT / AN	UT / AN	UT / AN								
10	Histidine–Hydrochloric acid							UT / AN	UT / AN	UT / AN	UT / AN	UT / AN						

(continued)

Table 11.3 *(continued)*

10 mM buffers		2.5	3.0	3.5	4.0	4.5	5.0	5.5	6.0	6.5	7.0	7.5	8.0	8.5	9.0	9.5	10.0	10.5
																		pH
11	Tris–Maleate							UT	UT	UT	UT	UT	*UT*					
								AN	*AN*	*AN*	*AN*	*AN*	*AN*					
12	Sodium phosphate monobasic – Sodium phosphate dibasic								UT	UT	UT	UT	UT					
									AN	*AN*	*AN*	*AN*	*AN*					
13	ADA–Hydrochloric acid									UT	UT	UT						
									AN	*AN*	*AN*	*AN*	*AN*					
14	MOPS–Potassium hydroxide									UT	UT	UT						
									AN	*AN*	*AN*							
15	HEPES–Sodium hydroxide										*UT*	UT	UT					
										AN	*AN*	*AN*	*AN*					
16	TES–Hydrochloric acid										UT	UT	UT					
										AN	*AN*	*AN*						
17	Tris–Hydrochloric acid											UT	UT	UT	UT			
											AN	*AN*	*AN*	*AN*	*AN*			
18	Glycine–Sodium hydroxide														UT	UT	UT	UT
														AN	*AN*	*AN*	*AN*
19	CAPSO–Sodium hydroxide														UT	UT	UT	UT
														AN	*AN*	*AN*	*AN*
20	Sodium carbonate–Sodium bicarbonate															UT	UT	*UT*
														AN	*AN*	*AN*	*AN*

buffer types in which protein concentration, turbidity measurement, intrinsic tyrosine fluorescence, Nile Red, and 1,8-ANS fluorescence indicated the chemical and physical stability of the molecule, and the presence or absence of aggregates.[22] The data in Table 11.3 show that the best stability of salmon calcitonin was found in 10 mM sodium acetate (Na-acetate) buffer, across a broad range of pH values. It is interesting to note that the observed stabilizing effect cannot be attributed to individual components of the buffer or to an "optimal pH", *i.e.* Table 11.3 shows that there are formulations with sodium 10 mM and with acetate 10 mM that are not stable. This is the general experience in our work; a "good" formulation is not the sum of individual components. The approach "find the best sugar, find the optimal pH, find the best detergent, then add up all the best conditions" will not lead to a good formulation. We never came across such a situation.

The fact that a stable formulation is not the result of the addition of individually determined "best stabilizers" was shown in the research field of thermophiles, organisms that can live only at temperatures between 60 °C and 120 °C and which die when brought to 20 °C.[23] Comparison of the amino acid sequences of the same proteins found in thermophiles and mesophiles (organisms that live between 20 °C and 40 °C) showed no major differences.[23] There are no unique protein stabilizers found in thermophiles. What was found is that "enhanced intrinsic stability in thermophiles is the cumulative effect of minute improvements of local interactions: higher packing efficiency (mainly through van der Waals interactions), networks of ion pairs and/or hydrogen bonds (including alpha-helix stabilization) and reduction of conformational strain (loop stabilization)".[23] The same is also true for the majority of biopharmaceuticals: stabilization of a protein is the result of cumulative effects of small improvements through local interactions with formulation ingredients.

Thus, as shown in Figure 11.8, the first screening of 240 formulations of a monoclonal antibody identified five "good" formulations. The best was formulation "4" which showed no changes after stress at both pH 6 and pH 7. Interestingly also at pH 5 this formulation "4" was the best, showing some changes only in one analytical method after stress. This was the first (or initial) HTF for this biological molecule. In a second step a formulation optimization around formulation "4" will likely result in a formulation that is stable across a wider range of pH values (as was observed for salmon calcitonin, see Table 11.3).

In general, in HTF experiments multiwell plates, for example 96-, 384-, and 1536-well plates, are used so that low-volume protein solutions are tested. The use of robotic systems in the preparation of formulations can be employed for liquid, suspensions or powder. Various formulations containing the biological under development can be pipetted or prepared in multiwell plates, see (I) in Figure 11.7. Then the formulations are either analysed directly (IV), lyophilized and stressed in plates (II) (in the case of a lyophilized formulation development), or stressed as liquids (III) (in the case of a stable liquid formulation development). Subsequently, for example from 200–300

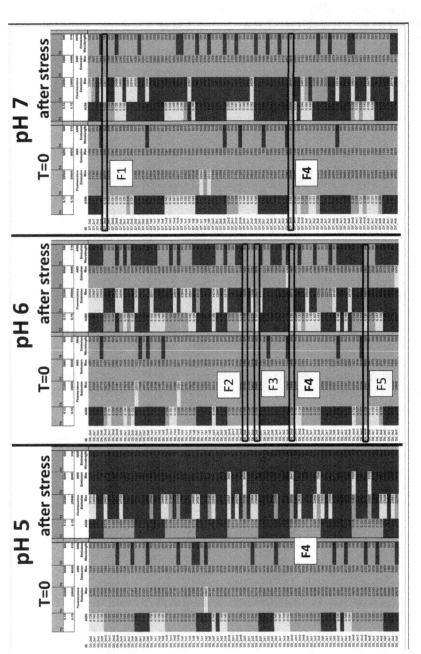

Figure 11.8 Example of data output from the high-throughput screening of 240 formulations at 100 mg ml⁻¹ of a monoclonal antibody. A total of 80 formulations were prepared, each at pH 5, pH 6 and pH 7. The formulations were analyzed in triplicate in 96-well plates. Each well contained 160 μl of mAb formulation. A total of 52 plates (including buffers alone) were prepared and analyzed. 58.5 g of mAb were used in these experiments. The samples were measured before and after a thermal stress with three methods: (i) UV absorption; (ii) intrinsic fluorescence emission; and (iii) 1,8-ANS fluorescence emission. The stability data for each method are shown in the columns using three different colors: white, grey or black. Grey indicates the formulation was stable; white indicates the formulation had acceptable stability; black indicates the analytical methods revealed degradation. The best five formulations found, which did not change strongly after the thermal stress, are indicated clearly in the figure by strong black outlines and are labelled F1–F5.

formulations, 20 may be selected. Stability testing of these 20 formulations in primary packaging, vials and syringes, leads to the selection of the best four. Plasma and blood compatibility testing may result in the selection of one single formulation to be further used in development, manufacturing and on the market.

11.6.2 HTA Methods

The analytical methods for formulations should detect small changes in the stability of the biopharmaceuticals after freeze–thaw cycles, thermal or shaking stress. In formulation studies, the focus is on inhibition of both aggregation in solution and chemical degradation of the molecule. Conformational change and aggregation have been correlated with reduced efficacy of some biologicals.[24,25] Particle formation has long been documented to constitute a high risk of anaphylaxis and anaphylaxis-like reactions upon intravenous administration to patients.[26] Chemical stability, *i.e.* inhibiting potential chemical degradation of the biomolecules (*e.g.* oxidation, deamidation) can also be improved by the formulation.

The sensitivity of the analytical methods is of essence in any formulation work. Using methods that do not detect changes in the molecule after stress will result in many "good" formulations, generating false hopes of soon finding a developable or marketable formulation and the expectation that the molecule is very stable. To induce changes in the stability of a biopharmaceutics drug one can perform mild or very intense stress experiments (*e.g.*, 1 day at 25 °C or 1 day at 37 °C or 1 day at 60 °C). In our experience, it is better to use mild stress conditions similar to those the drug will be expected to encounter during shelf-life. The use of harsh stress (*e.g.* temperatures above 40 °C, such as one day at 50 °C) may degrade the molecule by other or unique mechanisms, and the stable formulations found in this way may not be stable long term when stored at 4 to 8 °C. On the other hand, the use of mild stress conditions results, in general, in small changes to the molecule; thus there is a need to use many sensitive analytical methods to detect these small changes. No single method can be relied on to guarantee the absence of conformational change or protein aggregation in a drug product. However, if the results of several complementary, or orthogonal, methods are combined, a reasonably certain characterization of biological stability may be drawn. Methods suitable for HTA were discussed by Capelle *et al.*[20] and include UV–visible absorbance, 90° light-scatter, intrinsic fluorescence emission (tryptophan, tyrosine), fluorescence microscopy (*e.g.* employing dyes such as Nile Red, Thioflavine T, Congo Red), asymmetrical flow field-flow fractionation (FFF), and electron microscopy.

Examples of the first successes of the use of high-throughput methods were the development of stable formulations of human calcitonin (hCT, marketed as Cibacalcin®) and of stable recombinant hirudin (marketed as Revasc®). Human calcitonin is a naturally occurring polypeptide hormone involved in the regulation of calcium and bone metabolism and used to treat osteoporosis. Aggregation of hCT was a big problem in the manufacturing of the peptide, when approximately one in five manufacturing lots did not pass quality control. Aggregation of hCT results in a loss of biological activity. Extensive research was carried in the team of T. Arvinte to understand the mechanisms of aggregation of hCT. A double-nucleation hCT-aggregation model was developed that showed a linear dependence between the ln (natural log) of hCT concentration and the ln of fibrillation time.[24] Concomitant with the aggregation mechanism studies we performed a broad screen of ingredients, buffers and different solutions with the aim of developing a stable liquid formulation for nasal delivery. The hCT product had to be a stable liquid for up to 18 months at 1 or 2 mg ml^{-1} concentration. A new formulation screening strategy, based on the understanding of the aggregation mechanisms, was developed and used. The stress parameter for formulation screening was not high temperature or shaking; it was the use of highly concentrated protein solutions. Thus, to develop stable formulations at 1–2 mg ml^{-1} we prepared formulations between 50 and 200 mg ml^{-1} hCT. The formulations found to be stable at the high concentrations were expected, based on the double nucleation aggregation mechanism, to be also stable at lower concentrations. The linearity in the ln(hCT-concentration) *vs.* ln(fibrillation time) plots predicted by the double nucleation model was used to determine the stability of a solution at 2 mg ml^{-1} hCT. In this way, it was found that 0.001% acetic acid stabilizes hCT dramatically, much better than water or other acetic acid concentrations such as 0.1%, 1% or 60% acetic acid.[27] 0.001% acetic acid was used in a nasal formulation of hCT, which also contained other ingredients.[28] The extrapolated aggregation stability of this formulation, based on the ln(hCT-concentration) *vs.* ln(fibrillation time) plot, was 100 years: experimentally, no aggregation was observed after 2 years incubation at 4 °C.

Based on hCT experience, since the 1990s we have used high protein concentration as one stress parameter for developing aggregation-stable formulations. The best formulations identified at high concentrations in the majority of cases are stable also at lower concentrations. In a second step, different stable formulations found at high concentration have to be prepared at low concentrations and tested for stability: a good formulation is very likely to be found in this way.

The mechanisms of hCT aggregation, formation of large fibrils,[24] fibril bundles and cables proposed by the team in 1993–1994 were among the first contributions to the research emerging at that time on the protein aggregation mechanisms of beta-amyloid formation, a process implicated in Alzheimer's disease and "mad cow disease"/Creutzfeld–Jacob disease.

Recombinant hirudin (Revasc) is a potent anticoagulant, which was launched in 1997 by Ciba-Geigy/Novartis in a lyophilized formulation stable at room temperature: the formulation was developed in the group of T. Arvinte using HTF and HTA approaches.[29] Hirudin degrades chemically very rapidly both in liquid and lyophilized form. It was necessary that the market formulation be stable at room temperature. Despite sustained work over a couple of years using standard lyophilization approaches, it was not possible to obtain a hirudin lyophilized formulation that was stable at room temperature. We know now that the standard freezing and drying procedures which were applied, with focus upon the lyophilization cycle and standard formulations, induced changes in hirudin conformation which resulted in a loss of chemical stability during storage at 4 °C. This approach, focusing the optimization of the lyophilization cycle on obtaining a "nice looking" lyophilized cake may be suitable for small molecules, but was not good for hirudin, and, in our experience, it is not good for the majority of biopharmaceutics.

The stable hirudin lyophilized formulation was developed using HTA and HTF approaches based on screening hundreds of different formulations. The aim was to find ingredients (solution conditions) in which the hirudin molecule is stabilized: to find a "custom-made suit" for the molecule. It was found that $MgCl_2$ stabilizes hirudin at a broad range of concentrations.[29] The stabilization of hirudin by $MgCl_2$ and $CaCl_2$, but not $ZnCl_2$, was shown to be due to the interaction of Mg^{+2} and Ca^{+2} ions with the amino acids that are involved in the chemical degradation (cyclization of aspartic acid–glycine to cyclic imide).[30] The marketed hirudin product consists of a vial containing the lyophilized powder of 15.75 mg Hirudin, 1.31 mg anhydrous $MgCl_2$ and NaOH and a prefilled syringe containing 0.3% mannitol as diluent.[29,31]

The stable hirudin lyophilized formulation had no "nice looking" cake; it is a film with a small amount of powder. In this lyophilized formulation hirudin is stable for more than 2 years at room temperature; there are no vial-to-vial variations and the reconstitution is within 1–2 seconds. Furthermore, the formulation improves the chemical stability of hirudin by reverting the chemically degraded molecule prior to lyophilization to not-degraded species after lyophilization: the purity of hirudin also increases after storage in lyophilized form at 40 °C for one month.[29,30] The hirudin $MgCl_2$ formulation obtained using HTA and HTF approaches was used for the hirudin product launched in 1997 by Novartis: the first recombinant biotech product developed and launched by a Swiss pharmaceutical company.

The cake appearance is not an important quality attribute for good protein lyophilized formulations. A good lyophilized formulation is one in which, after prolonged storage of the vials (*e.g.* for hirudin during 2 years of storage at room temperature), the lyophilized powder has a fast reconstitution after addition of diluent and the solution obtained after reconstitution contains minimally degraded protein (compared with the protein purity before lyophilization). Thus, the first step in determining a good formulation is the development of analytical methods which will reliably show the concentration and molecular stability of the biological and, optimally, at least give an indication that its biological activity is not compromised. This will require a good understanding of the physical and chemical properties of the molecule and will often be tailor-made for the biological.

11.7   The Biopharmaceutical Industry Today

Biopharmaceuticals and especially recombinant proteins are at present the major products of the pharmaceutical industry regarding sales. In 2014, from the 15 top selling drugs, eight were recombinant biopharmaceuticals

Table 11.4   Example of prices for some biopharmaceuticals launched in 2014 and
2015. Adapted from ref. 37 and 40.

Recombinant monoclonal antibodies	Indication	Company	Price in US $
Blincyto (blinatumomab)	Acute lymphoblastic leukemia	Amgen	178 000 per patient
Keytruda (pembrolizumab)	Melanoma	Merck	150 000 per year
Opdivo (nivolumab)	Melanoma	Bristol-Myers Squibb	143 000 per year
Yervoy (ipilimumab)	Metastatic melanoma	Bristol-Myers Squibb	120 000 per year
Portrazza (necitumumab)	Lung cancer	Eli Lilly	137 000 per year

(Table 11.1). The prices of new biopharmaceutics are very high compared with small molecules, prices justified by the complex research, manufacturing and pharmaceutical and clinical development (Table 11.4). Starting in 1982 when recombinant insulin was launched, the number of biopharmaceuticals on the market has increased slowly, and in recent years the number of new biopharmaceuticals approved by the FDA in one year has been around 10, with a total of 114 new biopharmaceuticals approvals since 1993 (Figure 11.9). In the same period, 1993–2016, 579 new small molecules were approved (Figure 11.9).

In a recent presentation[32] the FDA gave an overview of the work their office had performed between 1993 and 2016, reflected in the number of FDA activities on Investigational New Drug (IND) applications,[33] approved New Drug Applications (NDAs, *i.e.* small molecules)[34] and approved Biologics License Applications (BLAs, *i.e.* biopharmaceuticals).[35] Using these data we compared the number of IND applications and the number of approved licenses for both biopharmaceuticals (Figure 11.10), and small molecules (Figure 11.11). The number of IND applications for biologics received at the FDA between 2004 and 2016 increased from 588 in 2004 to 1480 in 2016 (Figure 11.10). However, the success rate, *i.e.* the percentage of granted BLAs from the total biologics INDs, was very small, between 0.2% and 0.9% with a mean over the last 13 years of 0.62%. The number of small-molecule INDs on which the FDA worked over the same period was larger than for the biopharmaceuticals, increasing from 3773 in 2004 to 5261 in 2016 (Figure 11.11). However, the percentages granted NDAs from the total INDs received were also very small, between 0.2% and 0.8%: the mean over the last 12 years was 0.47%. The high failure rate of IND applications becoming approved drugs, 99.4% for BLAs and 99.5% for NDAs, may be reduced by ensuring that the molecules tested in clinical trials are in stable, robust formulations that provide good chemical and physical stability during transport and storage, and that no aggregation occurs after *in vivo* administration.

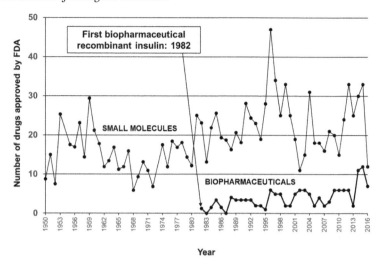

Figure 11.9   FDA approvals of small molecules and biopharmaceuticals between 1950 and 2016. The data for the number of approved drugs from 1950 to 1992 are from ref. 38. The data from 1993 to 2016 are from ref. 32. Multiple applications pertaining to a single new molecular/biological entity were only counted once. For the period from 1993 to 2016 approved BLAs and NDAs that did not contain a new active ingredient are excluded (*e.g.*, formulation changes of approved drugs are not counted).

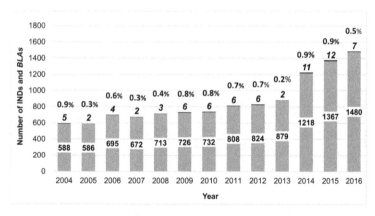

Figure 11.10   Biologic INDs[33] with FDA activities from 2004 to 2016 compared with the number of approved BLAs[35] issued in the same year (italic numbers). Multiple applications pertaining to a single new biologic entity are only counted once. The percentages of approved BLAs compared with biologics INDs received are shown above each bar. Figure adapted from data in ref. 32. The percentages of granted BLAs from the total biologics INDs were between 0.2% and 0.9%. The mean approval percentage was 0.62% over the last 13 year period.

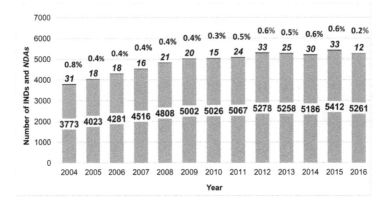

Figure 11.11   Drug (small molecules) INDs[33] with FDA activities from 2004 to 2016 compared with the number of approved NDAs[34] issued in the same year (italic numbers). Multiple applications pertaining to a single new biologic entity are only counted once. The percentages of approved NDAs compared with small-molecule INDs received are shown above each bar. Figure adapted from data in ref. 32. The percentages of granted NDAs from the total small molecules INDs were between 0.2% and 0.8%. The mean approval percentage was 0.47% over the last 13 year period.

The successful discovery, development and use of new medicines are and will remain important activities in health care efforts. In this context, biological products with optimized formulations will surely have an important contribution.

References

1. R. A. Rader, *BioExecutive International*, 2005 March, p. 60.
2. R. A. Rader, *BioExecutive International*, 2005 May, p. 42.
3. U S Food and Drug Administration, *What Are Biologics Questions and Answers*, e-accessed 15-Nov-2016 at: http://www.fda.gov/aboutfda/centersoffices/officeofmedicalproductsandtobacco/cber/ucm133077.htm.
4. Pharmaceutical Science Expert Advisory Panel: C. Bond, D. Craig and S. Denyer, et al., *New Medicines, Better Medicines, Better Use of Medicines, a Guide to the Science Underpinning Pharmaceutical Practice*, Royal Pharmaceutical Society, May 2014 2016, p. 25.
5. Pharmaceutical Science Expert Advisory Panel: C. Bond, D. Craig and S. Denyer, et al., *New Medicines, Better Medicines, Better Use of Medicines, a Guide to the Science Underpinning Pharmaceutical Practice*, Royal Pharmaceutical Society, May 2014 2016, p. 99.
6. www.fda.gov.
7. N. Ritter, *Analytical CMC Requirements for Biotech and Biosimilar Products: More Than Meets the Eye!* at: http://c.ymcdn.com/sites/casss.site-ym.com/resource/resmgr/BADG_Speaker_Slides/BADG_0215_Ritter_Slides.pdf.

8. G. Walsh, *Nat. Biotechnol.*, 2014, **32**, 992.
9. T. Arvinte, *BioWorld Europe*, 2007, 6–9.
10. J. M. McKoy, R. E Stonecash, D. Cournoyer, J. Rossert, A. R. Nisenson, D. W. Raisch, N. Casadevall and C. L. Bennett, *Transfusion*, 2008, **48**, 1754.
11. J. C. Ryff, *J. Interferon Cytokine Res.*, 1997, **17**(suppl. 1), S29.
12. J. Kotarek, C. Stuart, S. H. De Paoli, J. Simak, T. -L. Lin, Y. Gao, M. Ovanesov, Y. Liang, D. Scott, J. Brown, Y. Bai, D. D. Metcalfe, E. Marszal and J. A. Ragheb, *J. Pharm. Sci.*, 2016, **105**, 1023.
13. T. Arvinte, Concluding remarks: analytical methods for protein formulations, in *Methods for Structural Analysis of Protein Pharmaceuticals*, ed. W. Jiskoot and D. J. Crommelin, AAPS Press, 2005, pp. 661–666.
14. T. Arvinte, E. Poirier and C. Palais, in *Biobetters. Protein Engineering to Approach the Curative*, ed. A. Rosenberg and B. Demeule, Springer, New York, 2015, pp. 91–104.
15. T. Arvinte, C. Palais, E. Green-Trexler, S. Gregory, H. Mach, C. Narasimhan and M. Shameem, *mAbs*, 2013, **5**, 491.
16. S. Luo and B. Zhang, *mAbs*, 2015, **7**, 1094.
17. B. Demeule, C. Palais, G. Machaidze, R. Gurny and T. Arvinte, *mAbs*, 2009, **1**, 142.
18. T. Arvinte, C. Palais, E. Green-Trexler, S. Gregory, H. Mach, C. Narasimhan and M. Shameem, *mAbs*, 2013, **5**, 491.
19. FDA, *"*Important Drug Warning" Letter to Genentech on Avastin Side Effects*, January 5, 2005.
20. M. A. H. Capelle, R. Gurny and T. Arvinte, *Eur. J. Pharm. Biopharm.*, 2007, **65**, 131.
21. M. A. H. Capelle and T. Arvinte, *Drug Discovery Today: Technol.*, 2008, **5**, e71.
22. M. A. H. Capelle, R. Gurny and T. Arvinte, *Pharm. Res.*, 2009, **26**, 118.
23. R. Jaenicke, *J. Biotechnol.*, 2000, **79**, 193.
24. T. Arvinte, A. Cudd and A. F. Drake, *J. Biol. Chem.*, 1993, **268**, 6415.
25. N. Taschner, S. A. Mueller, V. R. Alumella, K. N. Goldie, A. F. Drake, U. Aebi and T. Arvinte, *J. Mol. Biol.*, 2001, **310**, 169.
26. J. M. Garvan and B. W. Gunner, *Med. J. Aust.*, 1964, **2**, 1.
27. B. Galli and T. Arvinte, Contribution to the Ciba-Geigy patent on human calcitonin: air-tight nasal applicator, European Patent EP 0 531 257 A1, 1993.
28. T. Arvinte and K. Ryman, Stabilisation of pharmaceutical compositions comprising calcitonin, European Patent, Publication Number 0490549B, 1995.
29. T. Arvinte, Stable dry powders containing hirudin, European Patent, Publication Number EP 0665019 B, 2000.
30. A. P. Nordmann, PhD thesis (hirudin), Birkbeck College, University of London, Sept 1996.
31. https://www.accessdata.fda.gov/drugsatfda_docs/label/2014/021271s006lbl.pdf.
32. J. K. Jenkins, *CDER New Drug Review: 2016 Update FDA/CMS Summit*, Dec 14 2016, at: http://www.fda.gov/downloads/AboutFDA/CentersOffices/OfficeofMedicalProductsandTobacco/CDER/UCM533192.pdf.

33. https://www.fda.gov/Drugs/DevelopmentApprovalProcess/HowDrug-sareDevelopedandApproved/ApprovalApplications/InvestigationalNew DrugINDApplication/.

34. https://www.fda.gov/drugs/developmentapprovalprocess/howdrug-saredevelopedandapproved/approvalapplications/newdrugapplica-tionnda/default.htm.

35. https://www.fda.gov/BiologicsBloodVaccines/DevelopmentApprovalPro-cess/BiologicsLicenseApplicationsBLAProcess/.

36. S. Stebbins, 24/7 Wall St, *The World's 15 Top Selling Drugs*, April 26 2016, at: http://247wallst.com/special-report/2016/04/26/top-selling-drugs-in-the-world/2/.

37. T. Staton, Nov 30 2015, http://www.fiercepharma.com/marketing/amgen-slaps-178K-price-on-rare-new-leukemia-drug-blincyto.

38. B. Munos, *Pharma's Innovation Challenge, Scientific American – Pathways: The Changing Science, Business & Experience of Health*, a custom collabo-ration with Quintiles, 2010, SAPathways@sciam.com.

39. E. Hochuli, Interferon immunogenicity: technical evaluation of interferon-α2A, *J. Int. Cytokine Res.*, 1997, **17**, S15–S21.

40. LIlly's Lung-Cancer Drug Portrazza to Cost $11,430 a Month, *First Word Pharma*, 11 December 2015, http://www.firstwordpharma.com/node/1340512#axzz4J5LlhXRS.

CHAPTER 12

Intellectual Property

DOMINIC ADAIR* AND CHLOE DICKSON

Bristows LLP, 100 Victoria Embankment, London, EC4Y 0DH, UK
*E-mail: Dominic.Adair@Bristows.com

12.1 Introduction to Intellectual Property

Unlike real property, intellectual property (IP) is intangible. Rather than existing as a physical item of property, intellectual property exists as a number of rights associated with something tangible. They are generally user rights, the effect of which is usually to create a time-limited negative monopoly, meaning that the owner of the right has the exclusive ability to use the tangible property with which the intellectual property right is associated. The rights exist above and beyond the tangible property, and are not limited to any particular item.

IP rights can be considered industrial in character (in fact, the old term for intellectual property was "industrial property"); the negative monopoly allows for exclusive commercial exploitation of tangible property made by industry. This fosters innovation because, by allowing the rights owner to commercialise the underlying tangible property without competition for a certain period, further creative development is incentivised.

Nowhere is this more true than in the pharmaceutical industry and in relation to patents in particular. Patents protect inventions. It is well known that the research and development activity necessary to create a new drug product is extremely expensive. A recent study by Tufts University has estimated

Drug Discovery Series No. 64
Pharmaceutical Formulation: The Science and Technology of Dosage Forms
Edited by Geoffrey D. Tovey
© The Royal Society of Chemistry 2018
Published by the Royal Society of Chemistry, www.rsc.org

the typical cost of developing a new drug at more than two billion dollars.[1] Each new drug product can be associated with a number of new inventions: process technology, drug substance, polymorphic form, formulation, dosage regimen, new combination, new indication *etc.* Patent protection allows the pharmaceutical company owning the inventions the opportunity to make exclusive sales of the new drug product and thus enables it to recoup its investment. Without patent protection, immediate market competition would prevent sufficient return on investment and the incentive to develop new drugs would be lost.

Patent protection is inherently technical in character. In contrast, other intellectual property rights such as trade marks, copyright or design rights largely protect the appearance of a product or its packaging. Unsurprisingly, therefore, patents are the most important intellectual property protection for pharmaceutical formulations. Branding and other aspects of the appearance of pharmaceutical products play a lesser role, largely because only a relatively small number of drug products are sold directly to consumers in a retail environment. Most are prescription-only medicines which are prescribed and dispensed without consumer choice. Accordingly, the bulk of this chapter focuses on patent rights, with the remainder addressing rights which relate to the appearance of a product.

12.2   Patents

Patents protect inventions. They give the owner the right to sue for infringement any person who makes, uses, sells or imports an article which makes use of the protected invention, without the patent owner's permission. This means the owner has a legal monopoly to use the invention set out in the patent. The monopoly lasts for 20 years from the time at which the patent application is filed. In return for this monopoly, the applicant must set out a description of the invention in sufficient detail that others can work the invention once the monopoly is over. Many good inventions have been made in the field of pharmaceutical formulation and there are several litigation cases concerning formulation patents available in the law reports. In order fully to illuminate these examples, it is necessary to start with a short introduction to the law.

Patents are designed to protect technical advances which are:

(i) Capable of being used in industry;
(ii) New; and
(iii) Inventive.

Whether an application for a patent clears these three hurdles is judged by viewing the application document, the patent "specification", through the eyes of the notional person to whom it is addressed, operating in the relevant technical field. This is the so-called "skilled addressee" or "person skilled in the art". This person is someone "likely to have a practical interest in the subject matter of the invention"[2] and "with practical knowledge and experience of the kind of work in which the invention was intended to be used".[2] The skilled addressee has the average level of skill in the technical area concerned and does not possess any inventive capacity; they can be thought of as an ordinary technician. The skilled addressee may also be a team, comprising individuals with different technical backgrounds. For a patent concerned with pharmaceutical formulations, the skilled addressee is likely to be a formulation scientist.

(i) Industrial application

A patentable invention must be capable of being used in industry and the law is not restrictive as to which types of industry are relevant. Furthermore, industry is construed very widely and should be:

"Understood in its broad sense meaning any physical activity of 'technical character' *i.e.* an activity which belongs to the useful or practical arts as opposed to the aesthetic arts."[3]

It is important to note that the industrial application does not have to be the final use of the invention. Thus, for example, in a case concerning certain antibodies of Neutrokine-α (a member of the TNF ligand superfamily) it was sufficient that the antibodies had some utility as a research tool in the development of therapies for certain diseases. The subject matter of the patent did not itself have to be an effective therapy for the diseases.[4]

(ii) Novelty

An invention is new if it does not form part of the "state of the art". The state of the art encompasses everything which has been made available to the public, in any way, before the patent application is filed.[†] If an invention is disclosed by any prior art then it is said to be "anticipated". However, in order to anticipate a patent the disclosure must also be "enabling". This means the skilled person described above must be able to put the invention into effect on the basis of what is disclosed.[5]

(iii) Inventive step

The invention must also involve an inventive step. This means that considered against the state of the art, it would not have been obvious to the skilled person to do what is claimed by the patent. Whilst this sounds simple enough, the question of obviousness is one of the most difficult, and frequently encountered, issues in patent law. A summary of the issue is contained in the case of *Generics vs. Lundbeck*:[6]

"The question of obviousness must be considered on the facts of each case. The court must consider the weight to be attached to any particular factor in the light of all the relevant circumstances. These may include

---

[†]An earlier date "the priority date" is used to assess the state of the art if the patent application claims a right to rely on an earlier (priority) application filed in the previous 12 months for the purpose of claiming the first filing of the invention.

such matters as the motive to find a solution to the problem the patent addresses, the number and extent of the possible avenues of research, the effort involved in pursuing them and the expectation of success."

While these are all useful considerations, none takes precedence and the question of whether the invention is obvious will, in the end, depend on all the circumstances. In order to develop the evidence on the issue, use of expert witnesses is essential.

12.2.1   The Patent System

Any person may apply for a patent,[7] but only the inventor (the deviser of the invention) may be granted a patent.[8] This is subject to the proviso that an invention made by a person employed in the UK is taken to belong to their employer if it was made in the course of their normal duties, or the employee has some special obligation to further the interest of the employer.[9]

An applicant for a United Kingdom patent has a choice between applying for:

(i) a national patent, for the United Kingdom only; or
(ii) a European patent, which designates the United Kingdom.

A European patent is generally chosen if the applicant wishes to obtain protection in multiple EU Member States, as the European Patent Office (EPO) will grant a bundle of equivalent national patents. This is usually a more cost-effective route than applying to each national office individually.

In the future, the options for obtaining a patent are likely also to include applying for a European patent with unitary effect, a "Unitary Patent", granted by the EPO with unitary effect in each of the EU Member States that have signed the enabling legislation.[‡]

For a UK patent, the application (consisting of a description of the invention, the claims, a summary or abstract of the invention and any drawings required to illustrate the invention) is filed at the UK Intellectual Property Office (UKIPO). A search request is then filed within the next six months and a patent examiner will gather information to determine whether the invention is new and inventive. Within 12 months, if they haven't already submitted them with the application, the applicant must file a set of claims which define the patented or "claimed" invention. 18 months after the application, the patent application will be published, the applicant then has a further six months to request examination of the patent and pay the

---

‡Following the outcome of the referendum on EU membership in the UK, the long-term future of the Unitary Patent in the UK is unclear. Although the UK will ratify the agreement creating the unitary patent system, at the time of writing the question of whether the UK will be able to continue to participate after leaving the EU is difficult to assess.

relevant fee, after which time the application will enter the examination phase. Approximately 12 months later, the UKIPO will issue an examination report and the patent is either granted, or the applicant must respond to concerns the examiner has (which may involve amending the claims). Following the applicant's response, a further examination report will then be issued. Examination continues until final refusal or grant of a patent. Once granted, renewal fees must be paid yearly until expiry of the patent, 20 years from the application date.

For an application to the EPO, the application (also consisting of a description of the invention, claims, an abstract and drawings) will be published six months from the application date, without a search report. The search report will be issued and published within six months from publishing the application and, within a further six months, the applicant must file a request for substantive examination. Following examination, the first examination report will be issued. As for UK patents, the applicant must then respond and further examinations continue until refusal or grant. A further step which must be taken with European patents drafted in English is translation of the claims into both French and German languages.

A comparative timeline for the UKIPO and EPO application procedure is set out in Figure 12.1.

As mentioned briefly above, and subject to the payment of annual renewal fees, the duration of patent protection in the UK and EU is 20 years from the date of filing the application.[10] In certain cases the period of protection may be extended by a Supplementary Protection Certificate (SPC). SPCs are discussed further under the heading *"Life Cycle Management"*.

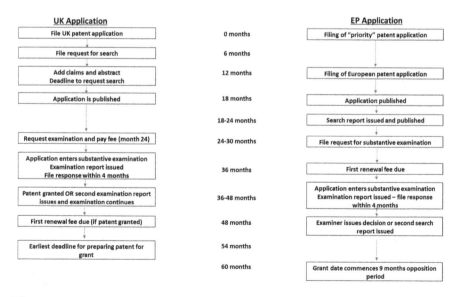

Figure 12.1 A Comparative timeline for the UKIPO and EPO patent application procedure.

12.2.2   Challenges to the Validity of a Patent

A 20 year monopoly can be a long time for a competitor to wait. Hence, it is common to find that attempts will be made to invalidate the patent by those seeking to bring a product to market for whom the patent represents an obstacle. There are two main types of challenge to a patent after grant: (i) central oppositions at the EPO within 9 months of grant (for European Patents only) and (ii) national revocation proceedings before national patent offices and courts at any time after grant.

A granted UK or European patent may be revoked under certain grounds and any member of the public may apply to have a patent revoked. It is not necessary to have a commercial interest in the invention.

The grounds on which a patent may be revoked are:

   (i)   The invention is not patentable (*i.e.* not new or inventive or capable of industrial application);

  (ii)   The patent was granted to a person not entitled to the patent;

 (iii)   The patent specification does not disclose the invention clearly and completely enough to enable a person skilled in the art to perform it ("insufficiency");

 (iv)   The matter disclosed in the patent specification extends beyond that disclosed in the application as filed ("added matter"); or

  (v)   The protection afforded by the patent has been extended by an amendment which should not have been allowed ("impermissible claim broadening").

12.2.3   Prior Disclosure

It will already be evident from the above that in order to satisfy the test for novelty, it is important to maintain the secrecy of the invention before the patent application is filed. If it is disclosed to just one person who is free to use the information, then the invention is no longer novel and hence cannot be patented. Limited exceptions exist[§] but pharmaceutical companies usually go to great lengths to prevent accidental disclosure prior to patent filing. However, there are certain danger areas of which to be aware.

12.2.4   Clinical Trials

One area in relation to which pharmaceutical companies must take particular care is clinical trials. Often patent applications are filed before clinical trials commence. However, if a clinical trial commences first, the invention might have been made available to the public if it is used in a trial where those concerned are not under an obligation of confidence. It is therefore

---

[§]In the UK there is a 6 month grace period which preserves novelty if the invention is disclosed in breach of confidence or at an officially recognised international exhibition.

important to ensure clinical trial agreements contain appropriate confidentiality obligations. This was highlighted in an appeal during opposition proceedings at the EPO in decision T 0007/07 (*Ethinylestradiol and drospirenone for use as a contraceptive*) concerning a micronized formulation of drospirenone in the oral contraceptive Yasmin®. The personnel conducting the trials, but not the women participating, had signed confidentiality agreements. The women knew the identity of the active substance in the trial but did not know that it was a micronized formulation. Furthermore, not all unused drugs had been returned at the end of the trial. As the patentee had effectively lost control of the unreturned drugs and the women appeared to be under no restrictions from disposing of them, all the tablets handed out were found to have been made publicly available. The Board of Appeal then considered whether the skilled person would have been able to analyse the formulations and determine that drospirenone was in a micronized state (it does not matter whether he would have had any motivation to do so). The evidence was that the micronized state could have been determined *via* Raman spectroscopy and so the formulation was held to lack novelty.

12.2.5 Life Cycle Management

Inventions in pharmaceuticals are not limited to the discovery of a new drug substance. On the contrary, many different aspects of pharmaceuticals can be used as the basis for a new invention, including drug substance combinations, dosage regimens, new treatment indications, polymorphs, and, of course, formulations. Each invention will gain its own 20 year monopoly and hence it is important for pharmaceutical companies to maximise the potential to protect all aspects of a drug in order to gain sufficient return on their investment. Typically, a new drug product costs over two billion dollars to develop and, given the regulatory delays in the process of getting a drug to market, peak sales are only achieved towards the end of the period of patent protection. As the patent which protects the active ingredient reaches expiry, so-called "secondary patents" are of increasing importance to ensure that pharmaceuticals companies can continue to protect their investments.

Some patent expiry dates after which, in theory, generic competition could lawfully appear are shown in Table 12.1.

Acknowledging this fact, pharmaceutical inventions in the European Union can benefit from an extension to patent protection under so-called Supplementary Protection Certificates (SPCs). The purpose of an SPC is to compensate the patent-holder for the delay in bringing a medicinal product to market until regulatory approval is obtained, in the form of a marketing authorisation (MA). The time associated with obtaining an MA may mean that the period of patent protection is not sufficiently long to recover the investment put into researching and developing the product. An SPC may be granted for a duration equal to the period elapsed between patent filing and MA grant, less five years and subject to a maximum duration of five years.

Table 12.1   Some example expiry dates.

Product INN[a] (*Originator brand*)	Compound patent expiry date
Valsartan (*Diovan®*)	2011
Atorvastatin (*Lipitor®*)	2012
Pregabalin (*Lyrica®*)	2013
Trastuzumab (*Herceptin®*)	2014
Pemetrexed (*Alimta®*)	2015
Imatinib (*Glivec®*)	2016
Adalimumab (*Humira®*)	2018

[a]INN, International non-proprietary name.

Note that while patent protection may be broad enough to cover multiple products, an SPC is limited to the product which has been authorised. The jurisprudence in SPC cases indicates that SPCs will only be awarded in respect of patents for new active ingredients or combinations of active ingredients, and not to new formulations of known active ingredients. However, at the time of writing, the question of whether a formulation patent can benefit from an SPC has been referred from the English Patents Court to the Court of Justice of the European Union.[11]

12.2.6   Case Law on Formulation Patents

As alluded to above, the most common basis on which patents for pharmaceutical formulations are challenged is lack of inventive step: that making the formulation is an exercise in the merely routine. This issue has come to the fore in a number of cases.

12.2.6.1   *Teva* vs. *Leo*

Teva vs. *Leo*[12] was a case concerning Leo's patents for an ointment formulation to alleviate the symptoms of psoriasis, an inflammatory disease. In patients with mild psoriasis, compliance with treatment tended to be a problem because treatments were perceived by patients to be inconvenient or causing unpleasant side effects. The patent claimed an ointment comprising:

- Two active ingredients (calcipotriol and betamethasone);
- A base; and
- A solvent called Arlamol E.

Each of the active ingredients was well-established for use as a sole active ingredient in psoriasis ointment or cream. Each active ingredient has a different beneficial effect and so patients were often prescribed both products. However, they could not be used at the same time. Each of the ingredients was unstable save in a narrow pH: alkali for calcipotriol and acid for betamethasone. The pHs for stability did not overlap and patients had to be warned not to apply the treatments together. A patient might use one in the morning and one in the afternoon.

The pH problem only arose because of the presence of water, without which there would be no pH at all. In theory, the simple solution would be to remove the water. However, using a non-aqueous solvent was not straightforward in practice. There was no evidence that any non-aqueous solvent would work; keeping water out was a well-known problem and in practice it was unlikely to be possible to remove all the water from the system. Even if a lab-based formulation was truly dry, some water was likely to be encountered during manufacture, processing and during the product's lifetime. This meant that just because a formulator identified the possibility of making an ointment using a non-aqueous solvent, they would not simply put their concerns about the pH to one side.

In the High Court, the Judge found that the use of the Arlamol E solvent to address the problem of formulating a combination formulation was obvious. The use of Arlamol E was disclosed in a document called Turi; Turi described a non-aqueous composition containing the solvent and betamethasone (one of the two active substances). The Judge went on to decide that the skilled formulator presented with Turi would proceed to test Arlamol E for the combination formulation and the result would be positive. A clinical study would then be carried out using the combination formulation and this would confirm it was an effective treatment. The invention was held to be obvious.

The Court of Appeal disagreed with the Judge's finding and held that while it was established that the skilled person would know that it would be necessary to use a non-aqueous solvent in order to formulate the two active ingredients, there was not a sufficient expectation of success that Arlamol E would work. The Court of Appeal held that finding a non-aqueous solvent which would work was a research project. There was nothing which pointed to Arlamol E as having better prospects than any other non-aqueous solvent.

In its judgment, the Court of Appeal also noted that there was a long-felt want for a combination product and no explanation of why that solution was not made before when it could have been. In short, this was "the classic sort of case where the Courts have found invention over the years".

12.2.6.2 *Hospira* vs. *Genentech*

The patent in suit in *Hospira vs. Genentech*[13] was primarily concerned with the role of trehalose as a stabiliser in lyophilised preparations of trastuzumab, an antibody treatment for breast cancer.

Hospira argued that Genentech's patent was obvious, relying on the existence of Phase II clinical trials for trastuzumab in the treatment of breast cancer. Hospira argued that the skilled person would be motivated to develop a lyophilised formulation and that, using their common general knowledge, they would reach the patented formulation.

The claims in this case were held to be obvious since all aspects of the claims, including the use of the trehalose, could "be reached by the application of nothing other than routine approaches applied to excipients which were part of [the skilled person's] common general knowledge". This decision was upheld in the Court of Appeal,[14] where it was held that the skilled person would have conducted a screening programme with the expectation that they would identify a suitable formulation. Despite the fact that the skilled formulator would not be able to tell in advance exactly which formulation in the screening programme would work, this did not prevent the patent being obvious. It was noted that the facts of this case were very different from those in *Teva vs. Leo* where the skilled person would not have had the same expectation of success.

The difference in result between the *Hospira vs. Genentech* and *Teva vs. Leo* cases is instructive. Both concern patented formulations characterised by the choice of certain excipients. In both cases the question was whether it would be obvious to reach the patented formulation by including those excipients in a screening programme. The cases illustrate that there are two requirements which must be met in order for the patent to survive a challenge to its inventive step. First, the choice of excipient must offer a benefit. If it is just one of a number of choices, any of which would do, then the choice is arbitrary and offers no technical contribution to the invention. If there is a lack of technical contribution, the patent will be held to be obvious *i.e.* there is no inventive step. The second requirement is that the skilled person would not have a reasonable expectation that the choice of excipient would succeed in the formulation. If there was a motivation to try it, and an understanding that it should work, that will also serve to make the selection of excipient obvious. The key is to have something like Arlamol E in Leo's ointment, which has a distinct benefit, but is one of a large number of alternatives to try, with no special pointer to its success.

12.2.6.3   *Gedeon Richter* vs. *Bayer Schering Pharma*

In this case,[15] Gedeon Richter sought revocation of two of Bayer's patents which related to formulation of the contraceptive tablet Yasmin®, which contains two steroidal hormones, drospirenone and ethinylestradiol. Drospirenone was known to be sparingly soluble and to isomerise under acid conditions and hydrolyse under alkaline conditions. To ensure good bioavailability, the drospirenone therefore needed to be provided in a form which promotes rapid dissolution.

The documents relied on by Gedeon Richter to attempt to show that the patent was obvious, crucially, contained little or no information on the formulation of the contraceptives and other medicaments concerned.

The judgment makes clear that if the invention relates to a formulation of a compound, then if it can be shown that a formulator would, as a matter of routine, find out certain information about the compound, *e.g.* solubility in water at various temperatures and various pHs, then it is legitimate to take that information into account when assessing the obviousness of a particular formulation.

Both parties agreed that, while it would have been obvious to carry out *in vitro* pre-formulation testing, such tests would be performed in ignorance of the outcome and of whether any specific formulation strategy would have a fair expectation of success. This led the Court to conclude that the patent was not obvious; the finding was upheld on appeal. The Court of Appeal stated that the step from the key piece of prior art to the invention would have been more of a "speculative jump in the dark than anything else." The Court of Appeal also relied on the absence of a clear explanation as to how the claimed formulation worked, this indicating that the success of the claimed formulation was unlikely to have been predictable.

Again, this illustrates that it will be harmful to the arguments in support of patent validity if reasons exist to believe that the patented formulation would work and hence the choice of excipients would be obvious to try. It is helpful if there is a certain amount of mystery around the success of the formulation, such that no motivation exists to pursue a line of enquiry. Another good example of this is the Omnipharm case, concerning "spot-on" veterinary formulations.

12.2.6.4 *Omnipharm* vs. *Merial*

This case[16] concerned solutions of fipronil as an anti-parasitic agent for protecting animals, such as small mammals and pets, from fleas. One of the patents concerned a ready-to-use solution which comprised an organic solvent and an organic co-solvent, both of specified boiling points and dielectric constants, as well as a compound to inhibit crystallisation. The claim, as amended, was for a "spot-on" solution for localised application (which spreads to treat the entire skin of the animal).

Omnipharm alleged the spot-on application was obvious in light of a spray which was marketed for all-over use on cats and dogs ("Frontline"). Omnipharm's argument was that it was obvious to formulate a spot-on application as this would increase ease of application. Furthermore, Omnipharm said it was obvious to formulate a non-systemic formulation (such as a spray) and to come up with the formulation claimed in the patent.

The Court disagreed with Omnipharm's arguments and found that the skilled team would not have had a sufficient expectation that the spot-on formulation would be successful. There was no technical basis for suggesting that fipronil would work as a spot-on solution, just because other compounds had worked well in such a formulation. As in the Gedeon Richter case, at the time of the invention, the mechanism by which spot-on treatments worked was not known. As the Court of Appeal put it:

> "it was generally understood that the application of one of these formulations to a single point on the animal's skin or along a line down the back of the animal led to its distribution over the whole of the animal's body without the formulation ever entering the animal's bloodstream. But how this distribution occurred was something of a mystery and, even today, is not fully understood"

In the absence of any common general knowledge theory as to how non-systemic spot-on formulations worked, the person skilled in the art did not have a sufficient expectation of success to render the invention obvious.

12.2.7   Conclusion

Although three out of four of the formulation patent cases described above resulted in the challenge to inventive step being rejected, these cases were selected to illustrate points of principle. A more balanced view is that formulation patents often struggle to survive validity challenges unless they offer genuinely surprising benefits. In more than half of the 14 decisions involving formulation patents at first instance and appeal between the years 2008 and 2015, the patent was held to be invalid.

Formulation patents can be an important part of the intellectual property protection around the life cycle of a drug product but care must be taken to frame the invention appropriately. Often having a good "invention story", describing the problems which exist in the state of the art, and how the patent solves them in a manner which is not foreseeable, is an important part of defending the allegation that the formulation is merely routine.

12.3   Protecting a Product's Appearance

At least in the UK, medicines regulation is such that prescription-only medicines are not advertised to consumers and consumers usually have little or no choice to exercise over the products they receive. Accordingly, intellectual property protecting a product's appearance is usually only meaningful in connection with pharmaceuticals sold "over-the-counter", such as the cough and cold remedies which can be purchased without prescription from pharmacies, supermarkets and other retail environments.

There are a number of intellectual property rights which are connected with a product's appearance and directed to the consumer at the point of sale.

12.3.1   Trade Marks

12.3.1.1   Trade Mark Protection

Trade marks act as a badge of origin or a guarantee of quality. Their purpose is to guarantee the trade origin of the goods or services supplied under the mark. Furthermore, in order for a trade mark to even be granted, it is a requirement under section 1 of the Trademarks Act 1994 (TMA) that it is capable of distinguishing the goods or services of the trade mark owner from those of another.

Unlike patents, which have a 20 year term, trade mark protection is capable of indefinite renewal, and therefore has the potential to create a perpetual monopoly. Trade marks are therefore a potentially powerful form of IP protection. But the breadth of protection is much narrower than patents—they do not protect the product or technology *per se*, only the branding under which it is put on the market.

The TMA defines a trade mark as[17]

"Any sign capable of being represented graphically which is capable of distinguishing goods or services of one undertaking from those of another.

A trade mark may, in particular, consist of words (including personal names), designs, letters, numerals or the shape of goods or their packaging."

The definition of a trade mark as a "sign" is potentially wide-ranging and extends to shapes, colours tastes and sounds, however the Courts in the UK and in the Court of Justice of the European Union (CJEU), have provided some guidance on the limits of the definition. For example, a sign cannot include a "mere property of the product concerned"[¶] nor does it include the use of colours without any limitation or spatial delimitation.[18]

The UK Act requires that the trade mark must also be capable of graphical representation, this means it must:[19]

"enable the sign to be represented visually, particularly by means of images, lines or characters, so that it can be precisely defined"

In *Sieckmann* a trade mark application for a fruity cinnamon smell was found not to meet the requirements on the basis of a verbal description, chemical formula and odour sample provided with the application. The Court added:

"As regards a chemical formula... few people would recognise in such a formula the odour in question. Such a formula is not sufficiently intelligible. In addition... a chemical formula does not represent the odour of a substance, but the substance as such, nor is it sufficiently clear and precise. It is therefore not a representation... In respect of the description of an odour, although it is graphic, it is not sufficiently clear, precise and objective. As to the deposit of an odour sample, it does not constitute a graphic representation... Moreover, an odour sample is not sufficiently stable or durable"

It is possible olfactory marks will become more readily available as the new EU Trade Marks Regulation[‖] has removed the requirement for a trade

---

[¶]Dyson *vs.* Registrar of Trade Marks (C-321/03) [2007] ETMR, 34, concerning Dyson's application to register transparent bins or collection chambers forming at least part of the external surface of a vacuum cleaner.

[‖]Regulation (EU) No 2015/2424 of the European Parliament and the Council amending the Community trade mark regulation entered into force on 23 March 2016.

mark to be capable of being represented graphically. Instead, the sign should be represented in any appropriate form using generally available technology, as long as the representation is clear, precise, self-contained, easily accessible, intelligible, durable and objective. However, in general, non-traditional marks, such as smell or taste marks, are likely to continue to be problematic. An application to register a strawberry taste for pharmaceutical products was refused in Eli Lilly and Co's Application.[20] Generally, it has also proved difficult to show trade origin in a colour or shape, resulting in many applications, such as tablet shapes, being refused.

12.3.1.2 *Applying for a Trade Mark*

As for patents, applications for trade marks can be made at the national or European level. In contrast to the European patents system, an EU trade mark (EU TM) covers all 28 EU Member States with a single right.

In order to file a trade mark application, the applicant must include the following information in their application:

- Their full name and address;
- The country and, if appropriate, the state of incorporation of the applicant (in the case of federal states such as the USA);
- Full details of the trade mark;
- A good representation of the logo or design, if appropriate; and
- An indication of the goods and services sold or to be sold under the trade mark (there are 45 classes of trade marks).

For the UKIPO, an online search is conducted to determine if the mark is free to use and register (although this is optional). The application is then filed, either on paper or electronically. Around one to two weeks later the UKIPO will issue an examination report. If objections are raised, these can be countered in writing.

The application will be advertised within around six months of the original filing. This is an opportunity for other brand owners to file any objections to the mark. The application is open to opposition for two months from the date of advertisement and may be extended to three months by any parties considering oppositions. If they do so, the application may fail completely or be delayed.

Many brand owners hire watching services that will automatically search every published application for potential conflicts.

If there are no oppositions, the UKIPO will issue the applicant or their trade mark attorney a certificate of registration around three to four months after the advertisement date.

Regarding the EUIPO application process, an online search is conducted to determine if the mark is free to use and register (although, again, this is optional). The application is then filed, either on paper or electronically.

Within around one month of the filing, the EUIPO provides comments on the formalities examination. This will raise any specific queries that they may have, for example regarding class choices, wording of the specification and objections to the distinctiveness of the mark (*i.e.* is the mark a term other traders need to use to describe their goods and services). If objections are raised, these can be countered in writing.

Within around four months of filing, the EUIPO will send Community and National Search Reports. These will list any trade marks, either granted or pending, that appear to conflict with the application. The application can be amended or withdrawn at this point, although the reports are only advisory.

The application will be published in the EUTM Bulletin on the EUIPO website within around six months of the original filing. The application is open to opposition for three months from the date of advertisement and, unlike for the UK application, the opposition period cannot be extended. If a formal opposition is filed, opposition procedures will commence which can last for two years or more.

If there are no oppositions, the EUIPO will issue the certificate of registration around six months after the advertisement date.

A comparative timeline for the UKIPO and EUIPO application procedures is set out in Figure 12.2.

12.3.1.3   Limitations

An application can be rejected under absolute grounds by the trade mark office or on opposition by a third party.

The absolute grounds for refusal under section 3 TMA focus on distinctiveness. The UKIPO will refuse an application for a mark that it considers to be devoid of distinctive character, customary or generic (*i.e.* in

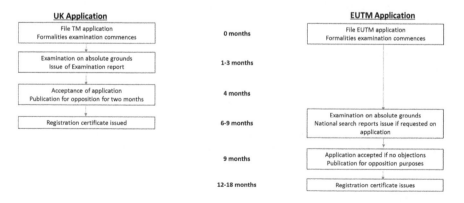

Figure 12.2   A comparative timeline for the UKIPO and EUIPO trade mark application procedures.

general use not specific to the goods in question). Applications that are descriptive of the kind, quality, quantity, intended purpose, value, geographical origin or time of production of the goods or services will also be rejected.

Relative grounds for refusal under section 5 arise where an existing registered or pending third-party trade mark conflicts with the mark applied for. In essence, a mark will not be registered if it is identical or confusingly similar with another mark. In addition, a trade mark will be infringed by a sign which is identical or confusingly similar with the trade mark. *Alcon Inc. vs. OHIM*[21] examined the likelihood of confusion in the context of pharmaceuticals. It concluded that the relevant consumer for pharmaceutical products is the health professional and the end consumer.

In practice, a trade mark search of a specialist pharmaceutical database, as well as a search for unregistered names such as drug indexes and the international non-proprietary names database, will minimise the risk of third-party oppositions.

12.3.1.4   Regulatory Approval

Applications for pharmaceutical trade marks must satisfy the requirements of the Medicines and Healthcare products Regulatory Agency (MHRA) guidelines which relate to public health. Some general requirements are that a product name must not:

- Be liable to cause confusion with the invented name of another medicinal product or with the product's common name;
- Convey a misleading message in relation to the therapeutic effect of the pharmaceutical;
- Be misleading as to the product's composition; or
- Convey a promotional message.

Centralised European marketing authorisations, effective in the United Kingdom, can also be obtained *via* the European Medicines Agency (EMA). A medicinal product authorised under the centralised procedure must have the same name in every member state and the name must be approved in advance of filing the MA application.

The EMA assesses proposed names through the Name Review Group, which cooperates with the relevant national authorities in member states and the World Health Organisation. Assessment is by reference to the "Guidelines on the Acceptability of Names for Human Medicinal Products Processed through the Centralised Procedure". The guidelines address whether the proposed name could cause confusion with another product, convey misleading pharmaceutical connotations or be misleading with respect to the composition of the product. Furthermore, an invented name should not be derived from its own international non-proprietary name.

12.3.1.5. International Non-proprietary Names

An international non-proprietary name (INN) is a unique name that is globally recognised for a particular pharmaceutical active substance and cannot be registered as a trade mark. This makes sense from a policy perspective: the name of the active substance must be free for all to use.

The MHRA has issued guidance for proposed product names, which covers the construction of pharmaceutical trade marks and the similarity of invented names to existing INNs.

12.3.1.6 Examples of Trade Marks for Formulations

The names of most branded pharmaceutical products are protected by trade marks. It is less common for a formulation *per se* to give rise to a trade mark of its own. However, trade mark registrations may allow the shape or colour of a tablet or patch to be protected, giving the owner a potentially indefinite monopoly over those aspects of the product's appearance.

Tiltab. Tiltab® was developed by SmithKline Beecham (now GlaxoSmithKline) to address the problem that elderly patients commonly have in taking daily medication, such as confusion when several products have to be taken in one day and difficulty handling the product, for example due to dexterity problems. The concept was originally developed for the antiarthritic product 'Ridaura®', which contains the gold compound auranofin. The tablet shape allows patients to pick up tablets from a flat surface by titling upwards, due to the projections on each side of the tablet, which improves handling.[22] Figure 12.3 shows one example of a Tiltab® and this shape has been registered as a trade mark.

Further examples of registered trade marks for tablets include Pfizer's well-known blue Viagra® diamonds, depicted in Figure 12.4.

There are some specific exclusions for marks such as shape marks. An example is the exclusion of shapes which result from the nature of the goods or which are necessary to obtain a technical function. This means that shapes such as standard tablet shapes or functional aspects are not protectable *via* trade marks. This reflects the fact that technical features are generally susceptible to patent protection. An example of refusal of a

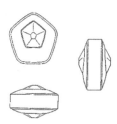

Figure 12.3 Glaxo Group Limited's trade mark registration for Tiltab®, application number 000719922.

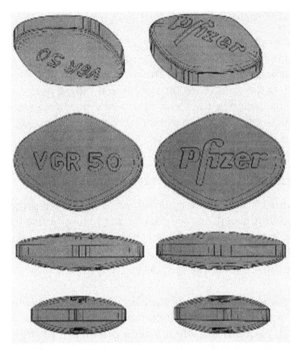

Figure 12.4   Pfizer Inc.'s trade mark registration for Viagra®, application number
000848812.

shape mark which was refused for lack of distinctive character where the
shape was the result of functional feature was demonstrated in decision
R0804/2008-4,[23] which concerned the following application for a shape
trade mark (Figure 12.5):

The registration was refused, since the grooves depicted are functional features which enable the tablet to be cut or broken into suitable portions.

Less successful attempts to register trade marks in the pharmaceutical
context include Eli Lilly's attempt to register the strawberry taste as a trade
mark, mentioned earlier.[20] Of relevance in that case was the public interest
in allowing competitors to use the flavour to mask unpleasant tastes in
this field. Other factors were the fact that consumers are unlikely to distinguish goods on the basis of this taste and also the vague description of
the mark.

Word marks which may be descriptive of one or more technical features
of a product are also unlikely to be registered. For example, MULTIPLA[24] was
refused registration by the EUIPO for "Pharmaceutical preparations and
products for the treatment of viral diseases; anti-viral agents; vaccines" on
the basis that it was descriptive of a vaccine that targets multiple diseases,
as opposed to a single disease. Similarly, PREDETECT[25] was refused registration for "diagnostic kits consisting of a test strip and buffer within a plastic
housing for testing of bodily fluids for use in detecting infectious diseases"
for being void of distinctive character.

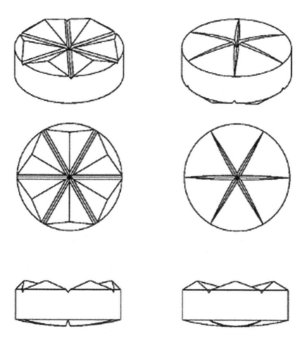

Figure 12.5 Abtei Pharma Vertriebs GmbH's trade mark application number 006093141.

Applications to register the shape of blister packaging for pharmaceutical preparations have also been rejected for lack of distinctive character, for example the EUIPO Board of Appeal held[26] that customers don't see the shape of blister packs as an indicator of origin, because they are always hidden in boxes.

12.3.2 Design Rights

12.3.2.1 Introduction to Design Rights

Design rights protect the appearance, look and shape of things, in contrast to trade marks which protect a brand and indicate trade origin, or patents which protect the way things work. Designs also protect articles which are made for mass market production. The utility of design rights in the pharmaceutical field is limited but they are sometimes used to protect tablet shapes and can be an important right in relation to product packaging.

12.3.2.2 Types of Design Right

Designs in the UK may be protected by:

(i) UK Registered Design Rights
(ii) UK Unregistered Design Rights

(iii) Registered Community Design Rights
(iv) Unregistered Community Design Rights
(v) Copyright

(i) UK Registered Design Rights
Registration of UK design rights affords 25 years of monopoly protection (*i.e.* the right-holder does not need to prove copying of the design) for the shape or surface design of an article. Unlike the application procedure for trade marks and patents, the registry will not search for prior designs which makes the system relatively cheap and quick. The entire registration process can be completed in around three to six months and makes the design easier to enforce than where an unregistered right is relied on. This is because in respect of unregistered rights the rights holder must prove copying in any infringement proceedings. Furthermore, the rights holder must prove ownership of their unregistered right.

The ease with which designs may be registered also means that they are frequently revoked in infringement proceedings, because the registry does not conduct an assessment of the validity of the design for which the application was made.

(ii) UK Unregistered Design Rights
Unregistered design rights protect the shape of articles and protect against copying of the design for the shorter of 15 years from the date of the design or 10 years from the first year of marketing the article. Unregistered design protects both functional and non-functional shape designs.

(iii) Registered Community Designs
A single Community Design can be registered centrally, offering protection throughout the EU and, like the UK registered design, it offers 25 years of monopoly protection. Community Registered Designs, again much like the UK Registered Design, protect both shape and surface designs.

In order to be eligible for a Community Design registration, the design must be novel (*i.e.* no identical design has been made available to the public) and must have individual character, meaning the overall impression it gives to a user must differ from any other design which has already been made available to the public.

There is a grace period of one year, during which the designer can market their product before deciding whether to register a Community Design.

(iv) Unregistered Community Design Rights
For a period of three years from first marketing a product in the EU, new and individual designs may also be protected without the need for an application to register the design. The Unregistered Community Design right arises automatically but is not a monopoly right. This means that in order to prove infringement of the unregistered right, the rights-holder must prove that the infringer copied their design.

Unlike the UK Unregistered Design Right, the Community Unregistered right also protects surface designs.

This right affords protection where the designer is taking advantage of the one year grace period for Community Registered Designs.

12.3.2.3 Interaction with Patents

Like a patent, *registered* design rights offer monopoly protection, however, they protect the appearance of the whole or part of a product, resulting from the features of its contour, texture, materials, colours and the ornamentation of the product.

Unlike patents, design rights are not intended to protect technical advances. This means that where features are dictated by a product's technical function, these features will not be protectable *via* design rights. Features which allow the product to be connected with or around another product so that the product works are also not protectable by design rights.

12.3.2.4 Interaction with Trade Marks

As discussed above, it is possible to obtain a trade mark for a shape, however, this is limited because trade marks cannot be obtained in respect of shapes which result from the shape of goods themselves or add substantial value to the goods. In circumstances where a shape mark cannot be obtained, design rights may be more relevant.

In contrast to trade marks, there is no formal system to notify potential infringers that a design is protected by a registered design right, unlike the ® symbol for registered trade marks. Furthermore, design protection will expire, at most, after 25 years, whereas trade marks may be renewed indefinitely.

12.3.2.5 Shape of Goods and Packaging

A number of possibilities exist for the protection of goods *via* design rights and tablet shapes are one example of the existence of design rights protecting pharmaceutical products. For example, as well as having trade mark protection, Pfizer's Viagra blue tablet was registered for design protection in 1998.[27]

In addition to the goods themselves, design rights can also protect the shape of their packaging. As well as the box or bottle itself, this could include the shape of devices for measuring liquid formulations such as Pfizer's UK design registration 2051614 for a liquid measuring vessel, shown in Figure 12.6.

12.3.3 Other Rights and Protection

12.3.3.1 Confidential Information/Trade Secrets

Confidential information can be incredibly valuable and has the potential to remain protected indefinitely, provided no breach of confidence occurs. Unlike IP rights expressed in a physical article or in a publication made available by being registered, protection as a trade secret could potentially last forever. The recipe for Coca-Cola is a famous example. However, it is generally

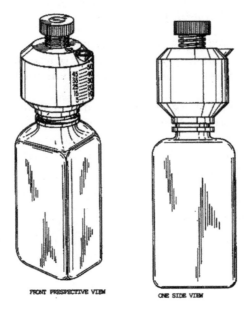

FRONT PRESPECTIVE VIEW   ONE SIDE VIEW

Figure 12.6   Pfizer's UK design registration 2051614 for a liquid measuring vessel.

unrealistic to rely indefinitely on confidential information in a pharmaceutical context since information will have to be disclosed to the regulator about the composition of the pharmaceutical product during the authorisation process. Product information will also have to be supplied to medical practitioners and patients. However, the obligations of confidence can be incredibly important in the early stages of pharmaceutical development, to preserve the ability to later file patents. Once the patent is filed, the information is necessarily made public—that is the bargain that the inventor makes with the state: a time-limited monopoly in return for teaching the world how to operate the invention.

Secret information can be protected by contractual restrictions but it may also be protected when no contract exists at all, *via* a court action for breach of confidence.

12.3.3.2   Copyright

Copyright protects original literary, dramatic or artistic works as well as sound recordings, films and broadcasts and typographical arrangements in publishing. The requirement that literary, dramatic and artistic works be original means that they must be the author's own intellectual creation. This means identical works can benefit from their own separate copyrights if they were created truly independently of each other.

Copyright works are protected as soon as they are "fixed", this means that the idea itself cannot be protected, only the expression of the idea, for example in writing or by a design drawing.

The applications of copyright law to pharmaceutical formulations are quite limited. For example, product information leaflets and the summary of product characteristics, which may be literary works, must for a generic pharmaceutical product be consistent in all relevant respects with that of the originator.[28] Where these documents are substantially the same, therefore, copyright infringement is unlikely to be a sound basis for action. One area of possible relevance is where there are drawings in the product information or summary of product characteristics to represent the product. This might be the case where the product is a plaster, patch or similar product and the shape is displayed in the documents.

12.3.3.3 *Databases*

Database rights are created from completion of the database and will last for 15 years from 1 January in the year following the date of completion. They are intended to protect the investment in storage and processing systems for the information.

For the purpose of the database right, databases are defined in the following manner:

"Collection of independent works, data or other materials; arranged in a systematic or methodical way; and individually accessible by electronic or other means."[29]

They are protected, provided that the maker can show either a substantial investment (whether human, technical or financial) in obtaining, verifying or presenting the content of the database. Note that the investment must be in the database itself, not in creating the data which goes into it.[30] This means that, for example, the investment in clinical trials in themselves would not warrant the protection of the data *via* the database right, only the investment in putting results into a database. This means that data such as clinical trial results *per se* will be protected as confidential information but not by database rights.

The maker must also be an EU national or a business having their principal place of business in an EU Member State or their registered office in and an economic link with the Member State, in order to benefit from the right. The rights in a database made by an employee will belong to their employer, subject to any other agreement.

12.3.3.4 *Customs Protection*

Many IP rights, including copyright, trade marks, patents and designs, can be registered with the EU Customs authorities. This means that when goods enter the EU for the first time they can be checked against recorded IP rights if they are suspected of infringing them. Goods which are suspected of infringing an IP right will be detained by EU Customs and, if found to infringe, may be destroyed.

Goods which infringe design and trade mark rights are particularly susceptible to detention by EU Customs as they are often easier to recognise than, for example, goods which infringe patent rights. Customs registration of IP rights is a powerful weapon in the fight against counterfeit medicines.

12.3.3.5   Protection Against Counterfeiting

Some reports indicate that pharmaceuticals are the most targeted counterfeit good in the online space.[31] An annual worldwide operation targeting illicit sales *via* the internet and spanning 115 countries, named "Operation Pangea", takes place in June each year. Between 9 and 16 June 2015, it resulted in the seizure of 20.7 million potentially dangerous drugs worth $81 million.[32] In the UK, 6.2 million doses of counterfeit and unlicensed drugs worth £15.8 million and 15 000 health devices were seized during the operation. These included slimming pills, treatments against erectile dysfunction, drugs for the treatment of anaemia and medicines treating sleep disorders. The UK operation also resulted in the closure of 1380 websites, 339 of which were local sites.[32]

The MHRA Enforcement Group, together with the Police and HM Customs and Revenue, investigates falsified and counterfeit medicines in the UK. Falsified medicines are products which pass themselves off as authorised medicines, but have not been evaluated to check their quality, safety and efficacy, as required by EU pharmaceuticals legislation. They may contain sub-standard ingredients, or the wrong dose of active substance, or no active substance at all. Counterfeit medicines are medicines which infringe intellectual property rights, most often by unauthorised use of trade marks. They are often also falsified medicines.

The enforcement agencies have a range of powers to investigate the trade in falsified and counterfeit medicines, which are largely contained in the Human Medicines Regulations 2012 and the Police and Criminal Evidence Act 1984. These include the right to:

- Search business and private premises.
- Seize and sample products.
- Inspect and copy documents.
- Interview witnesses.

In July 2015, MHRA Enforcement Officers carried out a raid on a personal dwelling and storage units containing 470 000 tablets of unlicensed erectile dysfunction medication, counterfeit medicines and the class C controlled drug tramadol. This ultimately led to the two individuals involved pleading guilty to illegally supplying medicines.[33]

The manufacture, distribution and supply of falsified medicines are criminal offences. Under the Human Medicines Regulations 2012, it is an offence to manufacture, distribute, act as a wholesale dealer, or broker the supply of medicines which do not have a marketing authorisation (or similar registration), subject to certain limited exceptions. Breach of these requirements is punishable with a fine of up to £5000 per offence if the case is dealt with by a Magistrates Court. If the case is tried in a Crown Court, an unlimited fine can be imposed and/or imprisonment for a term of up to two years.

Cases of counterfeiting involving unauthorised trade mark use can also be prosecuted under the Trade Marks Act 1994. The unauthorised use of a trade mark carries a maximum sentence of ten years imprisonment and/or an unlimited fine. Civil injunctions can also be obtained to prevent the supply of falsified or counterfeit medicines and to require the delivery up of such products.

In addition, proceedings can be brought under the Proceeds of Crime Act 2002 (POCA) to seek the confiscation or recovery of the proceeds of crime.

References

1. *Cost to Develop and Win Marketing Approval for a New Drug Is $2.6 Billion*, Tufts Center for the Study of Drug Development, November 18 2014, Retrieved from: http://csdd.tufts.edu/news/complete_story/pr_tufts_csdd_2014_cost_study.
2. Catnic Components Ltd v Hill & Smith Ltd (No. 1) [1982] RPC 183 at 242–3.
3. *Guidelines for Examination at the EPO*, November 2015, G-II, 1.
4. [2011] UKSC 51 at para 157.
5. Asahi Kasei Kogyo KK's Application [1990] RPC 485.
6. H. Lundbeck A/S, [2007] EWHC 101 (Pat) per Kitchin J at para 72.
7. PA 1977, s. 7(1).
8. PA 1977, s. 7(2).
9. PA 1977, s. 39(1).
10. PA 1977, s. 25(1); EPC 1973, Art 63(1).
11. Abraxis, Abraxis Bioscience LLC v The Comptroller General of Patents [2017] EWHC 14 (Pat).
12. Leo, [2014] EWHC 3096 (Pat); [2015] EWCA Civ 779.
13. Genentech, [2014] EWHC 3857 (Pat).
14. Genentech, [2016] EWCA Civ 780.
15. Bayer Schering Pharma, [2011] EWHC 583 (Pat).
16. Merial, [2011] EWHC 3393 (Pat).
17. Trade Marks Act 1994, s. 1(1).
18. Heidelberger Bauchemie C-49/02 [2004] ETMR, 99, CJEU and Libertel.
19. Sieckmann, Sieckmann v Deutsches Patent und Markenamt [2002] EUECJ, C-273/00, 12 December 2002.
20. Eli Lilly/The taste of artificial strawberry flavour, R120/2001-2, 4 August 2003.
21. Alcon v OHIM (Intellectual property) [2007] EUECJ, C-412/05, 26 April 2007.

22. G. D. Tovey, The Development of Tiltab Tablets, *Pharm. J.*, 1987, **239**, 363–364.

23. OHIM, *Fourth Board of Appeal*, November 19 2008, Case No. R 804/2008-4.

24. See Refusal MULTIPLA 015286248, Accord Healthcare Limited, Application reference: 26.14.M109611, EUIPO, 30 March 2016.

25. See Refusal PREDETECT 014354468, BBI Solutions OEM Limited, Application reference: 23642MAUPTCTMsn, EUIPO, 13 July 2015.

26. See Refusal 012856126, Novartis AG, Application reference: TMA035369-EM-CM, EUIPO, 7 May 2014.

27. UK registered design number 2078188, Pfizer Limited, 7 April 1998.

28. European Medicines Agency, *EMA/627621/2011, QRD General Principles Regarding the SmPC Information for a Generic/Hybrid/Biosimilar Product*, 2012.

29. Article 1(2) of Directive 96/9/EC of the European Parliament and of the Council of 11 March 1996.

30. BHB v William Hill [2004] EUECJ, C-203/02, 9 November 2009.

31. (a) K. Megget, *Pharma Most Counterfeited Category Online, Says Report*, 24 November 2016, Retrieved fromhttps://www.securingindustry.com/pharmaceuticals/pharma-most-counterfeited-category-online-says-report/s40/a3049/#.WKRtvW-LRpg, accessed 15 February 2017; (b) *Pangea viii: Over 20 Million of Illegal and Fake Drugs Seized in 115 Countries*, International Institute of Research Against Counterfeit Medicines, 23 June 2015, Retrieved from: http://www.iracm.com/en/2015/06/pangea-viii-over-20-million-of-illegal-and-fake-drugs-seized-in-115-countries/, accessed 15 February 2017.

32. *UK Leads the Way with £15.8 Million Seizure in Global Operation Targeting Counterfeit and Unlicensed Medicines and Devices*, Medicines and Healthcare products Regulatory Agency press release, 18 June 2015, Retrieved from: https://www.gov.uk/government/news/uk-leads-the-way-with-158-million-seizure-in-global-operation-targeting-counterfeit-and-unlicensed-medicines-and-devices, accessed 15 February 2017.

33. *Oxford Men Admit to Illegally Supplying Medicines*, Medicines and Healthcare products Regulatory Agency press release, 8 February 2017, Retrieved from: https://www.gov.uk/government/news/oxford-men-admit-to-illegally-supplying-medicines, accessed 15 February 2017.

CHAPTER 13

User-friendly Medicines

CLIVE WILSON*[a] AND MARTIN KOEBERLE[b]

[a]Strathclyde Institute of Pharmacy & Biomedical Sciences Glasgow, Scotland; [b]Hermes Pharma, A Division of Hermes Arzneimittel GmbH, Pullach, Germany
*E-mail: c.g.wilson@strath.ac.uk, koeberle@hermes-pharma.com

13.1 The Concept of User-friendly Medicine

The oral route remains the mainstay of methods of dosing with medicines on the grounds of cost, convenience and public familiarity. The tablet, soft gelatin and hard gelatin capsules are the most familiar oral medications that we encounter. Most developments are new embodiments of well-established formulations whose quality has steadily evolved through improved manufacturing practice and the efforts of regulatory authorities. Although alternative routes, for example skin and lung, exploiting developments in drug delivery technology have become popular through the last two decades the cost of goods is significant and many oral formulations can be produced at lower costs, allowing a wide range of generic alternatives.

A medicine for chronic conditions must be easy to take at convenient intervals during the day. Dosing once or twice per day can be coordinated with early morning and late evening activities and patient compliance maintained. The conditions of dosing may influence efficacy: for example, should the active pharmaceutical ingredient be taken with food and with water? For certain patients, mobility, motor and cognitive functions may be an issue and user-friendliness of the medicine a concern for both patient and carer.

Drug Discovery Series No. 64
Pharmaceutical Formulation: The Science and Technology of Dosage Forms
Edited by Geoffrey D. Tovey
© The Royal Society of Chemistry 2018
Published by the Royal Society of Chemistry, www.rsc.org

The concept of user-friendliness is interesting. It directs the public attention to the benefits of appropriate functionality and ergonomics in maximizing convenience and breadth of use. We have embraced this concept in electronics and household goods; now we have the opportunity to apply similar principles to the needs of those having the burden of illness and dependence on others.

It is recognized that pharmaceutical innovation might provide better solutions to access to healthcare systems, markets, generation of intellectual property and adjustment of regulatory requirements to advances in science and technology.[1] There are segments of the population that require dosage adjustment or alternative presentation; for example, elderly patients with dysphagia, children and those requiring lower doses by virtue of genetic predisposition. Innovations in the design of dosage forms might provide better patient convenience and acceptability, leading to increased compliance. A medicine that is consumed according to the intended schedule is better than the cheapest medicine, which may be disliked by the patient on the grounds of taste, difficulty in swallowing, *etc.*

The concept of user-friendly medicines is explored in this chapter. In particular, we consider the needs of two populations. Firstly, the elderly, who are the main consumers of medicines and represent an expanding proportion of society. They will increase the costs of provision of health care, particularly with regard to nursing care. Secondly, the young, who might need alteration of an existing medicine by the pharmacist or hospital. Increasingly, there is a call for medicines designed specifically for this segment of the population.

A wider range of alternative formulations allows physicians and patients to have more choice in selecting the presentation that they prefer, whilst preserving safety and efficacy.[2]

13.2 The Relevance of User-friendly Medicine to the Patient

We probably assume that user-friendliness applied to medicines is well understood by the profession and the public; however, there appear to be significant gaps in our knowledge base. For example, public health specialists in Utrecht have drawn attention to the fact that we know relatively little about the effects of pharmaceutical technology applied to paediatric oral drugs on patient-related outcomes. Their review of 95 articles concluded that there was dearth of evidence on the relationship between formulation factors and route of administration, specifically dosage form, route and frequency of administration, packaging and device characteristics on clinical efficacy and acceptance.[3] This establishes the need to begin discussion on the attributes of medicines that we might be able to improve to achieve true user friendliness.

Drugs often taste bitter or have unpleasant aftertaste and, in part, the unit dosage form helps disguise unfavourable organoleptic properties associated with granules, powders and solutions. There are advantages of easy transportability and clear identification of active pharmaceutical ingredient in tablets and capsules when packaged; however, the conventional unit lacks flexibility in terms of adjustment according to personal need. This, in part, stems for need to maintain formulation unit integrity to assure reproducible dosing and to avoid contamination.

Unit robustness is an attribute of good posology but in some situations, predispersion could be useful, especially where there is a perceived advantage for dosing. In addition, a predispersed dosage form skips one phase of the process, disintegration, converting the medicine to an absorbable dose. The fewer processes involved, the lower the number of complex interactions generating variability.

13.2.1 The Issue of Swallowing

Oral dosage forms generally need to be swallowed in order to provide the desired pharmaceutical effects. Although swallowing is almost an automatic physiological process it can be difficult for people: Swallowing requires a co-ordination of muscle actions and for saliva to be present as a lubricant. The mechanics of swallowing involve both a voluntary oral phase, when material is consciously sampled in the mouth by the tongue and then handed over to the pharynx where the initiation of automatic deglutition occurs. The time to swallow a bolus of water is usually around 1.5 s although tablet transit is slower, measured at 4–7 s by scintigraphy.[4] Tablets that take longer than 15 s to clear from the oesophagus are usually classified as adherent.[5] The incidence of swallowing difficulties, in elderly subjects measured in the upright position is typically 20% for capsules.[6,7] Uncoated tablets show a longer residence time than coated tablets, and a prolonged adhesion to the oesophageal mucosa is common.[8]

Although the process of swallowing is almost an automatic physiological manoeuvre in the young, adaptive cerebral changes in the co-ordination of the swallowing reflex are seen in the elderly, suggesting that the brain cortical region increases the time for pharyngeal triggering,[9] resulting in swallowing smaller volumes, accumulation of residue and a higher rate of laryngeal penetration.[10]

A recent survey by the independent market research firm Spiegel Institut Mannheim and Hermes Pharma[11] has shown that over 50% of people in the USA and Germany experience potentially serious problems when swallowing tablets and capsules. Many participants have interfered with their tablets or capsules in an attempt to overcome these issues: over a third reported breaking tablets before swallowing, 17% crushed and dissolved them in water in order to swallow them and 8% stopped taking their medication entirely. This problem has received little or no media attention but has potentially serious consequences for an individual's health.

The survey went on to make it clear that people would greatly prefer their medicines or supplements to be a positive experience. It highlighted several key criteria that people would like to see in their medicines or supplements (Figure 13.1).

QUESTION: If you had to take a medication or a food/dietary supplement short-term use/chronically i.e. over a number of years: What characteristics would then be particularly important to you? (At least one but not more than four characteristics)

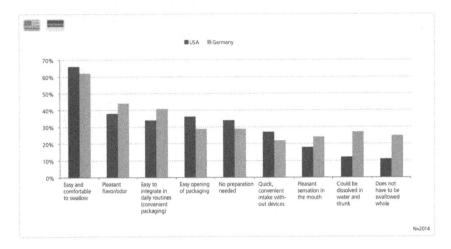

Figure 13.1   More than two thirds of the participants reported that products should be easy and comfortable to swallow. Around 40% said that a pleasant taste or odour was particularly important, 38% required products to be easy to integrate into their lives and 30% wanted packaging that was easy to open. Image courtesy of Hermes Pharma.

Designing or reformulating products to consider these factors can provide a major advantage when bringing new products to market, offering people precisely what they are looking for.

13.2.2   Organoleptic Assessments and Presentation

The process of deciding whether to swallow an object is associated with considering the textural properties of a material and the shape—the appearance of the object in our hand and the sampling of temperature, smell, texture and taste during the initiation of the manoeuvres to swallow the object. From previous experience, there is an expectation of what will happen when we attempt to swallow objects, even including the sound that we will hear, transmitted through the jaw. These properties, the organoleptic components of a medicine, are of great importance for oral and for topical delivery (creams and ointments).

The earlier use of organoleptic assessment for oral products arose from other industries, notably the meat and dairy industries. Meat, for example, is gauged on the parameters of flavour, juiciness, tenderness and general palatability as quality attributes.[12] Attention to organoleptics of medicines was recorded early in tribal knowledge. Etkin reports that the Kenyah Leppo'Ke tribe of Borneo classifies plant extracts that have strong chemosensory attributes to be more effective medicines. Bitter-tasting extracts are selected for treatment of medical conditions associated with fever, whereas astringent plants are used for the treatment of gastrointestinal disorders. Those with a high content of terpenoids, particularly those which are highly volatile with a citrus-like smell and taste, are identified as possessing a novel sensory quality *'nglidah'*.[13]

13.2.3 The Issue of Taste

The taste of a medication is one of the key attributes determining patient acceptability and compliance and will be important in market acceptance. In the past problems arose in the development of paediatric medicines, which were often liquids or suspensions due to the need to adapt medications designed for adults.[14] The employment of flash-dispersing formulations, taken without water has heightened awareness that taste masking is a critical attribute in formulation.[15] The use of an effervescent agent as a taste-masking principle is described in several compositions including chewing gums and in solution formulations for buccal, sublingual and peridontal application. A useful review of some of the established approaches used in the industry is provided by Sohi and colleagues.[16]

The sense of taste is mediated by taste buds, which are located around the four types of gustatory papillae – the small structures on the upper surface of the tongue, soft palate, upper oesophagus and epiglottis (Figure 13.2). It is estimated that there are about 10000 taste buds. This number decreases with advancing age, which results in altered taste sensations in old age. Taste buds contain the receptors for taste that are stimulated by substances dissolved in the mouth by saliva. A taste bud consists of about 50 taste receptor cells. Each taste bud has a small taste pore, through which fluids in the mouth interact with the surface of its receptor cells.

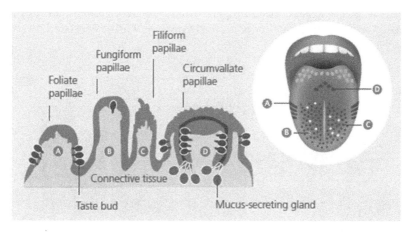

Figure 13.2 The tongue comprises three types of papillae (foliate papillae, fungiform papillae and circumvallate papillae) that contain taste buds, plus the filiform papillae that only detect the texture of food. Adapted by permission from Macmillan Publishers Ltd: *Nature* (ref. 17), copyright 2012.

Conventionally four basic taste sensations have been described:

- sweet
- salty
- sour
- bitter

Recently, the so called 'umami taste' has also been considered as a unique stimulus type (Figure 13.3). It is especially associated with the glutamate used in Asian cooking. All other sensations had been assumed to result from combinations of these five basic taste sensations. The 'physiology textbook' understanding is as follows: the area which is considered to be the most sensitive to sweet sensation is the tip of the tongue, whereas the front half of each side of the tongue is thought to be the area most sensitive to salty stimuli. The posterior half of each side of the tongue is the area considered to be most sensitive to sour sensation and receptors for bitter taste are mostly located at the base of the tongue. However, this division into five types may be over-simplistic. Receptors have been identified that probably integrate by a combination of exploration of texture, specifically viscosity and grittiness in fine discrimination and G-coupled proteins are thought to sample 'fattiness'.[18] Some materials, such as chilli, do not activate 'taste' receptors but have a direct on heat sensation through the VR-1 vanilloid receptor.[19]

While sweet taste receptor cells are most probably directly linked to positive hedonic centres of the brain, bitter taste is a major challenge in the pharmaceutical manufacturing process due to its aversive impact on ingestion.[20] From an evolutionary point of view, the bitter modality may protect humans against potentially toxic or harmful substances in nature.[21] Bitterness is widely distributed in nature and, in contrast to other taste qualities, there appears to be a wide structural range. From this we conclude that bitter taste seems to be a complex quality of all basic taste modalities assisted by smell.

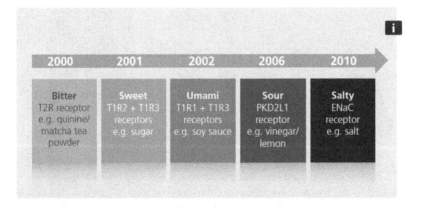

Figure 13.3 Timeline showing the characterisation of taste receptors with some typical examples. Based on data from ref. 17.

Taste perception is modulated by various factors including aging, gender, ethnicity, cigarette smoking, olfactory stimuli (as well as the temporary loss of smell sense during a cold), time of day and even psychological processes. Small and Prescott have concluded that the integration of smell and taste by the brain results in the individual perceiving that the whole sensation originates from the mouth, with ortho-nasal and retro-nasal stimuli contributing to the organoleptic experience.[22] As we get older, we lose the ability to detect salty and bitter substances, whereas the thresholds for sour or sweet stimuli appear to be unchanged up to the age of 90.[23] This change in the balance of taste perception is interesting because few subjects show a generalized loss.

13.2.4 Assessing Taste

Unfortunately, many active pharmaceutical ingredients (APIs) have an unpleasant taste—they may be bitter, salty or sour, or may cause an irritating mouth feeling—an astringent, metallic or spicy taste. The medicines *per se* may alter perception of flavour. For example, chlorhexidine gluconate, a commonly used anti-plaque agent in oral mouth washes, alters the taste perception to rinse challenges with salt or quinine hydrochloride but not citric acid.[24] Also, the disease or treatment may alter taste sensitivity thresholds. Thus in cancer it is well established that olfactory and gustatory sensitivity is altered and in Parkinson's disease[25] olfactory tests are regarded as providing early diagnosis.[26] Radiation therapy of head and neck directly destroys taste buds and the dysguesia associated with radiation therapy is an identified problem.[27] Children express a preference for antibiotic suspensions which, not surprisingly, correlates with after-taste.[28] The pharmaceutical industry has been heavily investing in technologies to mask unpleasant taste and odour which are considered to be main reasons for poor compliance.

There are various approaches available to check the effectiveness of flavouring efforts and taste-masking techniques, *in vivo* and *in vitro*. The most widely used method of assessing taste and other organoleptic properties of medicines is by the use of a panel who record their immediate impressions on a questionnaire. The pharmaceutical industry employs psychometric testing in which the characteristics of the medicine are described on intensity scales (Box 13.1) that are serially recorded to map the changes in intensity following exposure.

However, the human sense of taste is a highly developed mechanism accompanied by wide inter-individual variability. Sensory impressions remain subjective, even if members of a taste panel are trained and calibrated and, for ethical reasons, human taste panels comprise only healthy adults but no paediatric or geriatric patients.

In addition, human test panels are not acceptable for new drugs, because a complete toxicology profile will not be available. So, although taste trials are acceptable in the food industry, the fact that we are dealing with active medication (*e.g.* for children) causes obvious problems.

Box 13.1 Taste modalities used in medicines testing. Data from Fu *et al.*[15]

overall intensity
sweet
sour
salty
bitter
metallic
cooling
hot
spicy
burning
anaesthetic
astringent
medicinal
minty/menthol
warming
sharp
alcohol
painful
irritating
stinging
dry
peppery
paper

For a new chemical entity (NCE), laboratory animals are used and surrogate methods such as placing the substance in the drinking water at various concentrations and measuring the weight of the animal over a period of time assesses both aversion and metabolic effects.

Nowadays, instruments such as the "electronic tongue" are being explored as possibilities for biomimetic tasting sensing systems. Consisting of multiple sensors that evaluate different tastes, "electronic tongues" may offer an alternative as they enable the artificial assessment of taste and flavour of various liquids.[29,30] They can be qualified and validated and so this permits evaluations of taste in stability studies and formulation development.[31]

A review of this technology by Woertz and colleagues presents approaches used in pharmaceuticals, nutraceuticals and herbal medicines.[32] The systems described consist of a mixture of specific sensors whose voltage output is logarithmically proportional to signal intensity, in the same way that the human tongue responds. Two such detectors were commercially available at the time of the review, the Insent taste-sensing system and the Astree electronic tongue.

The Astree electronic tongue has been utilized to investigate the taste-masking abilities of numerous flavours in a paediatric suspension and solution.[33] The electronic tongue measured the taste perception changes of both a solution and suspension and found there to be improvement in taste perception using cherry and lemon flavours in the suspension tested while

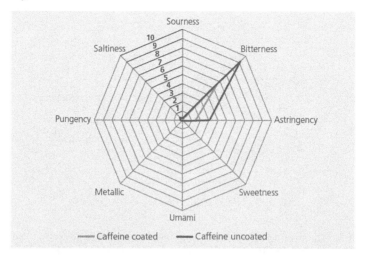

Figure 13.4 Radar plot showing a comparison of caffeine before and after taste-masking. A score of 0–10 denotes the strength of a particular characteristic. Hot-melt coating was used to significantly reduce the bitter and astringent taste of caffeine, to create intermediates that can be easily formulated into a pleasant-tasting product. Image courtesy of Hermes Pharma.

strawberry, vanilla and cherry were perceived by the electronic tongue to have an improved taste when measured in the test solution.

Using an electronic tongue, the data from the different taste sensors can be evaluated either graphically or mathematically:

- Graphically: The signals from the different sensors are depicted in a radar plot and the similarity or difference compared to the reference sample is evaluated visually (Figure 13.4)
- Mathematically: The signals from the different sensors are processed *via* multi-variate data analysis (MVDA) such as principle component analysis (Figure 13.5)

Using either approach, specific factors can be extracted from the complete data-set for comparison with other samples.

There are, however, some limitations of using electronic tongues. For example, only liquids can be analysed (which means that tablets need to be dissolved for testing), and it is a relative method rather than an absolute method. As such, a reference sample of an 'acceptable' taste is required for comparison. In addition, they do not monitor additional factors, including texture and olfactory sensations.

The disadvantage of electronic tongues is that they do not monitor additional factors including texture and olfactory sensations. An overview of the strengths and limitations of the electronic tongue systems is presented in Table 13.1.

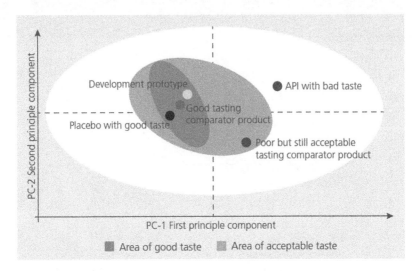

Figure 13.5 Illustration of the use of principle component analysis (PCA) to compare large, multifactorial data sets. Electronic tongue analysis demonstrates flavour as the combination of eight individual parameters. PCA is used to condense this data to two principle components, which might be an individual taste (*e.g.* bitterness) or an abstract mathematical term (*e.g.* sweetness^2 × pungency). Image courtesy of Hermes Pharma.

Table 13.1 Pros and cons of an electronic tongue.

Pros	Cons
No ingestion of samples required, ethically sound	Solid dosage forms need to be dissolved prior to measurement as instrument can only process liquids
Objective data and highly reproducible	Samples require preparation ahead of measurement (including preparation of reference and calibration samples)
Instrument and method can be qualified and validated	Only taste is measured (no olfactory input nor texture not the interplay of theses sensory impressions)
Faster taste assessment compared with a regulatory-approved assessment by a human tasting panel	Knowledge of data analysis and evaluation is required
Success of this approach proven in numerous published case studies	Taste sensors deteriorate with time and need to be replaced regularly and calibrated
Particularly suitable for patient groups that are difficult to access (including infants, children)	Instrument requires training and substantial upfront investment
Automated/high-throughput analysis approaches are in development	Interplay between taste, texture and smell (which is how humans instinctively evaluate) cannot be not taken into consideration
Sensitive (at least as sensitive as the human sense of taste) and capable of detecting a wide range of tastes	
A wide range of drug formulations can be screened in a short time	

13.2.5   The Interplay Between Taste, Texture and Smell

In addition to taste-related changes, there is evidence that nasal chemosensitivity is more strongly blunted, with less sensitivity to perceived intensity of odour and also reduced awareness of pungent or irritant vapour.[34] Scheiber and colleagues conclude that the taste system is generally robust with age; however the reductions in sensitivity coupled with a disordered neural system may yield abnormal stimulations. The sensory experiences can be distorted by certain drugs, a famous example being the restoration of normal taste with zinc sulphate after penicillin treatment.[35]

13.3   Palatability of Medicines: an Issue for Children

In Europe the WHO consultation "make medicines child size" accelerated action towards a model formulary in 2010.[36] The document highlighted areas for possible development with regards to paediatric medicines, including convenient and reliable administration, acceptability and palatability, minimum dosing frequency and end-user needs.

It is preferable for paediatric medicines to be administered in a ready to use format that does not require further handling by parents or carers; however, it is also a requirement that the formulation can provide the intended dose appropriate to the relevant target age group.[37]

In order to cover all age groups (Figure 13.6), it may be necessary for the design of a specific dose form to facilitate easy sub-division into smaller, uniform doses. In the case of liquid medicines, this can be done by accurate measurement of doses relevant to the specific age group.

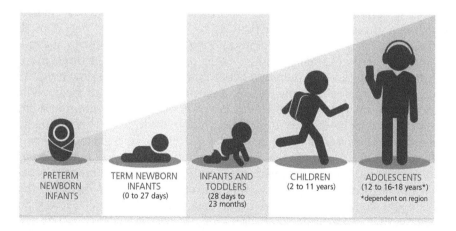

Figure 13.6   Stratification of children considered for paediatric medicine, based on physiological and pharmacokinetic differences, birth to adult. [from reference note for guidance on clinical investigation of medicinal products in the paediatric population (CPMP/ICH/2711/99)]. Image courtesy of Hermes Pharma.

Taste preference is of particular concern with regards to younger patients and taste-masking is one area which can be utilized in the design of paediatric medicine. Children with milk intolerance or having low bone mass may be prescribed supplements of calcium and vitamin D3. It has been reported that compliance is poor, which suggests that child-related organoleptic preferences might be important in the design of paediatric formulations. Bianchetti and colleagues conducted a comparison of lemon-flavored single-sachet and banana-flavoured suspensions in 40 Swiss children requiring supplementation. They noted that the younger children in the group 4–7 years preferred the banana flavoured suspension, but the older cohort (aged 8–11 years), measured by facial expression, favoured the lemon-flavoured formulation.[38]

Taste preferences and palatability are important factors for the chronically sick child because the parents have to crush formulations originally designed for adults. In a study of calcium channel blockers, Milani and colleagues compared the palatability of pulverized amlopidine besylate with lercanipidine in two cohorts of children aged 4–7 years and 8–11 years with arterial hypertension.[39] The neutral-tasting lercanipidine was strongly preferred compared with amlopidine.

Infants consistently prefer sweet, candy-like preparations with fruity flavours, whereas adults tend to prefer less sweet preparations.[40] As noted by empirical observation, citrus flavours are the most popular flavour for long-term use, although there are regional disparities in terms of taste preferences.

13.3.1 Dosing Issues in Paediatric Medicines

Minimum dosing frequency is of particular importance in the case of compliance with paediatric medicines[41] and there have been numerous studies, which have investigated the role that the dosing schedule can play in medication compliance.[42,43] These studies reveal that the simpler the medication regime, the greater the compliance. Twice-daily dosing at set times morning and evening can be adhered to better than a treatment regime requiring multiple doses across the day. This is of particular relevance to medications aimed at school age children where access to and administration of a particular drug may be difficult during the middle of the day. This highlights the consideration that the presented form of a paediatric dose might change across the entire paediatric age range (newborn to adolescent) due in part to enhanced handling ability with age. Small-volume liquids are most often used in the younger age groups while conventional wisdom has been that tablets and capsules allow greater acceptance in adolescents.[44] However, a recent study by Best and colleagues highlights key difficulties as some adolescents are unable to swallow adult sized tablets and crushing of tablets resulted in a clinically significant loss of bioavailability.[45] In another field, hypertension in children and adolescents, Flynn and Daniels lament the failure to develop suspension formulations of the most commonly used anti-hypertension agents.[46]

13.3.2   The Design of Paediatric Medicines

There are important physiological and metabolic differences between children and adults and simple extrapolation of data obtained from an adult medication dose cannot be achieved, especially where chronic disease is evident. For example, in chronic renal insufficiency, the predicted clearance of famotidine must be based on GFR (glomerular filtration rate) for sick children.[47] In addition, frequent feeding with milk may raise gastric pH, leading to changes in gastric pH and calcium–protein binding. Anatomical and biochemical changes during development contribute; the distribution rate of a drug is known to vary between adults and children due to relative proportions of body water, lean body mass and fat and also due to differences in metabolic enzymes and protein binding.[48]

Certain diseases have particular prominence in the paediatric age group and more attention is directed to early diagnosis. As an example, cystic fibrosis is an autosomal recessive genetic disorder that is now screened for at birth. Children are susceptible to chronic infections and are often prescribed long-term antibiotic therapy prophylactically. Patients in this disease group have been shown to have faster drug clearance[49] and, as such, the formulation and dose requires careful consideration. This illustrates that for some conditions, children present a situation that has to be considered separately from that of adults.

With regards to paediatric medicines, formulation design needs to be considered with respect to three broad categories of importance; Quality, the effect of the Biopharmaceutics Classification System (BCS) and the nature of the excipients included in the formulation.[50]

13.3.2.1   Quality and Quality Tests

During the development of paediatric medicines, companies should be aware of the current quality guidelines available, including those relevant to the development of generic products.[51] Attention should be given to any situation where the target market may not be exclusive to mature adults, since the quality test may have to be altered. For example, considering the higher gastric pH in younger children, it may be necessary to adapt the dissolution testing media.[52]

13.3.2.2   Biopharmaceutics Classification System (BCS)

The BCS assigns API's to Class I, II, III or IV dependent upon their aqueous solubility and intestinal permeability. The WHO paper concludes that aqueous solubility is not an issue for those compounds in Class I and Class III; however, for Class II and Class IV compounds the physicochemical properties that influence dissolution—particle size, polymorphism and added excipients—may be important. This limits the extrapolation of dosage adjustment from the adult dosage form—especially if the excipient used is different in the paediatric form.

13.3.2.3   Excipients

Although excipients are pharmacologically inactive substances used primarily as carriers of the active ingredients of a medication the WHO recommends that the use of excipients in paediatric medicines needs to be justified with consideration given to age, dosing frequency and treatment duration. It has been opined that more information is needed with regard to excipient safety in paediatric medicine, particularly as issues have been identified with the use of certain excipients including benzyl alcohol, propylene glycol, ethanol, azo dyes and parabens.[53,54]

Paediatric formulation design should bear several factors in mind with regards to the choice of excipient, including the safety profile of the excipient in children, the administration route, the treatment duration, possible alternatives and the regulatory status in the paediatric market.

13.4   The Elderly: Living Longer

Throughout the developed world, the improvements in standards of living and diet over the last 50 years have increased the proportion of the elderly in society. Previous reviews of health conducted in a province of Canada, 30 and 40 years ago established that the plight of the elderly should be of concern. Roos and colleagues estimated a 29% increase in the number of elderly white residents of Manitoba between 1971 and 1983, with a 73% increase in those who were in poor health.[55] Lin and colleagues report a Taiwanese government 2008 estimate, that in their country, 10.4% of the population was over 65.[56]

As we age, there is a noticeable slowing of behaviour with a decrease in reaction time, slowing of cognitive processes and a decline in somatosensory, visual and auditory sensitivity, coupled with muscle weakness and tendon stiffness. Decrease in swallowing function would be expected to be associated with Parkinson's disease and has been shown to diminish after treatment with leva-DOPA.[57]

In a review of swallowing dysfunction applied to oral drug therapy, Stegemann and colleagues also draw attention to the shifting demographic as the 'baby boomers' become aged, with a similar reporting of swallowing difficulties in the aged across three countries.[58]

The functional declines are often picked up as oesophageal dysfunction—especially an inability to swallow medications, but more severely, foreign objects lodged in the oesophagus. A Turkish study of 177 cases reported that the obstructions were varied: with coins (53% of cases) and bone and meat impaction (35%) being the most common objects retrieved by rigid endoscopy.[59]

Dysphagia is, in particular at higher age, a frequent problem. In Germany alone about five million people suffer from dysphagia. They are reported in every seventh to eighth of the over-sixties and in 30–55% of the persons in long-term care facilities.[60]

13.4.1   The Elderly and Flavour Preferences

The elderly suffer from disorders of smell and taste, although these sensory modalities receive much less attention than deficits of sight and hearing, which have obvious safety implications. Nevertheless, enjoyment of food is an important quality of life indicator and consumption has been reported to increase on addition of flavour enhancers to food with moderate weight gain in elderly patients. Mathey and colleagues compared monosodium glutamate-based flavour enhancers against control meals over a period of 16 weeks in a population of 67 elderly patients.[61] This study showed shifts in the elderly, with increased preference for salt and sugar, which is supported by a study by Mojet and colleagues.[62] Studies by Griep and colleagues cite evidence that flavour perception is the strongest determinant of food choice in the elderly, suggesting that flavour enhancement deserved attention.[63,64] A study by Ikeda and colleagues of a Japanese population noted that serum zinc was significantly lower in the elderly group compared with the younger population. Zinc supplementation raised serum zinc and improved taste sensation in 70% of the elderly cohort treated.[35]

Boyce and Shone have reviewed the effects of ageing on gustatory and olfactory senses and draw attention to the wide spectrum of diseases that affect the senses of taste and smell. This, in turn, leads to loss of appetite, weight loss and decreased immunity. In addition, a patient with these losses may become anxious. Flavour augmentation is not without risk: the authors comment that alteration to increase flavour sensation may result in additional sodium loads.[65] Additionally, it has been recently reported that increased salt intake is associated with frequent nocturia, which would be especially undesirable in the elderly.[66]

Elderly patients with cognitive disabilities suffer deterioration of their self-caring abilities. In addition, stroke impairs mobility, reinforcing a circle of social isolation. Institutionalized care has therefore become one of the principal choices for the elderly, in part as a decision by their families who may be unable to cope with the physical and emotional aspects of age, especially mental illness.

13.4.2   Community Homes Practice and Geriatric Medicine

Barnes and colleagues have drawn attention to the competing demands on nurses in the nursing home setting. The lack of time that might be needed to crush a medication or otherwise prepare a formulation might dissuade the carer from bothering, especially if the patient has issues such as dementia.[67] In the past, the practice of disguising medicines in food—covert dosing, particularly of patients with dementia—led to disciplinary action and secrecy surrounded the practice in the UK.[68] To obtain concordance with treatment, it would be better to design formulations as granules and administer with food where possible.

13.4.2.1   Crushing Tablets

Attempting to follow the physician's recommendation to cut or divide tablets is an intractable problem experienced by many elderly patients.[69] Nurses and other caretakers often split tablets not intended for breaking for patients and risk jeopardizing treatment as a result. Easily dividable formulations, however, provide an additional important condition to support user-friendliness and to empower personalized dosing, especially for patients who must follow complex regimens. Wening and Breitkreutz evaluated dosing approaches which could be applied to personalized medicine in an important European review.[70]

There are a large number of preparations that cannot be crushed prior to administration and alternative therapy has to be available. The principal classes of formulation include modified-release preparations, enteric coated preparations, sublingual or buccal systems cytotoxics and hormones.[71] In well-supervised clinical practice there are few reported problems associated with tablet crushing because alternatives can be provided, although in the past, there have been deaths.[72] The crushing device itself can be an issue, with contamination between one mortar and pestle-crushed formulation and the next. An old report surveying practice in ten nursing homes in Australia found that 17% of the medicines were altered, often with the medicines being crushed into one vessel.[73]

Stegeman has reflected on the issues surrounding the pharmaceutical industry and the clinical development program. He proposes that appropriate geriatric dosage forms will have to be provided with a simplified medication regimen and controls in place to ensure that drugs are taken appropriately and on time. In particularly he considers that oral dosage forms should be dispersible, that information leaflets should employ pictograms and caregivers should assist the patient by reinforcing situational cues as appropriate.[74]

The advantages of individualized, user-friendly dosage forms should also be considered for the elderly, who are often sensitive to adverse effects and whose adherence to complex tablet-taking regimens must be stricter.[75] As the growing population over 65 years of age is now the largest medication user group, the development of appropriate dosage forms for this patient population is absolutely essential.[76]

13.5   Mothers and Potential Mothers

The needs of the young and elderly have been mentioned, but there is a third group, mothers and potential mothers, that may need to be considered. The health and development of the baby is inevitably related to maternal nutrition, especially before and during the first trimester of pregnancy. A particular concern is low birth weight and pre-term delivery that compromises the development of the young. A study by Doyle and colleagues in 2000 showed that mothers of low birth weight babies did not appreciate that a small but apparently 'well' baby needed more nourishment.[77]

Folate deficiency was high in the cohort surveyed (47%) with low ferritin levels. Doyle *et al.* reported the estimate of Taylor *et al.* that, without supplementation, it takes up to two years for a mother to replace vitamins and minerals to the normal range.[78] Anaemia carried forward from previous pregnancies is a recognized problem in the developing world[79] and should be considered in recent migrations. The design of food additives for this group of patients with identifiable need therefore presents another opportunity in user-friendly medicine.

All of these innovations, attempting to improve user-friendliness of medicines, take place against a backdrop of tremendous changes in the structures of pharmaceutical companies and the healthcare systems that they supply.

13.6 Maximizing the Innovative Value of Pharmaceutical Products

The pharmaceutical industry faces significant challenges fuelled by patent protection issues, surviving patent expiration, rising R&D costs and increasing competition from generic products. Analysts of the pharmaceutical industry have long held that the conventional blockbuster business model is no longer sustainable.[80] Development and production cycles are becoming shorter, with the result that innovation itself cannot guarantee long-term profitability.[81]

Currently, in the pharmaceutical industry regulatory complexity is more pronounced than in most other branches. High costs in R&D and the long process of taking a new pharmaceutical product through the regulatory systems make it increasingly difficult to find novel APIs that can be manufactured at a reasonable price, as well as cost-effectively enough to meet reimbursement requirements.[81]

Frequent regulatory changes also play an important role in the identification of appropriate patent expiration strategies. One strategy, which has been confirmed by empirical research, is product-line extension that involves the innovative modification of pharmaceutical drugs into user-friendly dosage forms.[82] The potential benefits are many, ranging from enhanced product life-cycle management and extended intellectual property protection up to convenient dosing for patients with specific needs who would benefit from greater accessibility to a therapeutic agent.

Another useful patent expiration strategy is the development of a successor product, *i.e.*, development of single-pill combinations containing several APIs. The pairing of two or more components in one fixed-combination product can facilitate convenience, increased patient compliance and, from a therapeutic standpoint, better overall disease management. Moreover, combination products could provide a synergistic effect by targeting one or multiple diseases. Due to their many potential benefits, several pharmaceutical companies have developed double or triple combination products to match changing therapy needs.[83,84]

The case for innovation is strong, although health care providers may not have fully considered the benefits that new formulations may bring for improved therapy. Thinking about a product in terms of 'user-friendliness' helps identify primary users in the population and special requirements for those groups. Getting the right combination of drug and form of presentation should assist the patient to achieve concordance with treatment strategy.

13.6.1   Innovation Through User-friendly Formulations

Solid oral dosage forms, such as effervescent tablets and orally disintegrating granules (ODGs), have been developed to improve bioavailability and conduce to faster onset of action compared with conventional dosage forms. This is significant for BCS class II drugs (high permeability, low solubility) as the slow rate of dissolution limits bioavailability. The gastroduodenal area is an important region of the gastrointestinal tract because the low gastric pH provides an environment to increase the dissolution of basic drugs. Diseases and ageing alter secretion and motility, and disorders may increase, decrease or even cause retrograde movement.[85] The acidic flow into the duodenum, particularly following a meal, decreases the duodenal pH to between 5 and 6.[86] Further down the intestines, the pH rises to nearer to neutral and dissolved drugs may precipitate out of solution. Similarly, the dissolution of conventional tablet formulations may decrease, particularly if the high drug loading exerts a major influence on the dissolution of the formulation.

In this manner a "window of absorption" occurs and appropriate presentation improves bioavailability and site-specific presentation might provide more consistent absorption;[87] however, it was appreciated by many authors that the conditions in the gut were difficult to simulate in a single apparatus and, also, early systems based on CaCO2 cell culture had severe limitations. In the clinic, attempts at designing gastroretentive dosage forms using physical mechanisms including swelling, flotation or unfolding were pursued.[88] These were employed to exploit positioning above the area of maximum absorption, since transit through the duodenum occurs quickly. Additionally, the solubilisation provided if bile is released diminishes as the formulation transits through the length of the small intestine. A more robust approach may be to provide the formulation in a dispersed system. The higher surface area of a dispersed dosage form facilitates wetting and early release to dissolved and small suspended particles, which improves bioavailability. This is a special consideration in the pain control market, where a therapeutic effect is sought with some urgency. Moreover, innovations in dosage forms permit new routes of administration and dose delivery systems that offer substantial clinical advantages, including reduced dosing frequency and improved patient adherence to medication.[89] As a possible result, patients and consumers that seek modern dosage forms remain loyal to the brand and may also be prepared to accept higher prices.

However, often the new dosage forms permit a more cost-effective treatment altogether. Therapeutic success may be improved by the more specific use of the APIs and patient compliance increased by user-friendly application forms.[90]

13.6.1.1 Effervescent Formulations

A notable property of an effervescent formulation is that the drug is already in solution when administered and patient preference is usually to take a whole glass. Moreover, carbon dioxide generation in a drink contributes to taste masking, as identified in several well established patents. For the oral drug delivery of analgesic formulations and other drug compounds, where quick and effective onset of action is desirable, effervescent tablets are of potential benefit. Faster drug absorption associated with an effervescent formulation of acetylsalicylic acid when compared with a tablet formulation has been reported in a small number of patients.[91] In addition, an improvement in the onset of pain relief using the analgesic paracetamol in the treatment of postoperative dental pain has been demonstrated using an effervescent formulation compared with a tablet formulation. The median pain relief onset was reported to be 20 minutes for the effervescent formulation and this was significantly shorter than the tablet formulation pain relief onset at 45 minutes.[92]

The presence of carbon dioxide within a formulation may not only improve tablet disintegration but may also contribute to improved drug bioavailability. It has been demonstrated for a range of drug compounds that the presence of carbon dioxide in the intestine can improve drug permeability through the intestinal epithelium, affecting drug bioavailability.[93] The concept that carbon dioxide will disrupt the mucus layer, allowing access to the apical mucus layer, is important, but local higher concentrations have profound physiological actions. When the intestinal tissue is exposed to higher concentrations of carbon dioxide, the intestinal blood vessels dilate and the tissue becomes hyperaemic.[94] The muscular tone of the gut also decreases and rhythmic tone is abolished.

The usual agents in an effervescent mixture are sodium bicarbonate, combined with citric or tartaric acid. Dilute sodium salts produce an increased rate of gastric emptying, as first described by Hunt.[95] His work also refers to earlier studies by Lolli and colleagues, who reported that carbonation of drinks increased the rate of gastric emptying[96] although later work using MRI measuring the simultaneous emptying of gas and water indicated that carbonation may decrease the rate of gastric emptying and mainly contributes to variability.[97] Further studies showed that the main effect of carbonation was to alter the intragastric meal distribution rather than perturb emptying, the carbonated water causing the distension of the proximal stomach, holding solids above the gastric midband.[98]

Effervescent formulations can also be useful in improving the dissolution performance of difficult to deal with APIs. For example, an effervescent formulation has been shown to be useful in the oral delivery of benznidazole, a class IV drug compound associated with poor solubility and poor permeability, through improved drug dissolution.[99]

A classic example of a drug which continues to be a strong product despite competition from generic products is acetylsalicylic acid, commonly known by its trade name, Aspirin®. This brand owes its success not least to a product-line extension strategy which has resulted in the development of a multitude of formulations such as effervescent products and orally disintegrating granules.[100]

13.6.1.1.1 Advantages of Effervescent Dosage Forms. Oral dosage forms allowing dissolution or dispersion in water prior to administration can be formulated to be effervescent. In Germany, Austria and Switzerland effervescent dosage forms are widespread and offer another welcome alternative to swallowing tablets. Their main advantages include excellent bioavailability, rapid release of API and minimization of gastric irritation.[74]

Effervescent tablets are also suitable if large quantities of the API are required. The concentration of bicarbonate within effervescent tablets is normally high. As a result of this, once drug dissolution into solution has occurred the pH of solution delivered to the stomach is usually slightly alkaline.[101]

Epidemiologic studies have shown an association between the use of proton pump inhibitors (PPIs), low bone density and fractures. One of the proposed mechanisms of action is an increased gastric pH causing a decrease in absorption of calcium in patients who use PPIs long-term.[102-104] It is therefore recommended that persons who are at risk of osteoporosis should take adequate vitamin D and calcium supplements.[105] Taking this into consideration, calcium chloride and calcium citrate in the form of effervescent formulations are the best soluble calcium formulations for patients who have to use acid-inhibitory drugs: This strategy may achieve the maximal quantity of ionized calcium in the maximal volume of gastric fluid entering the small intestine from the stomach.[106] Related to this observation, patients with achlorhydria have difficulties absorbing weak, insoluble basic drugs such as ketoconazole. Omeprazole-induced achlorhydria results in the reduction of gastric acid volume, basal and stimulated acid output. Howden and colleagues reported a 92% reduction in acid secretion and a 59% reduction in basal volume six hours after a single dose of 60 mg given in the morning.[107] Administration of ketoconazole, with an acidic carbonated drink to normal subjects with an omeprazole-induced achlorhydria, resulted in a seven fold increase in AUC.[108] A similar effect has been reported for posaconazole suspension. Similar effects follow pH–pK_a properties of the compounds. For example buccal absorption of nicotine is increased when magnesium hydroxide is added to the formulation to make the mouth slightly alkaline on chewing.[109] The effects of proton pump inhibitors are often mediated through metabolic interactions with cytochrome P450 isoforms and therefore have larger effects than those predicted, due to pH effects on solubility and partition.

Additionally, it has been shown that effervescent dosage forms of orally administered bisphosphonates, which are used to inhibit bone resorption and to increase bone mineral density in osteoporosis, can provide an oral solution with a more favourable tolerability profile. Most oral bisphosphonate tablets are associated with gastro-oesophageal irritation because their exposure at pH values less than 3 is irritating to mucosal tissue. Effervescent formulations allow oral solutions buffered to a pH above 3, which may be associated with a better absorption profile and also a reduced risk of developing gastrointestinal lesions.[110,111] In an effervescent tablet, absorption of an API can be significantly quicker compared with a conventional tablet. The passage through the stomach is accelerated by the buffered solution and the API will finally reach the destination—the small intestine—earlier.[112] In addition, effervescent tablets do not need to disintegrate in the body as dissolution starts before intake.

Another benefit arising from effervescent dosage forms involves a classic example concerning a substance that is difficult to digest—calcium carbonate. Chewable calcium carbonate tablets, for example, are associated with the release of carbon dioxide during the reaction between calcium carbonate and gastric acid that usually causes erucation, bloating and other gastric adverse reactions. Taken in an effervescent formulation, insoluble calcium carbonate is converted into absorbable calcium citrate already in the glass and with no unwanted excessive carbon dioxide production.

13.6.1.2   Oral Liquids

Gastric emptying of liquids occurs much faster than that of solid objects, such as slowly disintegrating tablets, and shows interesting behaviour in the presence of food. Mixing efficiency is dependent on dispersion of food, which in turn is affected by components of a meal (especially bread) which affects heterogeneity. Thus liquids can move round a solid mass in the stomach and empty faster. Although this is affected by posture, the stomach maintains a pressure pump system, controlling emptying even when the subject is upside down![113] A non-nutrient, non-viscous liquid is therefore the least sensitive to emptying as a function of posture and emptying is usually mostly complete in 30 minutes. The residual volume is probably no more than 10–20 ml.

Oral liquids have advantages in paediatric dosing, particularly for the very young, as the wide range of bodyweight which is encountered requires unusual flexibility. There is also the danger of airway obstruction when swallowing tablets in children who have pre-existing obstruction. De Goede and colleagues describe the preparation of a stabilized clonidine solution as an adjunct to benzodiazepines after periods in intensive care[114] Liquid dosage forms do have disadvantages with regard to stability and bulk that may make them more expensive. In paediatric practice, dosing could become unreliable if the dosed volume is small. Syrups do not *per se* solve the problem of after-taste and for extremely bitter medicines taste masking may be easier with a simple tablet[115] or an ODG.

13.6.1.3   Chewable Vitamin and Micronutrient Products

Chewable formulations do have some advantages which are recognised in adult, paediatric dosage forms and the OTC (over the counter) market. For the patient who dislikes swallowing tablets, or has experienced previous difficulty doing so, the chewable system is convenient and a pleasant method of taking medication. Usually, the chewable tablet is consumed without additional liquid and in some locations where clean water is not immediately available—for example at work in unhygienic surroundings, where the mastication to reduce the size of the swallowed material and dilution in saliva is a significant patient convenience. For this reason, we encounter many embodiments of the chewable tablet, not only in formulations of antacids, vitamins, analgesics and laxatives but also in anticonvulsant and antibiotic preparations. In addition, the chewable formulation allows long contact within the mouth, which may facilitate buccal absorption, although most of the dose will be swallowed.

The uses in nutrient supplementation were early examples of employment of chewable dosage forms. In the days when vitamin D deficiencies were first noted, rickets and osteomalacia were regarded as hallmark indicators of the lack of this vitamin in the diet. More recently, non-skeletal disorders including auto-immune disease, cancer and metabolic syndromes such as diabetes have been identified as having links with low levels of circulating 25-OH vitamin D and there are suggestions that race and sunlight exposure should be taken into account in calculation of vitamin D requirements.[116,117] In Europe there are a large number of migrants from Asia who present with hypovitaminosis associated with lower levels of sunlight needed to produce sufficient vitamin D. This has significant clinical importance.

A concern has been the perceived choking hazard, although recent clinical experience in older children does not highlight this as an issue. For example, the WHO launched a large programme for de-infesting children of parasitic worms using this type of formulation. A recently published trial reported that chewable tablets of mebendazole were well tolerated and treatment-emergent events were drug- and not formulation-related.[118] Similarly, the IMPAACT P-1056 study outcomes reported in 2010, enrolled children in a young age group (less than 6 months) who were dosed with a chewable tablet formulation to deliver stavudine (7 mg), lamivudine (30 mg) and nevirapine (50 mg) commented on the safety and efficacy.[119] Children in this trial preferred the chewable tablet *versus* the large volumes required for the liquid formulations.

13.6.2   Enabling High-dose APIs and Combined Dosage Forms

In many respects, effervescent tablets and other user-friendly solid oral dosage forms represent an ideal basis for product line extensions targeted at all age groups and patient populations in order to ease drug intake and uphold a drug regimen, provided that the properties of the API are suitable for the respective dosage formulation. They may facilitate the incorporation of a wider range of dosage levels and even combined dosages, beyond what conventional formulations allow, so that a large amount of API can be taken in a single dose.

Complex drug regimens can be simplified by combining several agents in one formulation. Combining multiple drugs that have the same overall functional effects can also enable reduced dosage of each API through synergistic effects, and result in diminished side effects.[90] By adding specific excipients, combination products can also be used to enable a booster effect of active agent(s), while other excipients allow controlled rates of dissolution and correspondingly optimized drug release. Patients who are not required to take many pills tend to be more compliant, resulting in better health outcomes.[120]

From a practical point of view effervescent and other user-friendly solid oral dosage forms are also more portable than conventional liquid formulations.[121] The product can be individually packaged in convenient forms (*e.g.* stick packs, sachets) and therefore provide an added value. User-friendly dosage forms incorporating large amounts of API, which, therapy permitting, can be consumed all at once as a daily dose, rather than in multiple tablets over the course of the day, also offer patient benefit.

13.6.3   Packaging

When it comes to developing pharmaceuticals or even dietary or health supplements, the old blockbuster, one-size-fits-all approach has lost relevance: as consumers, people seek options to improve dosing. To accommodate these, developers need to start listening and putting people first, starting with convenience and packaging.

Packaging plays an essential dual role in the pharmaceutical industry: it has to maintain product integrity, protecting it from contamination and degradation. It also needs to be easy to use, attract the attention of consumers and be able to differentiate itself from the competition while remaining on-message. Crucial for successful packaging is its usability, or how user-friendly the product is. This has become of increasing importance in an age where convenience and choice are all-important.

13.7   Challenges in Developing and Manufacturing User-friendly Dosage Forms

13.7.1   Product Design and Formulation Development: Defining API Characteristics and Matching the Right Dosage Form

Understanding the characteristics and requirements of a drug substance is of critical importance for appropriate product design and performance. Before the formulation of an API in a dosage form, its chemical and physical attributes need to be characterised.

Physical properties, including particle size, particle size distribution and shape, may affect the flowing properties of solid matter during formulation, which itself has a direct effect on the performance characteristics of a drug product, including mouth feel—for example, large particles feel gritty and unpleasant in the mouth. Conversely if particles are too small, they tend to fully absorb the saliva available.

The age of the intended patient plays an important role in matching the API with an appropriate dosage form. Whereas liquids for infants and children aged less than 5 years are according to Allen the preferred delivery system,[40] the WHO[122] and Spomer and Colleagues[123] challenge this and also recommend granules and microtablets. Alternative dosage forms can be considered for developmentally handicapped patients and elderly patients who have difficulty swallowing solid dosage forms.[40]

To avoid dosage forms that may interfere with swallowing disorders, effervescent tablets may be useful. This dosage form is also recommendable to facilitate precise drug dosage[124] as with effervescent tablets the dose is standardized—even if the liquid quantity is variable.

Excipients are required to formulate an API for a final dosage form and need to be carefully selected and sourced. They range from filling materials such as sugar alcohols, fillers, flavours and sweeteners to colorants, to product appeal enhancements, to various coatings.

13.7.2 Taste-masking in Oral Pharmaceuticals

Effective taste masking is of great significance to reach an acceptable degree of palatability for orally administered drugs.[125,126] In paediatric formulations bitterness masking becomes essential to ensure patient adherence to medication, as young children show higher taste sensitivity to bitter-tasting substances.[127,128] Likewise for people on long-term medication and for those who need to regularly take high volumes of medicines, taste is an issue that matters.

It is important to consider that only the soluble portion of a drug can generate the sensation of taste when chemical molecules interact with taste receptors on the tongue after dissolving in saliva. Hence, in dosage forms that disintegrate rapidly in a glass of water or in the saliva, taste masking is of critical importance. Therefore, effervescent tablets and other user-friendly dosage forms, such as orally disintegrating granules, place a high demand on formulation scientists to improve taste.[125,129] With these dosage forms, taste masking of active ingredients is challenging because APIs often have a bitter taste which can diminish patient adherence to medication.

Pharmaceutical scientists use various technologies to improve the taste of a drug, each depending on the drug's nature and physicochemical properties. A variety of taste-masking formulation options are available, including incorporation of excipients such as flavours, sweeteners and amino acids.[130]

13.7.2.1 Taste-masking with Flavours

The flavouring of APIs is primarily used in solid dosage forms intended for oral administration; by adding flavouring excipients, the taste of APIs may be successfully covered. Excipients that could affect the stability of the API are preferably avoided, but the required flavours can still interact with the API due to the direct contact with it, which may degrade the API or even the flavour itself. It appears relevant to consider the targeted flavour at an early stage of drug development due to the potential of taste-masking technologies not only to change APIs but also to affect the approval process of the pharmaceutical drug.

To limit aroma in order to prevent API degradation, it is necessary to encapsulate volatile excipients such as flavouring agents. Encapsulation techniques make it possible to coat or entrap a flavour within another material or system.[131] By utilizing suitable technology, flavours may remain stable over years and the API release may remain unaffected.

Specialized manufacturers have patented procedures for developing and manufacturing custom flavours for specific dosage forms, such as effervescent tablets. During one procedure, for example, sugar alcohols, such as mannitol, sorbitol and glucono-delta-lactone (GDL), are melted until homogeneity. Subsequently, the liquid flavouring components are blended with a high-pressure blender so that micro droplets are formed. The melt is then deposited onto a conveyor belt and solidifies into an amorphous structure. After cooling, the non-crystalline melt is milled into particles of the desired particle size. When adding water (*e.g.* when putting an effervescent tablet in water), the matrix of the sugar alcohols and the GDL is dissolved and fine aroma droplets are released, suspended in water. Flavours produced according to this procedure have the benefits of long term stability and several advantages during manufacture.

13.7.2.2 Taste-masking with Physical Barriers (Coating)

For very bitter and highly water-soluble drugs taste-masking with ingredients such as (artificial) flavours and/or sweeteners alone can be insufficient.[130] Taste masking is of critical importance, especially in orally disintegrating drug delivery systems, because the drug must disperse in the saliva while maintaining a pleasant taste and mouth feeling. The use of various physical barriers, such as coating of APIs or lipid extrusion, can provide a feasible option to mask unpleasant taste. In addition, a sweetener and a flavouring agent may be added.

Chewable tablets, lozenges, orally disintegrating granules and tablets require very effective API taste-masking strategies as well, which usually can be achieved by the formation of drug granules that are coated before flavouring agents can be added.

Coatings are not only an essential consideration in the formulation of dosage forms to ensure aesthetic qualities including taste-masking, mouth feel, colour or texture, but they also offer physical and chemical protection for the API. Moreover, they can be used to modify drug release characteristics.[132]

Modern pharmaceutical coating was introduced in the 19th century with sugar coating, which was mainly used to improve the palatability of bitter tasting drugs. Due to the disadvantages of sugar coating, this method was replaced by film coating, which offered better reproducibility and process automation as well as the ability to apply it to a wider range of pharmaceutical dosage forms. However, film coatings applied by using organic solvents brought their own disadvantages and required long processing times, and the necessity for solvent-removal systems.[133]

Coating can be achieved by repeatedly exposing a particle to a spray containing solute and solvent. In fluid-bed coating, a stream of air disperses solid particles and each is coated when passing through the spray zone. The challenges in fluid-bed coating technology include:

- Optimization of process parameters (especially instrument temperatures, spray rate and air flow) for ideal product characteristics (release rate, taste-masking and further processability)
- Optimization of process parameters for best product stability (no chemical degradation during and after the coating process, no morphological changes, *e.g.* polymorphism)
- Optimization of process parameters for short production cycles and reduction of solvents—applicable only to solvent-based fluid-bed coating
- Long drying times to remove solvents and testing for residual solvents—applicable only to solvent-based fluid-bed coating
- Risk of twin formation (two particles sticking together because of their coating, caused by overly fast spray rate)
- Selection of robust process parameters that generate low batch-to-batch variability

Today there are several solvent-free coating techniques available, one of which is hot-melt coating.[134] Hot-melt coating is a unique, solvent-free process in which a molten lipid excipient is sprayed onto solid particles during fluidization in a fluid bed coating device (Figure 13.7).

The lipid solidifies upon cooling and coats the particles with a thin, homogenous film (Figures 13.8 and 13.9). The coated material can be used in solid dosage forms such as tablets, hard and soft gelatin capsules and ODGs. Hot-melt coating has been investigated for its use to improve stability, mask taste and achieve sustained as well as immediate release.

Lipid excipients must be GRAS (Generally Recognized As Safe) and well accepted by international authorities. The choice of lipid excipient is determined by the application. The amount of lipid excipient can range from 5% to 40% depending on the API characteristics. The type of lipid used must be selected to optimize the functional performance of the lipid film.

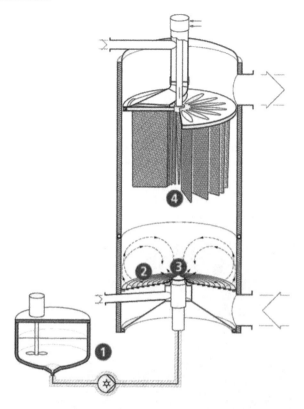

Figure 13.7 Hot-melt coating takes place in a fluid bed coater under controlled environmental conditions. The melting device (1) contains a heated tank for the molten coating materials, a stirrer and a heated tubing system to avoid solidification during transport of the molten mass into the fluid bed. The air layer gliding process (2) moves the particles into the spraying zone, creating a homogeneous temperature distribution as well as a spiral air circulation. The liquid is sprayed through a heated nozzle (3) and is atomized by compressed air into fine droplets, ensuring a precise coating. The dust filter system (4) works continuously during the process. The filter bags, of which there are several, are cleaned by blowing air through the filters, thus removing any dust stuck to them. This approach reduces downtime to a minimum. Image courtesy of Romaco Innojet.

Additionally, surface-active excipients can be added to modify the drug delivery properties of the coated API; this type of coating may also be applied for immediate-release formulations (Figure 13.10).

13.7.3 Manufacturing

The manufacture of pharmaceuticals is a very complex process. The formulation of the API into a drug product comprises many steps beginning with the sourcing of raw materials, then weighing and blending, selection of granulation technologies and tableting and filling processes.

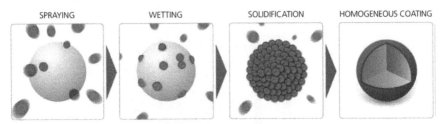

SPRAYING   WETTING   SOLIDIFICATION   HOMOGENEOUS COATING

Figure 13.8 The HMC process, from spraying through to the finished coating. In the first step, the coating constituents are heated up and melted. Following this, the coating droplets are sprayed onto the seed particle (API) and wetting occurs on the surface. As the seed particle is colder than the melting temperature of the coating mixture, the droplets solidify and form a homogeneous layer. Image courtesy of Hermes Pharma.

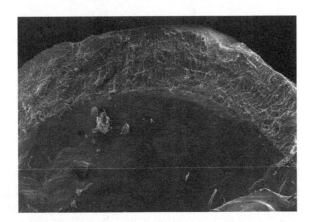

Figure 13.9 Scanning electron microscope (SEM) image of a bitter tasting, fast acting API coated with a lipid based mixture. The image shows a cross section of the crystalline structure of the API core covered with a very homogeneous coating layer. Image courtesy of Hermes Pharma.

Finally, the product has to be protected by proper packaging, labelled and delivered.

In many cases the development of user-friendly dosage forms introduces additional specific challenges. For example, the use of lubricants should be minimized when producing effervescent tablets, as they can lead to a final product with an unpleasant soapy taste that forms a cloudy solution upon dissolution. These problems can be avoided by using water-soluble lubricants such as high molecular weight polyethylene glycol. However, hydrophilic compounds are often poor lubricants and in most cases it is more effective to introduce external lubrication during the tableting process. External lubrication enables the tableting process to be carried out smoothly and at high speed. Furthermore, this ensures that the final product contains minimal traces of lubricant (less than 1 mg) and performs as expected when dissolved in water for administration, while minimizing the occurrence of defects on the surface of the final product.

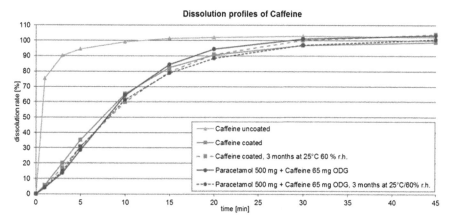

Figure 13.10 Dissolution profile of taste-masked caffeine using hot-melt coating over storage and in comparison with the uncoated raw material. Image courtesy of Hermes Pharma.

Due to their mode of administration, effervescent medications must also be manufactured in low-humidity environments to ensure maximal stability; unwanted moisture could accidentally trigger the effervescent reaction before the production process has been completed. It is also important to minimize the moisture taken up by ODGs during production to help increase the stability of APIs and excipients, for example phyto-extracts and vitamins. In some cases it may even be necessary to produce ODGs under inert-atmosphere conditions (*i.e.* in nitrogen), as is essential when working with oxygen-sensitive products, such as omega-3 fatty acids.

Factors such as moisture- and oxygen-sensitivity can influence every stage of the process—from formulation to manufacturing and packaging. Therefore, for sensitive products it is preferable to keep turnaround times as short as possible, minimizing the time between compounding and packaging to increase stability. In these cases, the final product should also be carefully packed and sealed to protect it from external elements.

13.7.3.1 Up-scaling and Granulation

Although the scale-up of manufacturing is never a simple task, this can be especially true when working with certain user-friendly dosage forms such as ODGs, which may be formulated using 'softer' compounds such as lipids. Such materials are prone to mass effects when stored in large containers, as the additional weight can cause crushing and compacting at the bottom of the container, affecting particle size and morphology. In such cases it is not uncommon to observe an asymmetric spread throughout the container, leading to variability in the formulation process and the appearance of structural artefacts. For example, particles of materials that have undergone a fluid-bed coating process, such as during hot-melt coating, are prone to caking in large containers. Therefore, such an effect needs to be considered already at the lab scale when the formulation is developed.

Scale-up challenges also extend beyond the storage of raw materials to the process itself. In many cases, it may be necessary to re-optimize parameters, such as drying time, to account for the larger amounts involved. If not carefully controlled, the final product may suffer from quality and/or stability issues that were not predicted or observed during earlier stages of the formulation process.

Granulation allows size enlargement by aggregating individual powder particles, usually of several different components, to form larger structured particles. For the production of effervescent tablets, various manufacturing methods are in use, including direct compression, dry granulation or fluid-bed granulation. However, before initiating the granulation process for the development of a new pharmaceutical application, the scale-up of each engineering process for future commercial purposes must be ensured. The up-scaling of granulation processes, in particular, is a great technical challenge due to the inherently heterogeneous nature of the materials used.[135] Batch size in the early-stage development of a solid dosage form is small, but the size of a batch in a later stage may be up to 100 times larger.[136]

The TOPO granulation technique (Figure 13.11) has been developed in Austria to prepare granules and coated particles under high vacuum (Box 13.2). This technology comprises synthetic granulation to enlarge the particles and

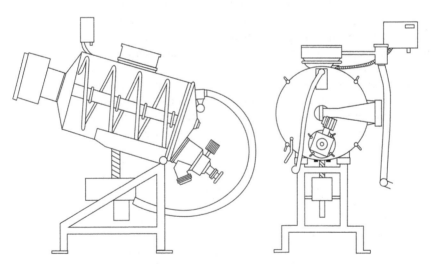

Figure 13.11   TOPO granulation technology. At the front is a suction hose where the raw materials are sucked into the vessel by the vacuum. At the front there is also a rotating sieve used to break up agglomerates at the end of the process. A spiralled stirrer, used to mix the granulate during the process, is located within the vessel. The vessel can be moved from a horizontal position into positions ±20° from horizontal. The up and down movement helps to achieve intense blending of the contents when stirring occurs at the same time. Image courtesy of Hermes Pharma.

> **Box 13.2** TOPO vacuum granulation.
>
> The TOPO technology is based on a surface modification of the citric acid applied in the effervescent mixture. During the process the reactive citric acid is coated with an alkaline carbonate and passivated, followed by development of sodium citrate on the acid surface. Depending on the duration of the granulation, about 20–30% of the citric acid converts into citrates during the reaction, so that the citric acid is coated with layers of citrates of only a few micrometres in thickness. The citrate coating facilitates increased stability *vis-à-vis* acid-sensitive agents. As also alkaline agents can be coated, this benefit is equally effective for agents sensitive to alkali.
>
> The proportion of the converted citric acid transformed into citrate can be identified *via* infrared analytics, titration of the citric acid or semi-quantitative determination of the carbon dioxide content after acidification. Only a small amount of water is added—that needed for granulation—during the TOPO process.
>
> Oscillating vacuum
> Additional water develops during the chain reaction from the conversion of citric acid with bi-carbonates or carbonates activated through moistening. To control and manage this chain reaction, a vacuum is applied repeatedly for certain periods of time during the granulation process to eliminate the reaction water. During this process step, the vacuum oscillates within the TOPO granulator between a maximum and a minimum value. The number of oscillating movements defines the reaction and the thickness of the moisture-resistant surfaces. This so-called "oscillating vacuum" is patented.

increase their stability, resulting in a granulate, which is easy to convert into tablets and extremely moisture-resistant. By granulating in a vacuum, an uncontrolled chain reaction of the acidic and alkaline components is prevented. This patented procedure can be utilized *e.g.* for the manufacture of effervescent granules containing at least one organic acid (*e.g.* citric acid) and at least one alkaline carbonate (*e.g.* sodium hydrogen carbonate).

When water is added, the acid dissolves on the surface, the sodium hydrogen carbonate starts a reaction, is then fixed and a granulate forms. Through the TOPO vacuum technology the reactive citric acid can be coated with an alkaline carbonate and passivated, followed by development of sodium citrate on the acid surface. With this citric acid passivating process even effervescent tablets with moisture sensitive APIs can be packed in polypropylene tubes at the final manufacturing step and remain stable for up to five years and even show three years or more in-use stability.

TOPO granulation provides further benefits compared with conventional procedures ranging from the creation of products that demonstrate a short disintegration time, to very high moisture resistance, to versatile use of the technology for a broad spectrum of purposes. The TOPO process can be applied to various dosage forms.

13.7.3.2 QbD and PAT

Quality has always been an important factor in the pharmaceutical industry to ensure that the medicines produced for patients are safe, reliable and effective. However, modern approaches are now starting to influence the

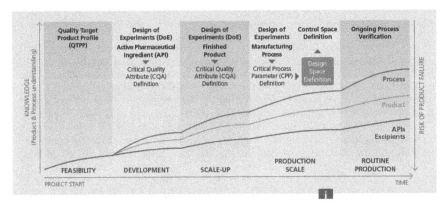

Figure 13.12   Overview of how knowledge about raw materials, product and involved processes increases with each QbD element from project start to routine production. At the same time the risk of product failure decreases significantly to an acceptable and manageable level. Image courtesy of Hermes Pharma.

industry, with advanced analytical methods becoming essential throughout every step of the pharmaceutical value chain, from early formulation development right through to commercial scale-up and ongoing production (Figure 13.12).

Driving this change are the twin approaches of Quality by Design (QbD) and Process Analytical Technology (PAT), which work in concert to maximize production efficiency and product quality, while reducing production cycle time and the costs associated with poor product quality. The advantages are a more robust, high-quality product development process, which is easy to upscale and provides a regulatory-friendly process aiding logical improvement through the development cycles.

13.7.3.2.1   Improved Product Development.   While QbD does increase the workload during the product development stage, it protects against variability later on, reducing risk and saving time. Extensive characterization of the API and excipients is performed in order to define critical quality attributes (CQAs). These are the chemical, physical, biological and microbiological attributes that can be defined, measured and continually monitored to ensure the final products remain within acceptable quality limits. To account for variability, multiple batches of API and excipients from multiple suppliers must be tested. This 'scientific approach' to determining the CQAs enables the more accurate identification of critical process parameters (CPP) and is more thorough, which reduces the number of 'dead ends' caused by the production of substandard batches. Not only does this reduce the overall time and costs associated with failed batches, but it also reduces the wastage of APIs and other important resources, such as potentially costly excipients.

13.7.3.2.2 Easier to Upscale. Moving from formulation-development scale to mass-production scale is a significant task. Fewer adjustments are generally required during the scale-up when following a QbD approach, since the CQAs and CPPs are well understood and it is easier to develop a product that will scale effectively. This makes the process more efficient and reduces the risk of any unexpected problems that could negatively affect production throughput and timescales.

13.7.3.2.3 Improved and Easier Regulatory Compliance. Another benefit of having a good understanding of the CQAs and CPPs is establishing a formulation and process design space that ensures compliance with the quality target product profile (QTPP). In particular, this prevents the need to register later adjustments with regulatory bodies after large-scale production has begun, which is important as such holdups can lead to significant delays when it comes to getting a product to market. Using a data-driven approach such as QbD also allows companies to manage risk more effectively and make better decisions on exactly which product changes will require additional regulatory submissions.

The QbD/PAT approach creates the potential for real-time release testing (RTRT). This not only gives better assurances when it comes to product quality but also enables a faster and more informed response when problems arise during the manufacturing process.

13.7.3.2.4 Providing a Framework for Continuous Improvement. Continuous process verification (as opposed to assessing a number of discrete batches) can reduce sampling effort and increases quality assurance. Furthermore, it keeps pushing processes towards optimization and maintains manufacturing parameters within acceptable limits. This improves quality and enables manufacturers to take action at the earliest opportunity even before problems have arisen, in order to ensure that quality issues are rectified faster.

13.7.3.3 Product Protection

Pharmaceutical packaging involves designing a system capable not only of ensuring that the drug product remains safe and efficacious, but also of ensuring that the container closure system is suitable for the intended use. The packaging must be compatible with the dosage form, and must protect the dosage form adequately; it must also ensure that the materials that are used are safe for the route of administration.[137]

Age-associated conditions, such as blindness, muscle weakness and cognitive changes, strongly affect the ability of the elderly to comply with prescription regimes. The manual dexterity needed to open medication packaging to remove tablets or capsules from the container has been identified as an issue in geriatric medicine. In a study in which 120 elderly patients admitted to a geriatric treatment centre, 94 patients of the cohort were unable to break the tablet or open one or more of the containers.[138] Blister packaging is useful for the preservation of hygroscopic and photo-labile drugs. It also offers some degree of child protection. In an elderly population, it might be expected that these types of packages might cause problems but there are scant data available to assess the scale of the problem if indeed it does exist. Older studies, specifically addressing the problems of patients with rheumatoid arthritis, noted that patients had problems with click-lock container closures and for this group small containers provided special issues. The mean grip strength appeared to correlate with the ability to open containers. Surprisingly, blister packs were not found to be a specific problem, since the patient could resort to the use of fingernails or scissors to pierce the film so that the tablet could fall out. Self administration in the home using blister packs and child-proof bottles has been reported to be problematic in several Australian studies.[139]

As the mean age of our population increases, it will be important to provide safe medicines which cannot be abused or accidentally ingested by children but which will be accessible to patients with neuromuscular degeneration, sight difficulties and a disease burden.

ODGs are filled into small 'stick packs' with size depending on the needs of the product and customer. The filling process requires several specific pieces of manufacturing equipment that includes an inline weight control for each stick pack, with a feedback loop to the dosing system that adjusts itself automatically. Precise screw dosing methods place up to a maximum of 4000 mg of product into each stick pack, with a weight variability of less than 2%. ODGs remain easy to swallow, even when large amounts of API need to be included. Larger doses or combinations of APIs in a single pack can simplify dosing regimens and improve overall convenience.

For the primary packaging material, laminates consisting of PET–aluminium–PE and paper–aluminium–PE are the most commonly used materials. These laminates offer not only protection from light, oxygen and moisture, but also allow ease-of-processing and almost unlimited options for their graphical design. PET laminates can be used for creating childproof packaging.

Adding tear notches to the stick pack is a simple method to improve ease-of-opening, making the intake more user-friendly. When creating the tear notch, an additional area is sealed and then cut. The size of this area needs to be well defined and the cutting has to be precise in order to achieve easy opening and emptying, while avoiding creating a literal bottleneck or compromising tightness (Figure 13.13).

Secondary packaging in the form of cardboard boxes is convenient and provides sufficient space for relevant information, such as ingredients and instructions. From a marketing perspective, it can be designed to assure a high recognition value and differentiation from competitor products. Furthermore, in order to prohibit counterfeit medicines, serialization will soon become mandatory in the pharma industry. QR codes and/or special tamper-proof labels can be added onto the secondary packaging to differentiate original from counterfeit products.

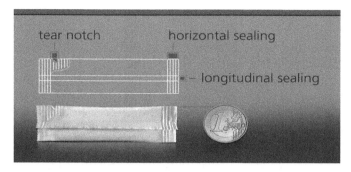

Figure 13.13　Typical stick-pack using paper–aluminium–PE laminate and incorporating a tear-notch. Image courtesy of Hermes Pharma.

Box 13.3　Packaging

Using a vacuum conveyor technique, the granules are transported to the tube filling/stick/sachet line, where they can be dosed to the individual stick/sachet. If the formulation contains several APIs in very different ratios (*e.g.* API 1 > 1g, API 2 < 50 mg) it is also possible that they can be delivered to the stick/sachet line from two different containers. The APIs are then individually dosed into the stick or sachet and thus dosing accuracy and homogeneity are increased.

While these stick or sachet lines are optimized for high throughput (up to ten sticks or sachets in parallel), they are also flexible to allow quick and easy product changes and product-specific adjustments (*e.g.* stick or sachet size, cardboard quality and printing).

Modern tableting equipment, immediate packaging of the tablets into primary packaging inline and selection of the most suitable packaging material can minimize the risk of degradation through moisture and damage through mechanical handling (Box 13.3).

13.8　Concluding Remarks

The term adherence is used to describe dosing history and various other measures including persistence, the time elapsed between a drug prescription and intake, discontinuation, or the extent of alignment between a patient's actual drug intake and the prescribed regimen.[140] Medical non-adherence has been identified as a major public health problem that induces considerable health burden and related economic drawbacks.[141] Overall, up to 70% of patients do not take their prescribed medication properly.[142] Lack of medication adherence leads to unnecessary disease progression, disease complications, reduced functional abilities, a lower quality of life and premature death. In the USA poor adherence has been estimated to cost approximately $177 billion annually in total and indirect healthcare expenditure.[143] Poor adherence can undermine the effectiveness of care at many steps in the treatment process.[144] The likelihood of non-adherence increases with the complexity of the drug regimen and is of particular concern in elderly patients, who often experience complex health problems compounded by multimorbidity and polypharmacy.[145]

Experience garnered from various indication areas has shown that advanced delivery systems can reduce many of the barriers related to taking medication properly including complex dosage regimens, undesirable side effects or physical characteristics (*e.g.* size of tablet, aftertaste, gastrointestinal stress).[90] Improved adherence associated with new dosage forms has been reported by studies concerning the treatment of patients, *e.g.* with human immunodeficiency virus (HIV), osteoporosis, diabetes, hypertension, the need for contraception and overactive bladder.[90,146-148] Elderly patients often benefit from user-friendly dosage forms due to their reduced side effects and simplified regimen.[149,150]

Orally disintegrating formulations have gained popularity because they are easy to administer and may lead to better patient compliance.[151,152] They may alleviate administration to patients who refuse or are unable to swallow a tablet, such as paediatric and geriatric patients. Good mouth feel and taste-masking techniques may change the perception of medication as a bitter pill met with resistance, particularly in paediatric patients.[153]

Moreover, they may offer improved biopharmaceutical properties, improved efficacy and better safety compared with conventional oral dosage forms. Fast orally disintegrating products dissolve in the mouth in seconds and that can enhance the clinical effect of a drug because of pre-gastric absorption in the mouth, pharynx and oesophagus.[153] In such cases, by avoiding first-pass hepatic metabolism, bioavailability of a drug might be significantly increased compared with conventional tablets.

In some indications, such as migraine or insomnia, the faster onset of action associated with orally disintegrating forms of medication can increase patient satisfaction and adherence to the treatment.[154-156] Although advanced formulations may be more expensive than conventional dosage forms, user-friendly dosage forms often have a more favourable pharmacological profiles and therefore the potential to reduce overall treatment cost through reduced burden on healthcare services.[90]

Another aspect, which is critical for certain patient populations—especially the elderly coping with more than one disease—is the need to take multiple drugs. Moreover, various conditions, including cancer or HIV/AIDS, are treated with combination therapies. Patients who take fewer pills tend to show a higher compliance, which results in better health outcomes. The development of fixed-dose combination products may lead to fewer prescriptions and a better cost–benefit-ratio than prescribing medications separately.[120]

In conclusion, patient convenience is a decisive factor in medication compliance and improved outcome. The utility of a medicine can be primarily achieved by designing and developing appropriate formulation options tailored to the needs of specific patient groups.[120] Consistent use of medications can lead to a reduced reliance on hospital and other medical services and improved treatment outcomes. It should not be forgotten that the healthcare system benefits financially from user-friendly dosage forms. Oral delivery systems save costs because they can be produced in large quantities within short production times maintaining consistent quality.[157] Modern, optimized oral delivery systems, of which effervescent formulations are but one example, are a worthwhile undertaking that enables the healthcare system to promote well being and save costs. The pharmaceutical industry benefits from expanding product lines and growing brands—revealing new revenue opportunities.

References

1. The Pharmaceutical Innovation Platform: Sustaining Better Health For Patients Worldwide October 2004-International Federation of Pharmaceutical Manufacturers Associations IFPMA.
2. SCRIP Intelligence Sept 16th 2011, interview with Dr Thomas Hein & Dr Carsten Enssle.
3. D. A. van Riet-Nales, A. F. Schobben, T. C. Egberts and C. M. Rademaker, *Clin. Therap.*, 2010, **32**(5), 924.
4. A. C. Perkins, C. G. Wilson, R. M. Vincent, M. Frier, P. E. Blackshaw, R. J. Dansereau, K. D. Juhlin, P. J. Bekker and R. C. Spiller, *Int. J. Pharm.*, 1999, **186**, 169.
5. A. C. Perkins, C. G. Wilson, M. Frier, P. E. Blackshaw, D. Juan, R. J. Dansereau, S. Hathaway, Z. Li, P. Long and R. C. Spiller, *Aliment. Pharmacol. Ther.*, 2001, **15**, 115.
6. K. T. Evans and G. W. Roberts, *J. Clin. Hosp. Pharm.*, 1981, **6**, 207–208.
7. A. C. Perkins, C. G. Wilson, P. E. Blackshaw, R. M. Vincent, R. J. Dansereau, K. D. Juhlin, P. J. Bekker and R. C. Spiller, *Gut*, 1994, **35**, 1363.
8. A. C. Perkins, C. G. Wilson, M. Frier, P. E. Blackshaw, R. J. Dansereau, R. Vincent, D. Wenderoth, S. Hathaway, Z. Li and R. C. Spiller, *Int. J. Pharm.*, 2001, **222**, 295.
9. I. K. Teismann, O. Steinstraeter, W. Schwindt, E. B. Ringelstein, C. Pantev and R. Dziewas, *Neurobiol. Aging*, 2010, **31**(6), 1044.
10. M. Yoshikawa, M. Yoshida, T. Nagasaki, K. Tanimoto, K. Tsuga, Y. Akagawa and T. Komatsu, *J. Gerontol., Ser. A*, 2005, **60**, 506.
11. Hermes Pharma and Spiegel Institut Mannheim, *A Hard Truth to Swallow?* http://www.swallowingtablets.com.
12. R. G. Sunki, R. Annapureddy and D. R. Rao, *J. Anim. Sci.*, 1978, **46**, 584.
13. E. L. Etkin, *Edible Medicines: An Ethnopharmacology of Food*, University of Arizona Press, 2006.
14. S. S. Schiffman, J. Zervakis, M. S. Suggs, K. C. Budd and L. Iuga, *Pharmacol., Biochem. Behav.*, 2000, **65**(4), 599.
15. Y. Fu, S. Yang, S. H. Jeong, S. Kimura and K. Park, *Crit. Rev. Ther. Drug Carrier Syst.*, 2004, **21**(6), 433.

16. S. Harmik, Y. Sultana and R. K. Khar, *Drug Dev. Ind. Pharm.*, 2004, **30**(5), 429.
17. B. P. Trivedi, *Nat. Rev.*, 2012, **486**(7403), S7–S9.
18. M. M. Galindo, N. Voigt, J. Stein, J. van Lengerich, J. D. Raguse, T. Hofmann, W. Meyerhof and M. Behrens, *Chem. Senses*, 2011, **37**, 123.
19. J. B. Davis, J. Gray, M. J. Gunthorpe, J. P. Hatcher, P. T. Davey, P. Overend, M. H. Harries, J. Latcham, C. Clapham, K. Atkinson and S. A. Hughes, *Nature*, 2000, **405**(6783), 183.
20. A. Drewnoswki, *Nutr. Rev.*, 2001, **59**, 163.
21. J. A. Mennella and G. K. Beauchamp, *Clin. Ther.*, 2008, **30**, 2120.
22. D. M. Small and J. Prescott, *Exp. Brain Res.*, 2005, **166**, 345.
23. J. Mojet, E. Christ-Hazelhof and J. Heidema, *Chem. Senses*, 2001, **26**, 845.
24. J. A. Helms, M. A. Della-Fera, A. E. Mott and M. E. Frank, *Arch. Oral Biol.*, 1995, **40**, 913.
25. R. L. Doty, M. B. Stern, C. Pfieffer, S. M. Gollomp and H. I. Hurtig, *J. Neurol., Neurosurg. Psychiatry*, 1992, **55**, 138.
26. A. Takeda, A. Kikuchi, M. Matsuzaki-Kobayashi, N. Sugeno and Y. Itoyama, *J. Neurol.*, 2007, **254**, IV2.
27. S. R. Porter, S. Fedele and K. M. Habbab, *Oral Oncol.*, 2010, **46**, 457.
28. R. W. Steele, B. Estrada, B. E. Begue, A. Mirza, D. A. Travillion and M. P. Thomas, *Clin. Pediatr.*, 1997, **36**(4), 193.
29. http://www.alpha-mos.com/analytical-instruments/astree-electronic-tongue.
30. http://www.insent.co.jp/en/products/ts5000z_index.html.
31. World Health Organization, *Development of Paediatric Medicines: Points to Consider in Pharmaceutical Development*, 2011, Working Document QAS/08.257/Rev.3.
32. K. Woertz, C. Tissen, P. Kleinbudde and J. Breitkreutz, *Int. J. Pharm.*, 2011, **417**, 256.
33. G. A. Campbell, J. A. Charles, K. Roberts-Skilton, M. Tsundupalli, K. C. Oh, A. Weinecke, R. Wagner and D. Franz, *Powder Technol.*, 2012, **224**, 109.
34. F. Scheiber, J. L. Fozzard, S. Gordon-Salant and J. M. Weiffenbach, *Int. J. Ind. Ergon.*, 1991, **7**, 133.
35. M. Ikeda, A. Ikui, A. Komiyama, D. Kobayashi and M. Tanaka, *J. Laryngol. Otol.*, 2008, **122**, 155.
36. World Health Organization, *Development of Paediatric Medicines: Points to Consider in Pharmaceutical Development*, 2011, Working Document QAS/08.257/Rev.3.
37. T. B. Ernest, J. Craig, A. Nunn, S. Salunka, C. Tuleu, J. Breitkreutz, A. Rainer and J. Hempinstall, *Int. J. Pharm.*, 2012, **435**, 124.
38. A. A. Bianchetti, S. A. Lava, A. Bettineli, M. Rizzi and G. D. Simonetti, *Clin. Ther.*, 2010, **32**, 1083.
39. G. Milani, M. Ragazzi, G. D. Simonetti, G. P. Ramelli, M. Rizzi, M. G. Bianchetti and E. F. Fossali, *Br. J. Clin. Pharmacol.*, 2009, **69**, 204.

40. L. V. Allen, *Clin. Ther.*, 2008, **30**, 2102.
41. S. Winnick, D. O. Lucas, A. L. Hartman and D. Toll, *Pediatrics*, 2005, **115**, e718.
42. J. A. Coutts, N. A. Gibson and J. Y. Paton, *Arch. Dis. Child.*, 1992, **67**, 332.
43. N. A. Gibson, A. E. Ferguson, T. C. Aitchison and J. Y. Paton, *Thorax*, 1995, **50**, 1274.
44. T. Nunn and J. Williams, *Br. J. Clin. Pharmacol.*, 2005, **59**, 674.
45. B. M. Best, E. V. Capparelli, H. Diep, S. S. Rossi, M. J. Farrell, E. Williams, G. Lee, J. N. van den Anker and N. Rakhmanina, *J. Acquired Immune Defic. Syndr.*, 2011, **58**, 385.
46. J. T. Flynn and S. R. Daniels, *J. Pediatr.*, 2006, **149**, 746.
47. H. D. Maples, L. P. James, C. D. Stowe, D. P. Jones, E. B. Hak, J. L. Blumer, B. Vogt, J. T. Wilson, G. L. Kearns and T. G. Wells, *J. Clin. Pharmacol.*, 2003, **43**, 7.
48. M. Strolin Benedetti and E. L. Bates, *Fundam. Clin. Pharmacol.*, 2002, **17**, 281.
49. G. L. Kearns, S. M. Abdel-Rahman, S. W. Alander, D. L. Blowey, J. S. Leeder and R. E. Kauffman, *N. Engl. J. Med.*, 2003, **349**, 1157.
50. C. Tuleu and J. Brietkreutz, *Eur. J. Pediatr.*, 2012, **172**, 717.
51. J. Pogány, *Pharmaceutical Development for Multisource (Generic) Pharmaceutical Products*, Working document QA/0.8.251/Rev.1, 2007.
52. M. Siewert, J. Dressman, C. K. Brown, V. P. Shah, J.-M. Aiche, N. Aoyagi, D. Bashaw, C. Brown, W. Brown, D. Burgess and J. Crison, *AAPS PharmSciTech*, 2003, **4**, 43.
53. J. Breitkreutz, European Perspectives on Pediatric Formulations, *Clin. Therap.*, 2008, **30**, 2146.
54. V. Fabiano, C. Marmeli and G. V. Zuccotti, *Pharmacol. Res.*, 2011, **63**, 362.
55. N. P. Roos, B. Havens and C. Black, *Soc. Sci. Med.*, 1993, **36**, 273.
56. H. C. Lin, C. J. Chen, H. H. Lin, J. T. Huang and M. J. Chen, *Int. J. Gerontol.*, 2013, **7**, 35.
57. J. L. Fuh, R. C. Lee, S. J. Wang, C. H. Lin, P. N. Wang, J. H. Chiang and H. S. Liu, *J. Clin. Neurol. Neurosurg.*, 1997, **99**, 106.
58. S. Stegemann, M. Gosch and J. Brietkreutz, *Int. J. Pharm.*, 2012, **43**, 197.
59. A. Nadir, E. Sahin, I. Nadir, S. Karadayi and M. Kaptanoglu, *Dis. Esophagus*, 2011, **24**, 6.
60. K. P. Schaps, O. Kessler and U. Fetzner, *Das Zweite Kompackt. Querschnittsbereiche*, Springer Medizin Verlag, 2008.
61. A. M. M.-F. Mathey, E. Siebelink, C. de Graaf and W. A. Van Staveren, *J. Gerontol.*, 2001, **56**, M200.
62. J. Mojet, E. Christ-Hazelhof and J. Heidema, *Chem. Senses*, 2001, **26**, 845.
63. M. L. Griep, T. F. Mets and D. L. Massart, *Food Qual. Prefer.*, 1997, **8**, 151.
64. M. L. Griep, T. F. Mets and D. L. Massart, *Br. J. Nutr.*, 2000, **83**, 105.
65. J. M. Boyce and G. R. Shone, *Postgrad. Med. J.*, 2006, **82**, 239.
66. European Association of Urology, *Night-time Urination Reduced by Cutting Salt in Diet*, ScienceDaily, http://www.sciencedaily.com/releases/2017/03/170327083711.htm.

67. L. Barnes, J. Cheek, R. L. Nation, A. Gilbert, L. Paradiso and A. Ballantyne, *J. Adv. Nurs.*, 2006, **56**, 190.

68. A. Treloar, B. Beats and M. Philpot, *J. R. Soc. Med.*, 2000, **93**, 408.

69. N. Rodenhuis, P. A. G. M. De Smet and D. M. Barends, *Eur. J. Pharm. Sci.*, 2004, **21**, 305.

70. K. Wening and J. Breitkreutz, *Int. J. Pharm.*, 2011, **404**, 1.

71. J. Stubbs, C. Haw and G. Dickens, *Int. Psychogeriatr.*, 2008, **20**, 616.

72. J. G. Schier, M. A. Howland, R. S. Hoffman and L. S. Nelson, *Ann. Pharmacother.*, 2003, **37**, 1420.

73. L. M. Paradiso, E. E. Roughead, A. L. Gilbert, D. Cosh, R. L. Nation, L. Barnes, J. Cheek and A. Ballantyne, *Aust. J. Ageing*, 2002, **21**, 123.

74. S. Stegemann, F. Ecker, M. Maio, P. Kraahs, R. Wohlfart, J. Breitkreutz, A. Zimmer, D. Bar-Shalom, P. Hettrich and B. Broegmann, *Ageing Res. Rev.*, 2010, **9**, 384.

75. D. Shukla, S. Chakraborty, S. Singh and B. Mishra, *Sci. Pharmacol.*, 2009, **76**, 309.

76. R. G. Strickley, Q. Iwata, S. Wu and T. C. Dahi, *J. Pharm. Sci*, 2008, **97**, 1731.

77. W. Doyle, A. Srivastava, M. A. Crawford, R. Bhatti, Z. Brooke and K. L. Costeloe, *Br. J. Nutr.*, 2001, **86**, 81.

78. D. J. Taylor, C. Mallen, N. McDougall and T. Lind, *Br. J. Obstet. Gynaecol.*, 1982, **89**, 1011.

79. S. N. Massawe, E. N. Urassa, L. Nystrom and G. Lindmark, *East Afr. Med. J.*, 2002, **79**, 461.

80. J. Garnier, *Harv. Bus. Rev.*, 2008, **86**, 68.

81. H. W. Chesbrough, *MIT Sloan Manage. Rev.*, 2007, **48**, 22.

82. J. Mittra, *Int. J. Biotechnol.*, 2008, **10**, 416.

83. A. Galbraith, S. Bullock, E. Manias and B. Hunt, *Fundamentals of Pharmacology*, Pearson Education, 2007.

84. E. A. Zannou, P. Li and W. Tong, Product lifecycle management (LCM), *Developing Solid Dosage Forms: Pharmaceutical Theory and Practice*, Elsevier, 1st edn, 2009, p. 911.

85. G. Vantrappen, J. Janssens, G. Coremans and R. Jian, *Dig. Dis. Sci.*, 1986, **31**, 5S.

86. J. R. Malagelada and V. L. W. Go, *Dig. Dis. Sci.*, 1979, **24**, 101.

87. N. Rouge, P. Buri and E. Doelker, *Int. J. Pharm.*, 1996, **136**, 117.

88. M. D. Burke and C. G. Wilson, *Drug Delivery*, 2006, **6**, 26.

89. C. Raasch and O. Schöffski, Management des Patenablaufs, in *Pharmabetriebslehre, Springer Medizin Der Patentauslauf von Pharmazeutika als Herausforderung beim Management des Produktlebenszyklus*, ed. O. Schöffski, *et al.*, Gabler Verlag/Springer Fachmedien, 2010.

90. A. I. Wertheimer, T. M. Santella, A. J. Finestone and R. A. Levy, *Adv. Ther.*, 2005, **22**, 559.

91. W. Mason and N. Winer, *J. Pharm. Sci.*, 1981, **70**, 265.

92. P. L. Møller, S. E. Nørholt, H. E. Ganry, J. H. Insuasty, F. G. Vincent, L. A. Skoglund and S. Sindet-Pedersen, *Clin. Pharmacol.*, 2000, **40**, 370.

93. J. Eichman and J. Robinson, *Pharm. Res.*, 1998, **15**, 925.
94. D. R. Hooker, *Am. J. Physiol.*, 1912, **31**, 47.
95. J. N. Hunt, *Am. J. Dis.*, 1963, **8**, 885.
96. G. Lolli, L. A. Greenberg and A. Lester, *N. Engl. J. Med.*, 1952, **235**, 490.
97. L. Ploutz-Snyder, J. Foley, R. Ploutz-Snyder, J. Kanaley, K. Sagendorf and R. Meyer, *Eur. J. Appl. Physiol. Occup. Physiol.*, 1999, **79**, 212.
98. P. Pouderoux, N. Friedman, P. Shirazi, J. G. Ringelstein and A. Keshavarzian, *Dig. Dis. Sci.*, 1997, **42**, 34.
99. F. P. Maximiano, G. H. Y. Costa, L. Lira de Sá Barreto, M. T. Bahia and M. S. S. Cunha-Filho, *J. Pharm. Pharmacol.*, 2011, **63**, 786.
100. F. Harms and M. Drüner, *Pharmamarketing – Innovations management im 21, Jahrhundert*, Lucius & Lucius, Stuttgar, 2003.
101. M. E. Aulton and K. Taylor, *Pharmaceutics: The Science of Dosage Form Design*, Churchill Livingston, 2nd edn, 2001, p. 102.
102. G. S. Banker, in *Drug Products: Their Role in the Treatment of Disease, Their Quality, and Their Status and Future as Drug-delivery Systems*, ed. G. S. Banker and C. T. Rhodes, Modern Pharmaceutics, Taylor & Francis e-Library, 2006.
103. T. Ito and R. T. Jensen, *Curr. Gastroenterol. Rep.*, 2010, **12**, 448.
104. T. Ali, *et al.*, Long-term safety concerns with proton pump inhibitors, *Am. J. Med.*, 2009, **122**, 896.
105. FDA Drug Safety Communication 3/23/2011.
106. P. Sipponen and M. Kärkönen, *Scand. J. Gastroenterol.*, 2010, **45**, 133.
107. C. W. Howden, J. A. H. Forrest and J. L. Reid, *Gut*, 1984, **25**, 707.
108. T. W. Chin, M. Loeb and I. W. Fong, *Antimicrob. Agents Chemother.*, 1995, **39**, 1671.
109. G. Ikinci, S. Senel, C. G. Wilson and M. Sumnu, *Int. J. Pharm.*, 2004, **277**, 173.
110. C. P. Peter, M. V. Kindt and J. A. Majka, *Dig. Dis. Sci.*, 1998, **43**, 1009.
111. L. A. Hodges, S. M. Connolly, J. Winter, T. Schmidt, H. N. E. Stevens, M. Hayward and C. G. Wilson, *Int. J. Pharm.*, 2012, **432**, 57.
112. B. Nurnberger and K. Brune, *Biopharm. Drug Dispos.*, 1989, **10**, 377–387.
113. A. Steingoetter, M. Fox, R. Treier, D. Weishaupt, B. Marincek, P. Boesiger, M. Fried and W. Schwize, *Scand. J. Gastroenterol.*, 2006, **41**, 1155.
114. A. L. De Goede, R. R. Boedhram, M. Eckhardt, I. M. Hanff, B. C. P. Kock, C. H. Vermaat and A. Vermes, *Int. J. Pharm.*, 2012, **433**, 119.
115. E. K. Ansah, J. O. Gyapong, I. A. Agyepong and D. B. Evans, *Trop. Med. Int. Health*, 2001, **6**, 495.
116. M. S. Calvo, S. J. Whiting and C. N. Barton, *J. Nutr.*, 2005, **135**, 310.
117. A. B. Murphy, B. Kelley, Y. A. Nyame, I. K. Martin, D. J. Smith, L. Castaneda and G. J. Zagaja, *Am. J. Men's Health*, 2012, **6**, 420.
118. A. J. Friedman, S. M. Ali and M. Albonico, *J. Trop. Med.*, 2012, **2012**, 590463.
119. N. Vanprapar, T. R. Cressey, K. Chokephaibulkit, P. Muresan, N. Plipat, V. Sirisanthana, W. Prasitsuebsai, S. Hongsiriwan, T. Chotpitayasunondh, A. Eksaengsri, M. Toye, M. E. Smith, K. McIntosh, E. Capparelli and R. Yogev, *Paediatr. Infect. Dis. J.*, 2010, **29**, 940.

120. A. Shahiwala, *Expert Opin. Drug Delivery*, 2011, **8**, 1521.
121. EMEA/CHMP/PEG/194810/2005.
122. Headquarters, W. H. O, *Report of the Informal Expert Meeting on Dosage Forms of Medicines for Children*, Geneva, Switzerland, 2008.
123. N. Spomer, V. Klingmann, I. Stoltenberg, C. Lerch, T. Meissner and J. Breitkreutz, *Arch. Dis. Child.*, 2012, **97**, 283.
124. E. Dejaeger, in *Swallowing Disorders and Medication in the Elderly*, ed. S. Jackson, P. Jansen and A. Mangoni, Prescribing for elderly patients, Wiley-Blackwell, 2009.
125. D. Douroumis, *Expert Opin. Drug Delivery*, 2007, **4**, 417.
126. O. Rachid, F. E. R. Simons, M. Rawas-Qalaji and K. J. Simons, *AAPS PharmSciTech*, 2010, **11**, 550.
127. V. D. Wagh and S. V. Ghadlinge, *J. Pharm. Res.*, 2009, **2**, 1049.
128. K. Woertz, C. Tissen, P. Kleinebudde and J. Breitkreutz, *Int. J. Pharm.*, 2010, **400**, 114.
129. B. P. Badgujar and A. S. Mundada, *Acta Pharm.*, 2011, **61**, 117.
130. H. Sohi, Y. Sultana and R. K. Khar, *Drug Dev. Ind. Pharm.*, 2004, **30**, 429.
131. A. Madene, M. Jacquot, J. Scher and S. Desobry, *Int. J. Food Sci. Technol.*, 2006, **41**, 1.
132. S. Bose and R. H. Bogner, *Pharm. Dev. Technol.*, 2007, **12**, 115.
133. S. C. Porter and J. E. Hogan, *Pharm. Int.*, 1984, **5**, 122.
134. K. Becker, E. M. Saurugger, D. Kienberger, D. Lopes, D. Haack, M. Köberle, M. Stehr, D. Lochmann, A. Zimmer and S. Salar-Behzadi, *Int. J. Pharm.*, 2016, **497**, 136.
135. Y. He, J. D. Litster and L. X. Liu, Scale-up considerations in granulation, *Handbook of Pharmaceutical Granulation Technology*, CRC Press, 2nd edn, 2005, pp. 459–490.
136. H. Leuenberger, Scale-up in the field of granulation and drying, in *Pharmaceutical Process Scale-up*, ed. M. Levin, Informa Healthcare, 2001.
137. E. O. Akala, Effect of packaging and stability of drugs and drug products, in *Pharmaceutical Manufacturing Handbook. Regulations and Quality*, ed. S. C. Gad, Wiley & Sons, 2008.
138. P. A. Atkin, T. P. Finnegan, S. J. Ogle and G. M. Shenfield, *Age Ageing*, 1994, **23**, 113.
139. R. A. Elliot, *J. Pharm. Pract. Res.*, 2006, **36**, 58.
140. B. Vrijens and J. Urquhart, *J. Antimicrob. Chemother.*, 2005, **55**, 616.
141. *WHO Adherence to Long-term Therapies: Evidence for Action*, World Health Organization, 2003.
142. S. Ellis, S. Shumaker, W. Sieber, C. Rand, R. Anderson, M. Benes-Malone, S. Cohen, G. Crowley, B. Dugan, D. Farmer and N. P. Betty Kreuziger, *Controlled Clin. Trials*, 2000, **21**, 218.
143. National Council on Patient Information and Education (NCPIE), *Enhancing Prescription Medicine Adherence: A National Action Plan*, August 2007.
144. R. B. Haynes, H. P. McDonald and A. X. Garg, *JAMA*, 2002, **288**, 2880.
145. M. Hite, *Pharmatech*, 2004, 1–4.

146. S. D. Portsmouth, J. Osorio, K. McCormick, B. G. Gazzard and G. J. Moyle, *HIV Med.*, 2005, **6**, 185.
147. C. Melikian, T. J. White, A. Vanderplas, C. M. Dezii and E. Chang, *Clin. Ther.*, 2002, **24**, 460.
148. C. M. Dezii, H. Kawabata and M. Tran, *South. Med. J.*, 2002, **95**, 68.
149. F. M. Feinsod, K. P. Prochoda, A. L. Anneberg and W. Solomon, *Ann. Longterm Care*, 2000, **8**, 43.
150. J. D. McCue, *Consultant*, 1997, **37**, 2135.
151. K. Deepak, *Tablets Capsules*, 2004, **7**, 30.
152. D. Brown, *Drug Delivery Technol.*, 2001, **3**, 58–61.
153. N. Neeta, *Int. J. Pharm. Sci.*, 2012, **1**, 228.
154. B. Charlesworth and A. J. Dowson, *Expert Opin. Pharmacother.*, 2002, **3**, 993.
155. R. Englert, G. Fontanesi, P. Muller, H. Ott, L. Rehn and H. Silva, *Clin. Ther.*, 1996, **18**, 843.
156. W. Siegmund, C. Hoffmann, M. Zschiesche, V. W. Steinijans, R. Sauter, W. D. Krueger and F. Diedrich, *Int. J. Clin. Pharmacol. Ther.*, 1998, **36**, 133.
157. F. Gabor, C. Fillafer, L. Neutsch, G. Ratzinger and M. Wirth, Improving oral delivery, in *Drug Delivery*, ed. M. Schäfer-Korting, Springer, 2010.

Subject Index